Atomic Physics

Atomic Physics

Edited by
Titus Mason

Larsen & Keller
www.larsen-keller.com

Atomic Physics
Edited by Titus Mason
ISBN: 978-1-63549-036-7 (Hardback)

Larsen & Keller
Published by Larsen and Keller Education,
5 Penn Plaza,
19th Floor,
New York, NY 10001, USA

Cataloging-in-Publication Data

Atomic physics / edited by Titus Mason.
p. cm.
Includes bibliographical references and index.
ISBN 978-1-63549-036-7
1. Nuclear physics. 2. Atomic theory. I. Mason, Titus.
QC776 .A86 2017
539.7--dc23

Printed and bound in China.

For more information regarding Larsen and Keller Education and its products, please visit the publisher's website www.larsen-keller.com

Table of Contents

Preface

Atomic physics as a part of physics deals with the study of atoms and its primary components, which are neutrons, electrons and protons. The field is primarily focused on the presence of electrons in the atom, their arrangement and the processes, which can change this arrangement. This field is also closely related to molecular, atomic and optical physics. The aim of this book is to present these topics in the most easy to comprehend language. It attempts to understand the multiple branches that fall under the discipline of atomic physics and how such concepts have practical applications. This textbook is an essential guide for both students and those who wish to pursue this discipline further.

A detailed account of the significant topics covered in this book is provided below:

Chapter 1- An atom consists of a specific number of electrons revolving in notational shells around a positively charged nucleus. Electromagnetic force keeps the particles together. Atomic physics examines this atomic structure and the means by which this arrangement undergoes modification. This chapter provides an extensive initiation to this field of physics.

Chapter 2- Atom is the smallest unit of any matter. Solid, liquid, gas and plasma are made up of neutral or ionized atoms. These atoms can differ in their size and have no well-defined boundaries. The aspects explained in the following are atomic number, atomic orbital, atomic mass, atomic radius etc. The chapter on atoms offers an insightful focus, keeping in mind the complex subject matter.

Chapter 3- As early as the 19th century, it was recognized that all matter consisted of smaller units called atoms. Experiments in electromagnetism and radioactivity paved way for a better understanding of the atomic structure and the previously thought indivisible units were found to be composed of even smaller components like quarks, bosons, alpha and beta particles. This chapter chronologically illustrates atomic models like plum pudding model, Rutherford model, Bohr model and theories like atomic theory and atomism. A section of the chapter outlines atomic orbital and its classification.

Chapter 4- The atomic nucleus comprises of neutrons and protons, making it positively charged on the whole. The components of the nucleus are held together by a strong nuclear force. This chapter provides comprehensive information on the nuclear makeup of an atom and analyses the nuclear structure and nuclear shell model. This chapter also discusses the semi-empirical mass formula which explains the atomic masses of various elements and is based on the liquid drop model.

Chapter 5- Spectral lines help identify molecules and atoms as each atom has its own unique spectral line. This enables one to identify the chemical composition of various materials and is of particular use in identifying the atomic and molecular structure of stars and other planets. This chapter studies spectral lines, emission spectrum, Doppler broadening, Zeeman effect, Stark effect, forbidden mechanism, selection rule among others.

Chapter 6- There are principally four quantum numbers for each electron orbital and the characteristic shape of each orbital reflects the electron's motion in that orbital. Angular momentum emulates the angular velocity at which an electron is moving in its orbit. This chapter investigates concepts like spin, spin quantum number, electron magnetic moment, spin-orbit interaction, J-coupling, fine structure and hyperfine structure.

Chapter 7- The deviation or reflection of radiation from its original path due to any deviation in the medium like the introduction of particles, density fluctuations, bubbles etc. The chapter examines scattering and its types like Raman scattering, Rayleigh scattering, Rutherford scattering, Compton scattering, small-angle scattering etc. The topics discussed in the chapter are of great importance to broaden the existing knowledge on atomic physics.

It gives me an immense pleasure to thank our entire team for their efforts. Finally in the end, I would like to thank my family and colleagues who have been a great source of inspiration and support.

Editor

1

Introduction to Atomic Physics

An atom consists of a specific number of electrons revolving in notational shells around a positively charged nucleus. Electromagnetic force keeps the particles together. Atomic physics examines this atomic structure and the means by which this arrangement undergoes modification. This chapter provides an extensive initiation to this field of physics.

Atomic Physics

Atomic physics is the field of physics that studies atoms as an isolated system of electrons and an atomic nucleus. It is primarily concerned with the arrangement of electrons around the nucleus and the processes by which these arrangements change. This comprises ions, neutral atoms and, unless otherwise stated, it can be assumed that the term *atom* includes ions.

The term *atomic physics* can be associated with nuclear power and nuclear weapons, due to the synonymous use of *atomic* and *nuclear* in standard English. Physicists distinguish between atomic physics — which deals with the atom as a system consisting of a nucleus and electrons — and nuclear physics, which considers atomic nuclei alone.

As with many scientific fields, strict delineation can be highly contrived and atomic physics is often considered in the wider context of *atomic, molecular, and optical physics*. Physics research groups are usually so classified.

Isolated Atoms

Atomic physics primarily considers atoms in isolation. Atomic models will consist of a single nucleus that may be surrounded by one or more bound electrons. It is not concerned with the formation of molecules (although much of the physics is identical), nor does it examine atoms in a solid state as condensed matter. It is concerned with processes such as ionization and excitation by photons or collisions with atomic particles.

While modelling atoms in isolation may not seem realistic, if one considers atoms in a gas or plasma then the time-scales for atom-atom interactions are huge in comparison to the atomic processes that are generally considered. This means that the individual atoms can be treated as if each were in isolation, as the vast majority of the time they are. By this consideration atomic physics provides the underlying theory in plasma physics and atmospheric physics, even though both deal with very large numbers of atoms.

Electronic Configuration

Electrons form notional shells around the nucleus. These are normally in a ground state but can be excited by the absorption of energy from light (photons), magnetic fields, or interaction with a colliding particle (typically ions or other electrons).

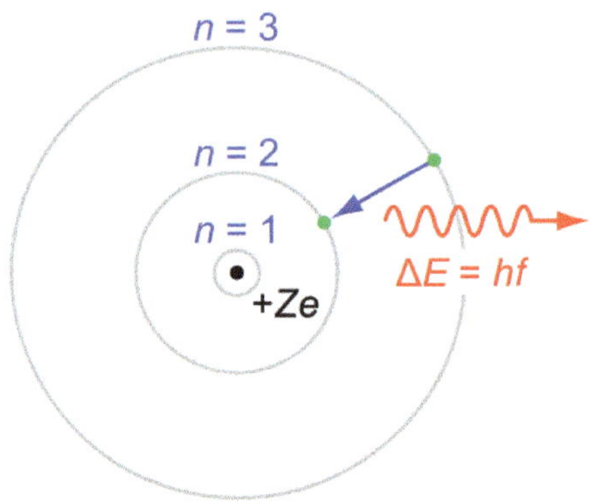

In the Bohr model, the transition of an electron with n=3 to the shell n=2 is shown, where a photon is emitted. An electron from shell (n=2) must have been removed beforehand by ionization

Electrons that populate a shell are said to be in a bound state. The energy necessary to remove an electron from its shell (taking it to infinity) is called the binding energy. Any quantity of energy absorbed by the electron in excess of this amount is converted to kinetic energy according to the conservation of energy. The atom is said to have undergone the process of ionization.

If the electron absorbs a quantity of energy less than the binding energy, it will be transferred to an excited state. After a certain time, the electron in an excited state will "jump" (undergo a transition) to a lower state. In a neutral atom, the system will emit a photon of the difference in energy, since energy is conserved.

If an inner electron has absorbed more than the binding energy (so that the atom ionizes), then a more outer electron may undergo a transition to fill the inner orbital. In this case, a visible photon or a characteristic x-ray is emitted, or a phenomenon known as the Auger effect may take place, where the released energy is transferred to another bound electron, causing it to go into the continuum. The Auger effect allows one to multiply ionize an atom with a single photon.

There are rather strict selection rules as to the electronic configurations that can be reached by excitation by light — however there are no such rules for excitation by collision processes.

History and Developments

The majority of fields in physics can be divided between theoretical work and experimental work, and atomic physics is no exception. It is usually the case, but not always, that progress goes in alternate cycles from an experimental observation, through to a theoretical explanation followed by some predictions that may or may not be confirmed by experiment, and so on. Of course, the current state of technology at any given time can put limitations on what can be achieved experimentally and theoretically so it may take considerable time for theory to be refined.

One of the earliest steps towards atomic physics was the recognition that matter was composed of *atoms*. It forms a part of the texts written in 6th century BC to 2nd century BC such as those of Democritus or Vaisheshika Sutra written by Kanad. This theory was later developed in the modern sense of the basic unit of a chemical element by the British chemist and physicist John Dalton in the 18th century. At this stage, it wasn't clear what atoms were although they could be described and classified by their properties (in bulk). The invention of the periodic system of elements by Mendeleev was another great step forward.

The true beginning of atomic physics is marked by the discovery of spectral lines and attempts to describe the phenomenon, most notably by Joseph von Fraunhofer. The study of these lines led to the Bohr atom model and to the birth of quantum mechanics. In seeking to explain atomic spectra an entirely new mathematical model of matter was revealed. As far as atoms and their electron shells were concerned, not only did this yield a better overall description, i.e. the atomic orbital model, but it also provided a new theoretical basis for chemistry (quantum chemistry) and spectroscopy.

Since the Second World War, both theoretical and experimental fields have advanced at a rapid pace. This can be attributed to progress in computing technology, which has allowed larger and more sophisticated models of atomic structure and associated collision processes. Similar technological advances in accelerators, detectors, magnetic field generation and lasers have greatly assisted experimental work.

2

Atom: An Integrated Study

Atom is the smallest unit of any matter. Solid, liquid, gas and plasma are made up of neutral or ionized atoms. These atoms can differ in their size and have no well-defined boundaries. The aspects explained in the following are atomic number, atomic orbital, atomic mass, atomic radius etc. The chapter on atoms offers an insightful focus, keeping in mind the complex subject matter.

Atom

An atom is the smallest constituent unit of ordinary matter that has the properties of a chemical element. Every solid, liquid, gas, and plasma is composed of neutral or ionized atoms. Atoms are very small; typical sizes are around 100 pm (a ten-billionth of a meter, in the short scale). However, atoms do not have well-defined boundaries, and there are different ways to define their size that give different but close values.

Atoms are small enough that attempting to predict their behavior using classical physics - as if they were billiard balls, for example - gives noticeably incorrect predictions due to quantum effects. Through the development of physics, atomic models have incorporated quantum principles to better explain and predict the behavior.

Every atom is composed of a nucleus and one or more electrons bound to the nucleus. The nucleus is made of one or more protons and typically a similar number of neutrons. Protons and neutrons are called nucleons. More than 99.94% of an atom's mass is in the nucleus. The protons have a positive electric charge, the electrons have a negative electric charge, and the neutrons have no electric charge. If the number of protons and electrons are equal, that atom is electrically neutral. If an atom has more or fewer electrons than protons, then it has an overall negative or positive charge, respectively, and it is called an ion.

The electrons of an atom are attracted to the protons in an atomic nucleus by this electromagnetic force. The protons and neutrons in the nucleus are attracted to each other by a different force, the nuclear force, which is usually stronger than the electromagnetic force repelling the positively charged protons from one another. Under certain circumstances the repelling electromagnetic force becomes stronger than the nuclear force, and nucleons can be ejected from the nucleus, leaving behind a different element: nuclear decay resulting in nuclear transmutation.

The number of protons in the nucleus defines to what chemical element the atom belongs: for example, all copper atoms contain 29 protons. The number of neutrons defines the isotope of the element. The number of electrons influences the magnetic properties of an atom. Atoms can attach to one or more other atoms by chemical bonds to form chemical compounds such as molecules. The ability of atoms to associate and dissociate is responsible for most of the physical changes observed in nature, and is the subject of the discipline of chemistry.

History of Atomic Theory

Atoms in Philosophy

The idea that matter is made up of discrete units is a very old idea, appearing in many ancient cultures such as Greece and India. The word "atom" was coined by ancient Greek philosophers. However, these ideas were founded in philosophical and theological reasoning rather than evidence and experimentation. As a result, their views on what atoms look like and how they behave were incorrect. They also could not convince everybody, so atomism was but one of a number of competing theories on the nature of matter. It was not until the 19th century that the idea was embraced and refined by scientists, when the blossoming science of chemistry produced discoveries that only the concept of atoms could explain.

First evidence-based theory

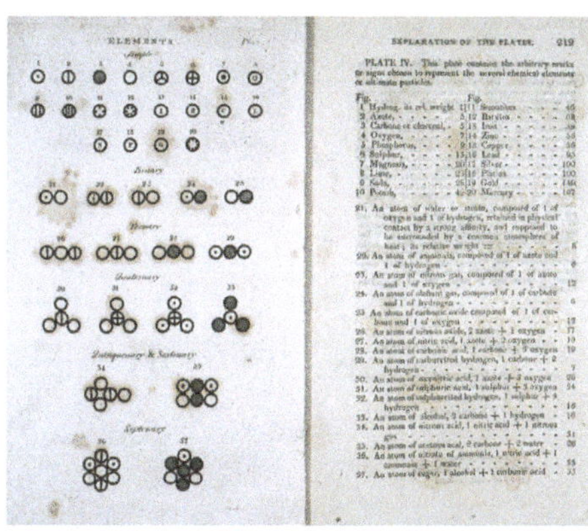

Various atoms and molecules as depicted in John Dalton's *A New System of Chemical Philosophy* (1808).

In the early 1800s, John Dalton used the concept of atoms to explain why elements always react in ratios of small whole numbers (the law of multiple proportions). For instance, there are two types of tin oxide: one is 88.1% tin and 11.9% oxygen and the other is 78.7% tin and 21.3% oxygen (tin(II) oxide and tin dioxide respectively). This means that 100g of tin will combine either with 13.5g or 27g of oxygen. 13.5 and 27 form a ratio of 1:2, a ratio of small whole numbers. This common pattern in chemistry suggested to Dalton that elements react in whole number multiples of discrete units—in other words, atoms. In the case of tin oxides, one tin atom will combine with either one or two oxygen atoms.

Dalton also believed atomic theory could explain why water absorbs different gases in different proportions. For example, he found that water absorbs carbon dioxide far better than it absorbs nitrogen. Dalton hypothesized this was due to the differences between the masses and configurations of the gases' respective particles, and carbon dioxide molecules (CO_2) are heavier and larger than nitrogen molecules (N_2).

Brownian Motion

In 1827, botanist Robert Brown used a microscope to look at dust grains floating in water and discovered that they moved about erratically, a phenomenon that became known as "Brownian motion". This was thought to be caused by water molecules knocking the grains about. In 1905 Albert Einstein proved the reality of these molecules and their motions by producing the first Statistical physics analysis of Brownian motion. French physicist Jean Perrin used Einstein's work to experimentally determine the mass and dimensions of atoms, thereby conclusively verifying Dalton's atomic theory.

Discovery of the electron

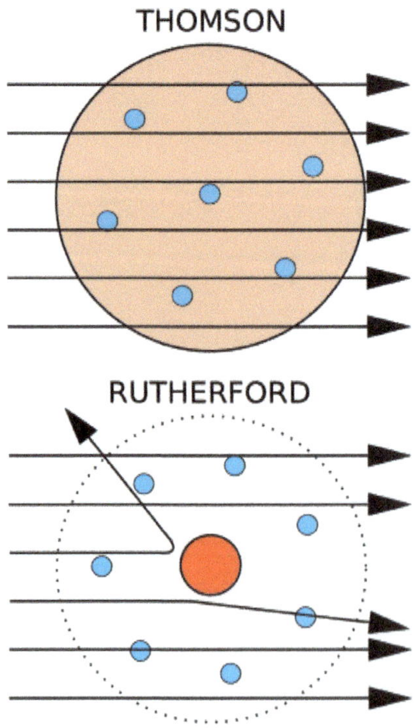

The Geiger–Marsden experiment *Top:* Expected results: alpha particles passing through the plum pudding model of the atom with negligible deflection. *Bottom:* Observed results: a small portion of the particles were deflected by the concentrated positive charge of the nucleus.

The physicist J. J. Thomson measured the mass of cathode rays, showing they were made of particles, but were around 1800 times lighter than the lightest atom, hydrogen. Therefore, they were not atoms, but a new particle, the first *subatomic* particle to be discovered, which he originally called "*corpuscle*" but was later named *electron*, after particles postulated by George Johnstone Stoney in 1874. He also showed they were identical to particles given off by photoelectric and radioactive materials. It was quickly recognized that they are the particles that carry electric currents in metal wires, and carry the negative electric charge within atoms. Thomson was given the 1906 Nobel Prize in Physics for this work. Thus he overturned the belief that atoms are the indivisible, ultimate particles of matter. Thomson also incorrectly postulated that the low mass, negatively charged electrons were distributed throughout the atom in a uniform sea of positive charge. This became known as the plum pudding model.

Discovery of the Nucleus

In 1909, Hans Geiger and Ernest Marsden, under the direction of Ernest Rutherford, bombarded a metal foil with alpha particles to observe how they scattered. They expected all the alpha particles to pass straight through with little deflection, because Thomson's model said that the charges in the atom are so diffuse that their electric fields could not affect the alpha particles much. However, Geiger and Marsden spotted alpha particles being deflected by angles greater than 90°, which was supposed to be impossible according to Thomson's model. To explain this, Rutherford proposed that the positive charge of the atom is concentrated in a tiny nucleus at the center of the atom.

Discovery of Isotopes

While experimenting with the products of radioactive decay, in 1913 radiochemist Frederick Soddy discovered that there appeared to be more than one type of atom at each position on the periodic table. The term isotope was coined by Margaret Todd as a suitable name for different atoms that belong to the same element. J.J. Thomson created a technique for separating atom types through his work on ionized gases, which subsequently led to the discovery of stable isotopes.

Bohr Model

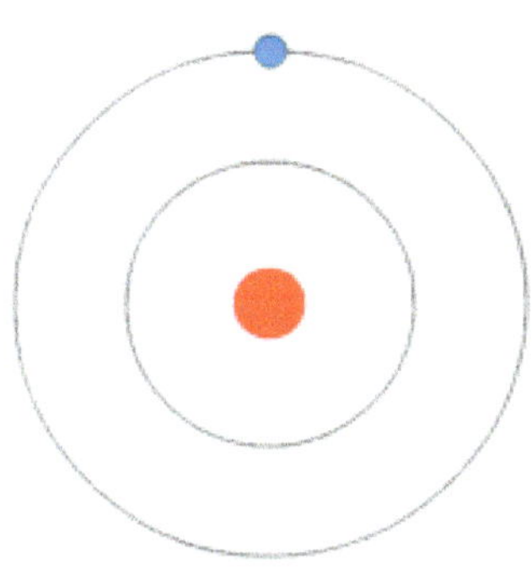

The Bohr model of the atom, with an electron making instantaneous "quantum leaps" from one orbit to another. This model is obsolete.

In 1913 the physicist Niels Bohr proposed a model in which the electrons of an atom were assumed to orbit the nucleus but could only do so in a finite set of orbits, and could jump between these orbits only in discrete changes of energy corresponding to absorption or radiation of a photon. This quantization was used to explain why the electrons orbits are stable (given that normally, charges in acceleration, including circular motion, lose kinetic energy which is emitted as electromagnetic radiation, and why elements absorb and emit electromagnetic radiation in discrete spectra.

Later in the same year Henry Moseley provided additional experimental evidence in favor of Niels Bohr's theory. These results refined Ernest Rutherford's and Antonius Van den Broek's model, which proposed that the atom contains in its nucleus a number of positive nuclear charges that is equal to its (atomic) number in the periodic table. Until these experiments, atomic number was not known to be a physical and experimental quantity. That it is equal to the atomic nuclear charge remains the accepted atomic model today.

Chemical Bonding Explained

Chemical bonds between atoms were now explained, by Gilbert Newton Lewis in 1916, as the interactions between their constituent electrons. As the chemical properties of the elements were known to largely repeat themselves according to the periodic law, in 1919 the American chemist Irving Langmuir suggested that this could be explained if the electrons in an atom were connected or clustered in some manner. Groups of electrons were thought to occupy a set of electron shells about the nucleus.

Further Developments in Quantum Physics

The Stern–Gerlach experiment of 1922 provided further evidence of the quantum nature of the atom. When a beam of silver atoms was passed through a specially shaped magnetic field, the beam was split based on the direction of an atom's angular momentum, or spin. As this direction is random, the beam could be expected to spread into a line. Instead, the beam was split into two parts, depending on whether the atomic spin was oriented up or down.

In 1924, Louis de Broglie proposed that all particles behave to an extent like waves. In 1926, Erwin Schrödinger used this idea to develop a mathematical model of the atom that described the electrons as three-dimensional waveforms rather than point particles. A consequence of using waveforms to describe particles is that it is mathematically impossible to obtain precise values for both the position and momentum of a particle at a given point in time; this became known as the uncertainty principle, formulated by Werner Heisenberg in 1926. In this concept, for a given accuracy in measuring a position one could only obtain a range of probable values for momentum, and vice versa. This model was able to explain observations of atomic behavior that previous models could not, such as certain structural and spectral patterns of atoms larger than hydrogen. Thus, the planetary model of the atom was discarded in favor of one that described atomic orbital zones around the nucleus where a given electron is most likely to be observed.

Discovery of the Neutron

The development of the mass spectrometer allowed the mass of atoms to be measured with increased accuracy. The device uses a magnet to bend the trajectory of a beam of ions, and the amount of deflection is determined by the ratio of an atom's mass to its charge. The chemist Francis William Aston used this instrument to show that isotopes had different masses. The atomic mass of these isotopes varied by integer amounts, called the whole number rule. The explanation for these different isotopes awaited the discovery of the neutron, an uncharged particle with a mass similar to the proton, by the physicist James Chadwick in 1932. Isotopes were then explained as elements with the same number of protons, but different numbers of neutrons within the nucleus.

Fission, High-energy Physics and Condensed Matter

In 1938, the German chemist Otto Hahn, a student of Rutherford, directed neutrons onto uranium atoms expecting to get transuranium elements. Instead, his chemical experiments showed barium as a product. A year later, Lise Meitner and her nephew Otto Frisch verified that Hahn's result were the first experimental *nuclear fission*. In 1944, Hahn received the Nobel prize in chemistry. Despite Hahn's efforts, the contributions of Meitner and Frisch were not recognized.

In the 1950s, the development of improved particle accelerators and particle detectors allowed scientists to study the impacts of atoms moving at high energies. Neutrons and protons were found to be hadrons, or composites of smaller particles called quarks. The standard model of particle physics was developed that so far has successfully explained the properties of the nucleus in terms of these sub-atomic particles and the forces that govern their interactions.

Structure

Subatomic Particles

Though the word *atom* originally denoted a particle that cannot be cut into smaller particles, in modern scientific usage the atom is composed of various subatomic particles. The constituent particles of an atom are the electron, the proton and the neutron; all three are fermions. However, the hydrogen-1 atom has no neutrons and the hydron ion has no electrons.

The electron is by far the least massive of these particles at 9.11×10^{-31} kg, with a negative electrical charge and a size that is too small to be measured using available techniques. It is the lightest particle with a positive rest mass measured. Under ordinary conditions, electrons are bound to the positively charged nucleus by the attraction created from opposite electric charges. If an atom has more or fewer electrons than its atomic number, then it becomes respectively negatively or positively charged as a whole; a charged atom is called an ion. Electrons have been known since the late 19th century, mostly thanks to J.J. Thomson.

Protons have a positive charge and a mass 1,836 times that of the electron, at 1.6726×10^{-27} kg. The number of protons in an atom is called its atomic number. Ernest Rutherford (1919) observed that nitrogen under alpha-particle bombardment ejects what appeared to be hydrogen nuclei. By 1920 he had accepted that the hydrogen nucleus is a distinct particle within the atom and named it proton.

Neutrons have no electrical charge and have a free mass of 1,839 times the mass of the electron, or 1.6929×10^{-27} kg, the heaviest of the three constituent particles, but it can be reduced by the nuclear binding energy. Neutrons and protons (collectively known as nucleons) have comparable dimensions—on the order of 2.5×10^{-15} m—although the 'surface' of these particles is not sharply defined. The neutron was discovered in 1932 by the English physicist James Chadwick.

In the Standard Model of physics, electrons are truly elementary particles with no internal structure. However, both protons and neutrons are composite particles composed of elementary particles called quarks. There are two types of quarks in atoms, each having a fractional electric charge. Protons are composed of two up quarks (each with charge +2/3) and one down quark (with a charge of −1/3). Neutrons consist of one up quark and two down quarks. This distinction accounts for the difference in mass and charge between the two particles.

The quarks are held together by the strong interaction (or strong force), which is mediated by gluons. The protons and neutrons, in turn, are held to each other in the nucleus by the nuclear force, which is a residuum of the strong force that has somewhat different range-properties. The gluon is a member of the family of gauge bosons, which are elementary particles that mediate physical forces.

Nucleus

All the bound protons and neutrons in an atom make up a tiny atomic nucleus, and are collectively called nucleons. The radius of a nucleus is approximately equal to $1.07\,\sqrt[3]{A}$ fm, where A is the total number of nucleons. This is much smaller than the radius of the atom, which is on the order of 10^5 fm. The nucleons are bound together by a short-ranged attractive potential called the residual strong force. At distances smaller than 2.5 fm this force is much more powerful than the electrostatic force that causes positively charged protons to repel each other.

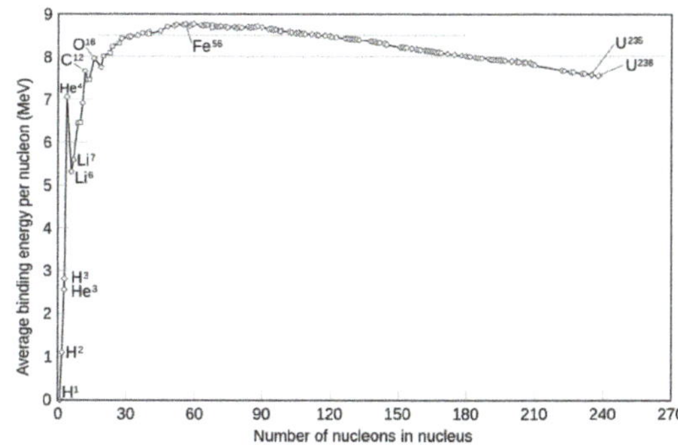

The binding energy needed for a nucleon to escape the nucleus, for various isotopes

Atoms of the same element have the same number of protons, called the atomic number. Within a single element, the number of neutrons may vary, determining the isotope of that element. The total number of protons and neutrons determine the nuclide. The number of neutrons relative to the protons determines the stability of the nucleus, with certain isotopes undergoing radioactive decay.

The proton, the electron, and the neutron are classified as fermions. Fermions obey the Pauli exclusion principle which prohibits *identical* fermions, such as multiple protons, from occupying the same quantum state at the same time. Thus, every proton in the nucleus must occupy a quantum state different from all other protons, and the same applies to all neutrons of the nucleus and to all electrons of the electron cloud. However, a proton and a neutron are allowed to occupy the same quantum state.

For atoms with low atomic numbers, a nucleus that has more neutrons than protons tends to drop to a lower energy state through radioactive decay so that the neutron–proton ratio is closer to one. However, as the atomic number increases, a higher proportion of neutrons is required to offset the mutual repulsion of the protons. Thus, there are no stable nuclei with equal proton and neutron numbers above atomic number Z = 20 (calcium) and as Z increases, the neutron–proton ratio of stable isotopes increases. The stable isotope with the highest proton–neutron ratio is lead-208 (about 1.5).

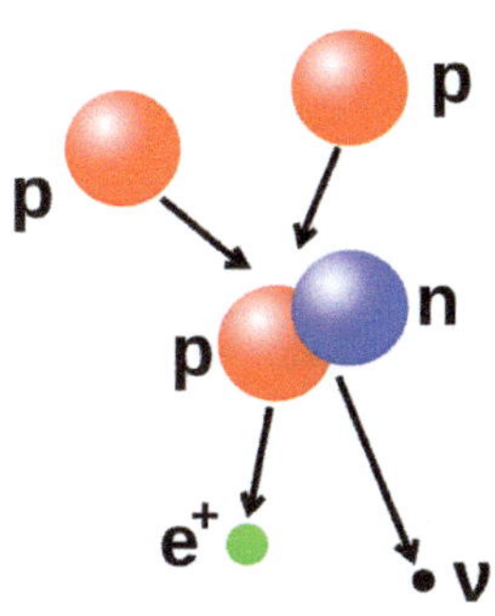

Illustration of a nuclear fusion process that forms a deuterium nucleus, consisting of a proton and a neutron, from two protons. A positron (e+)—an antimatter electron—is emitted along with an electron neutrino.

The number of protons and neutrons in the atomic nucleus can be modified, although this can require very high energies because of the strong force. Nuclear fusion occurs when multiple atomic particles join to form a heavier nucleus, such as through the energetic collision of two nuclei. For example, at the core of the Sun protons require energies of 3–10 keV to overcome their mutual repulsion—the coulomb barrier—and fuse together into a single nucleus. Nuclear fission is the opposite process, causing a nucleus to split into two smaller nuclei—usually through radioactive decay. The nucleus can also be modified through bombardment by high energy subatomic particles or photons. If this modifies the number of protons in a nucleus, the atom changes to a different chemical element.

If the mass of the nucleus following a fusion reaction is less than the sum of the masses of the separate particles, then the difference between these two values can be emitted as a type of usable energy (such as a gamma ray, or the kinetic energy of a beta particle), as described by Albert Einstein's mass–energy equivalence formula, $E = mc^2$, where m is the mass loss and c is the speed of light. This deficit is part of the binding energy of the new nucleus, and it is the non-recoverable loss of the energy that causes the fused particles to remain together in a state that requires this energy to separate.

The fusion of two nuclei that create larger nuclei with lower atomic numbers than iron and nickel—a total nucleon number of about 60—is usually an exothermic process that releases more energy than is required to bring them together. It is this energy-releasing process that makes nuclear fusion in stars a self-sustaining reaction. For heavier nuclei, the binding energy per nucleon in the nucleus begins to decrease. That means fusion processes producing nuclei that have atomic numbers higher than about 26, and atomic masses higher than about 60, is an endothermic process. These more massive nuclei can not undergo an energy-producing fusion reaction that can sustain the hydrostatic equilibrium of a star.

Electron Cloud

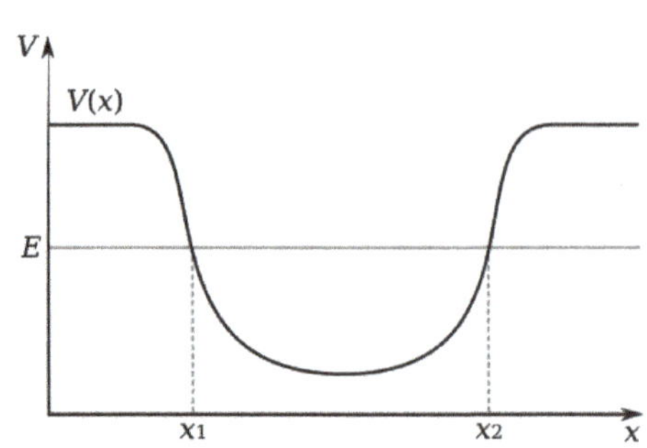

A potential well, showing, according to classical mechanics, the minimum energy V(x) needed toreach each position x. Classically, a particle with energy E is constrained to a range of positions between x1 and x2.

The electrons in an atom are attracted to the protons in the nucleus by the electromagnetic force. This force binds the electrons inside an electrostatic potential well surrounding the smaller nucleus, which means that an external source of energy is needed for the electron to escape. The closer an electron is to the nucleus, the greater the attractive force. Hence electrons bound near the center of the potential well require more energy to escape than those at greater separations.

Electrons, like other particles, have properties of both a particle and a wave. The electron cloud is a region inside the potential well where each electron forms a type of three-dimensional standing wave—a wave form that does not move relative to the nucleus. This behavior is defined by an atomic orbital, a mathematical function that characterises the probability that an electron appears to be at a particular location when its position is measured. Only a discrete (or quantized) set of these orbitals exist around the nucleus, as other possible wave patterns rapidly decay into a more stable form. Orbitals can have one or more ring or node structures, and differ from each other in size, shape and orientation.

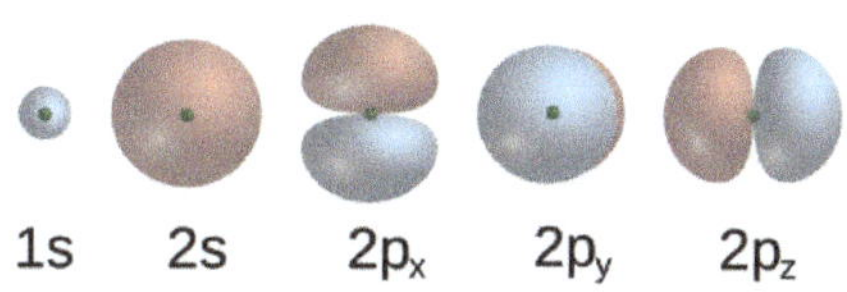

Wave functions of the first five atomic orbitals. The three 2p orbitals each display a single angular node that has an orientation and a minimum at the center.

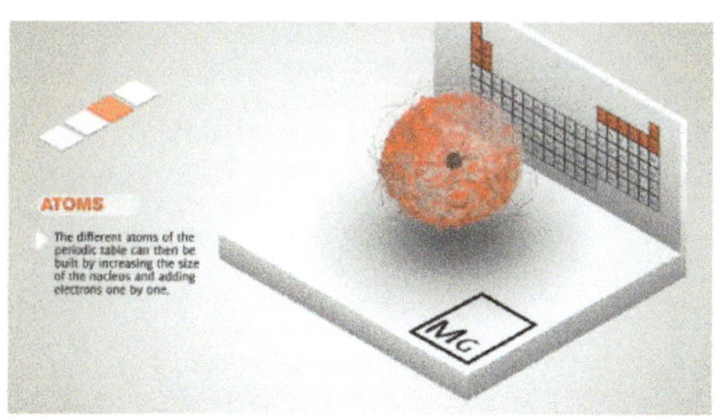

How atoms are constructed from electron orbitals and link to the periodic table

Each atomic orbital corresponds to a particular energy level of the electron. The electron can change its state to a higher energy level by absorbing a photon with sufficient energy to boost it into the new quantum state. Likewise, through spontaneous emission, an electron in a higher energy state can drop to a lower energy state while radiating the excess energy as a photon. These characteristic energy values, defined by the differences in the energies of the quantum states, are responsible for atomic spectral lines.

The amount of energy needed to remove or add an electron—the electron binding energy—is far less than the binding energy of nucleons. For example, it requires only 13.6 eV to strip a ground-state electron from a hydrogen atom, compared to 2.23 *million* eV for splitting a deuterium nucleus. Atoms are electrically neutral if they have an equal number of protons and electrons. Atoms that have either a deficit or a surplus of electrons are called ions. Electrons that are farthest from the nucleus may be transferred to other nearby atoms or shared between atoms. By this mechanism, atoms are able to bond into molecules and other types of chemical compounds like ionic and covalent network crystals.

Properties

Nuclear Properties

By definition, any two atoms with an identical number of *protons* in their nuclei belong to the same chemical element. Atoms with equal numbers of protons but a different number of *neutrons* are different isotopes of the same element. For example, all hydrogen atoms admit exactly one proton, but isotopes exist with no neutrons (hydrogen-1, by far the most common form, also called protium), one neutron (deuterium), two neutrons (tritium) and more than two neutrons. The known elements form a set of atomic numbers, from the single proton element hydrogen up to the 118-proton element ununoctium. All known isotopes of elements with atomic numbers greater than 82 are radioactive.

About 339 nuclides occur naturally on Earth, of which 254 (about 75%) have not been observed to decay, and are referred to as "stable isotopes". However, only 90 of these nuclides are stable to all decay, even in theory. Another 164 (bringing the total to 254) have not been observed to decay, even though in theory it is energetically possible. These are also formally classified as "stable". An additional 34 radioactive nuclides have half-lives longer than 80 million years, and are long-lived enough to be present from the birth of the solar system. This collection of 288 nuclides are known as primordial nuclides. Finally, an additional 51 short-lived nuclides are known to occur naturally, as daughter products of primordial nuclide decay (such as radium from uranium), or else as products of natural energetic processes on Earth, such as cosmic ray bombardment (for example, carbon-14).

For 80 of the chemical elements, at least one stable isotope exists. As a rule, there is only a handful of stable isotopes for each of these elements, the average being 3.2 stable isotopes per element. Twenty-six elements have only a single stable isotope, while the largest number of stable isotopes observed for any element is ten, for the element tin. Elements 43, 61, and all elements numbered 83 or higher have no stable isotopes.

Stability of isotopes is affected by the ratio of protons to neutrons, and also by the presence of certain "magic numbers" of neutrons or protons that represent closed and filled quantum shells. These quantum shells correspond to a set of energy levels within the shell model of the nucleus; filled shells, such as the filled shell of 50 protons for tin, confers unusual stability on the nuclide. Of the 254 known stable nuclides, only four have both an odd number of protons *and* odd number of neutrons: hydrogen-2 (deuterium), lithium-6, boron-10 and nitrogen-14. Also, only four naturally occurring, radioactive odd–odd nuclides have a half-life over a billion years: potassium-40, vanadium-50, lanthanum-138 and tantalum-180m. Most odd–odd nuclei are highly unstable with respect to beta decay, because the decay products are even–even, and are therefore more strongly bound, due to nuclear pairing effects.

Mass

The large majority of an atom's mass comes from the protons and neutrons that make it up. The total number of these particles (called "nucleons") in a given atom is called the mass number. It is a positive integer and dimensionless (instead of having dimension of mass), because it expresses a count. An example of use of a mass number is "carbon-12," which has 12 nucleons (six protons and six neutrons).

The actual mass of an atom at rest is often expressed using the unified atomic mass unit (u), also called dalton (Da). This unit is defined as a twelfth of the mass of a free neutral atom of carbon-12, which is approximately 1.66×10^{-27} kg. Hydrogen-1 (the lightest isotope of hydrogen which is also the nuclide with the lowest mass) has an atomic weight of 1.007825 u. The value of this number is called the atomic mass. A given atom has an atomic mass approximately equal (within 1%) to its mass number times the atomic mass unit (for example the mass of a nitrogen-14 is roughly 14 u). However, this number will not be exactly an integer except in the case of carbon-12 (see below). The heaviest stable atom is lead-208, with a mass of 207.9766521 u.

As even the most massive atoms are far too light to work with directly, chemists instead use the unit of moles. One mole of atoms of any element always has the same number of atoms (about 6.022×10^{23}). This number was chosen so that if an element has an atomic mass of 1 u, a mole of atoms of that element has a mass close to one gram. Because of the definition of the unified atomic mass unit, each carbon-12 atom has an atomic mass of exactly 12 u, and so a mole of carbon-12 atoms weighs exactly 0.012 kg.

Shape and Size

Atoms lack a well-defined outer boundary, so their dimensions are usually described in terms of an atomic radius. This is a measure of the distance out to which the electron cloud extends from the nucleus. However, this assumes the atom to exhibit a spherical shape, which is only obeyed for atoms in vacuum or free space. Atomic radii may be derived from the distances between two nuclei when the two atoms are joined in a chemical bond. The radius varies with the location of an atom on the atomic chart, the type of chemical bond, the number of neighboring atoms (coordination number) and a quantum mechanical property known as spin. On the periodic table of the elements, atom size tends to increase when moving down columns, but decrease when moving across rows (left to right). Consequently, the smallest atom is helium with a radius of 32 pm, while one of the largest is caesium at 225 pm.

When subjected to external forces, like electrical fields, the shape of an atom may deviate from spherical symmetry. The deformation depends on the field magnitude and the orbital type of outer shell electrons, as shown by group-theoretical considerations. Aspherical deviations might be elicited for instance in crystals, where large crystal-electrical fields may occur at low-symmetry lattice sites. Significant ellipsoidal deformations have recently been shown to occur for sulfur ions and chalcogen ions in pyrite-type compounds.

Atomic dimensions are thousands of times smaller than the wavelengths of light (400–700 nm) so they cannot be viewed using an optical microscope. However, individual atoms can be observed using a scanning tunneling microscope. To visualize the minuteness of the atom, consider that a typical human hair is about 1 million carbon atoms in width. A single drop of water contains about 2 sextillion (2×10^{21}) atoms of oxygen, and twice the number of hydrogen atoms. A single carat diamond with a mass of 2×10^{-4} kg contains about 10 sextillion (10^{22}) atoms of carbon. If an apple were magnified to the size of the Earth, then the atoms in the apple would be approximately the size of the original apple.

Radioactive Decay

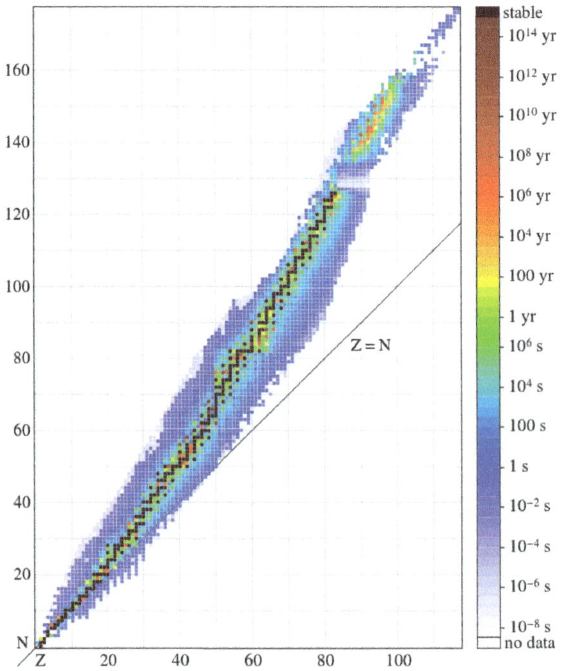

This diagram shows the half-life ($T_{1/2}$) of various isotopes with Z protons and N neutrons.

Every element has one or more isotopes that have unstable nuclei that are subject to radioactive decay, causing the nucleus to emit particles or electromagnetic radiation. Radioactivity can occur when the radius of a nucleus is large compared with the radius of the strong force, which only acts over distances on the order of 1 fm.

The most common forms of radioactive decay are:

- Alpha decay: this process is caused when the nucleus emits an alpha particle, which is a helium nucleus consisting of two protons and two neutrons. The result of the emission is a new element with a lower atomic number.

- Beta decay (and electron capture): these processes are regulated by the weak force, and result from a transformation of a neutron into a proton, or a proton into a neutron. The neutron to proton transition is accompanied by the emission of an electron and an antineutrino, while proton to neutron transition (except in electron capture) causes the emission of a positron and a neutrino. The electron or positron emissions are called beta particles. Beta decay either increases or decreases the atomic number of the nucleus by one. Electron capture is more common than positron emission, because it requires less energy. In this type of decay, an electron is absorbed by the nucleus, rather than a positron emitted from the nucleus. A neutrino is still emitted in this process, and a proton changes to a neutron.

- Gamma decay: this process results from a change in the energy level of the nucleus to a lower state, resulting in the emission of electromagnetic radiation. The excited state of a nucleus which results in gamma emission usually occurs following the emission of an alpha or a beta particle. Thus, gamma decay usually follows alpha or beta decay.

Other more rare types of radioactive decay include ejection of neutrons or protons or clusters of nucleons from a nucleus, or more than one beta particle. An analog of gamma emission which allows excited nuclei to lose energy in a different way, is internal conversion— a process that produces high-speed electrons that are not beta rays, followed by production of high-energy photons that are not gamma rays. A few large nuclei explode into two or more charged fragments of varying masses plus several neutrons, in a decay called spontaneous nuclear fission.

Each radioactive isotope has a characteristic decay time period—the half-life—that is determined by the amount of time needed for half of a sample to decay. This is an exponential decay process that steadily decreases the proportion of the remaining isotope by 50% every half-life. Hence after two half-lives have passed only 25% of the isotope is present, and so forth.

Magnetic Moment

Elementary particles possess an intrinsic quantum mechanical property known as spin. This is analogous to the angular momentum of an object that is spinning around its center of mass, although strictly speaking these particles are believed to be point-like and cannot be said to be rotating. Spin is measured in units of the reduced Planck constant (ħ), with electrons, protons and neutrons all having spin ½ ħ, or "spin-½". In an atom, electrons in motion around the nucleus possess orbital angular momentum in addition to their spin, while the nucleus itself possesses angular momentum due to its nuclear spin.

The magnetic field produced by an atom—its magnetic moment—is determined by these various forms of angular momentum, just as a rotating charged object classically produces a magnetic field. However, the most dominant contribution comes from electron spin. Due to the nature of electrons to obey the Pauli exclusion principle, in which no two electrons may be found in the same quantum state, bound electrons pair up with each other, with one member of each pair in a spin up state and the other in the opposite, spin down state. Thus these spins cancel each other out, reducing the total magnetic dipole moment to zero in some atoms with even number of electrons.

In ferromagnetic elements such as iron, cobalt and nickel, an odd number of electrons leads to an unpaired electron and a net overall magnetic moment. The orbitals of neighboring atoms overlap and a lower energy state is achieved when the spins of unpaired electrons are aligned with each other, a spontaneous process known as an exchange interaction. When the magnetic moments of ferromagnetic atoms are lined up, the material can produce a measurable macroscopic field. Paramagnetic materials have atoms with magnetic moments that line up in random directions when no magnetic field is present, but the magnetic moments of the individual atoms line up in the presence of a field.

The nucleus of an atom will have no spin when it has even numbers of both neutrons and protons, but for other cases of odd numbers, the nucleus may have a spin. Normally nuclei with spin are aligned in random directions because of thermal equilibrium. However, for certain elements (such as xenon-129) it is possible to polarize a significant proportion of the nuclear spin states so that they are aligned in the same direction—a condition called hyperpolarization. This has important applications in magnetic resonance imaging.

Energy Levels

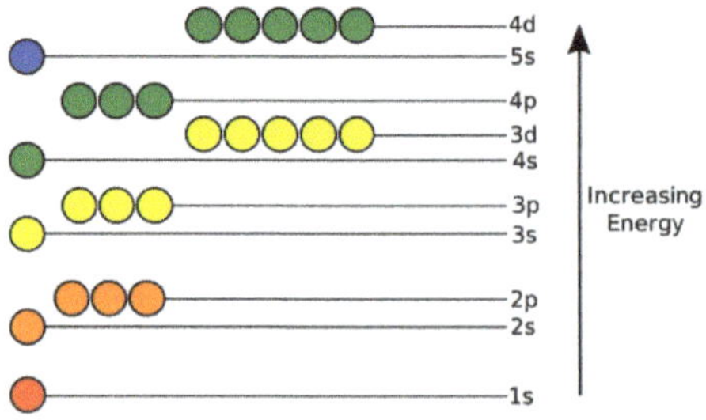

These electron's energy levels (not to scale) are sufficient for ground states of atoms up to cadmium (5s2 4d10) inclusively. Do not forget that even the top of the diagram is lower than an unbound electron state.

The potential energy of an electron in an atom is negative, its dependence of its position reaches the minimum (the most absolute value) inside the nucleus, and vanishes when the distance from the nucleus goes to infinity, roughly in an inverse proportion to the distance. In the quantum-mechanical model, a bound electron can only occupy a set of states centered on the nucleus, and each state corresponds to a specific energy level; see time-independent Schrödinger equation for theoretical explanation. An energy level can be measured by the amount of energy needed to unbind the electron from the atom, and is usually given in units of electronvolts (eV). The lowest energy state of a bound electron is called the ground state, i.e. stationary state, while an electron transition to a higher level results in an excited state. The electron's energy raises when n increases because the (average) distance to the nucleus increases. Dependence of the energy on ℓ is caused not by electrostatic potential of the nucleus, but by interaction between electrons.

For an electron to transition between two different states, e.g. grounded state to first excited level (ionization), it must absorb or emit a photon at an energy matching the difference in the potential energy of those levels, according to Niels Bohr model, what can be precisely calculated by the Schrödinger equation. Electrons jump between orbitals in a particle-like fashion. For example, if a single photon strikes the electrons, only a single electron changes states in response to the photon; see Electron properties.

The energy of an emitted photon is proportional to its frequency, so these specific energy levels appear as distinct bands in the electromagnetic spectrum. Each element has a characteristic spectrum that can depend on the nuclear charge, subshells filled by electrons, the electromagnetic interactions between the electrons and other factors.

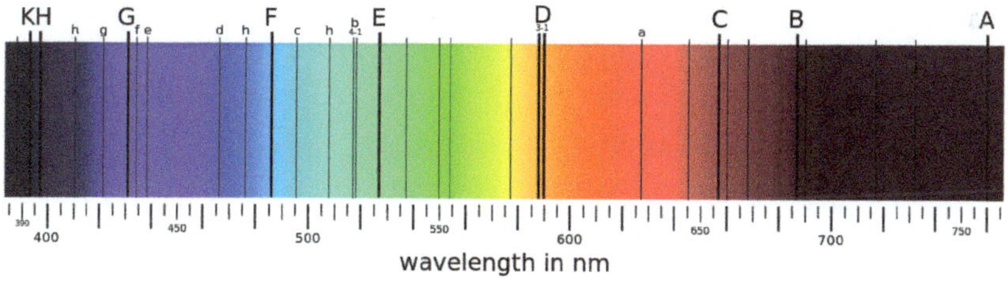

An example of absorption lines in a spectrum

When a continuous spectrum of energy is passed through a gas or plasma, some of the photons are absorbed by atoms, causing electrons to change their energy level. Those excited electrons that remain bound to their atom spontaneously emit this energy as a photon, traveling in a random direction, and so drop back to lower energy levels. Thus the atoms behave like a filter that forms a series of dark absorption bands in the energy output. (An observer viewing the atoms from a view that does not include the continuous spectrum in the background, instead sees a series of emission lines from the photons emitted by the atoms.) Spectroscopic measurements of the strength and width of atomic spectral lines allow the composition and physical properties of a substance to be determined.

Close examination of the spectral lines reveals that some display a fine structure splitting. This occurs because of spin–orbit coupling, which is an interaction between the spin and motion of the outermost electron. When an atom is in an external magnetic field, spectral lines become split into three or more components; a phenomenon called the Zeeman effect. This is caused by the interaction of the magnetic field with the magnetic moment of the atom and its electrons. Some atoms can have multiple electron configurations with the same energy level, which thus appear as a single spectral line. The interaction of the magnetic field with the atom shifts these electron configurations to slightly different energy levels, resulting in multiple spectral lines. The presence of an external electric field can cause a comparable splitting and shifting of spectral lines by modifying the electron energy levels, a phenomenon called the Stark effect.

If a bound electron is in an excited state, an interacting photon with the proper energy can cause stimulated emission of a photon with a matching energy level. For this to occur, the electron must drop to a lower energy state that has an energy difference matching the energy of the interacting photon. The emitted photon and the interacting photon then move off in parallel and with matching phases. That is, the wave patterns of the two photons are synchronized. This physical property is used to make lasers, which can emit a coherent beam of light energy in a narrow frequency band.

Valence and Bonding Behavior

Valency is the combining power of an element. It is equal to number of hydrogen atoms that atom can combine or displace in forming compounds. The outermost electron shell of an atom in its uncombined state is known as the valence shell, and the electrons in that shell are called valence electrons. The number of valence electrons determines the bonding behavior with other atoms. Atoms tend to chemically react with each other in a manner that fills (or empties) their outer valence shells. For example, a transfer of a single electron between atoms is a useful approximation for bonds that form between atoms with one-electron more than a filled shell, and others that are one-electron short of a full shell, such as occurs in the compound sodium chloride and other chemical ionic salts. However, many elements display multiple valences, or tendencies to share differing numbers of electrons in different compounds. Thus, chemical bonding between these elements takes many forms of electron-sharing that are more than simple electron transfers. Examples include the element carbon and the organic compounds.

The chemical elements are often displayed in a periodic table that is laid out to display recurring chemical properties, and elements with the same number of valence electrons form a group that is aligned in the same column of the table. (The horizontal rows correspond to the filling of a quantum shell of electrons.) The elements at the far right of the table have their outer shell completely filled with electrons, which results in chemically inert elements known as the noble gases.

States

Quantities of atoms are found in different states of matter that depend on the physical conditions, such as temperature and pressure. By varying the conditions, materials can transition between solids, liquids, gases and plasmas. Within a state, a material can also exist in different allotropes. An example of this is solid carbon, which can exist as graphite or diamond. Gaseous allotropes exist as well, such as dioxygen and ozone.

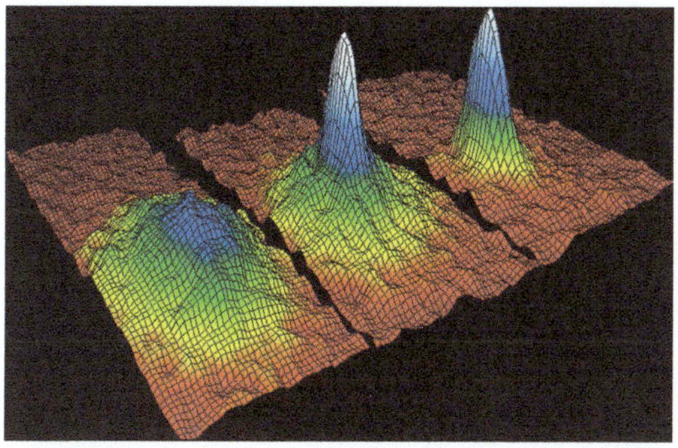

Snapshots illustrating the formation of a Bose–Einstein condensate

At temperatures close to absolute zero, atoms can form a Bose–Einstein condensate, at which point quantum mechanical effects, which are normally only observed at the atomic scale, become apparent on a macroscopic scale. This super-cooled collection of atoms then behaves as a single super atom, which may allow fundamental checks of quantum mechanical behavior.

Identification

Scanning tunneling microscope image showing the individual atoms making up this gold (100) surface. The surface atoms deviate from the bulk crystal structure and arrange in columns several atoms wide with pits between them (See surface reconstruction).

The scanning tunneling microscope is a device for viewing surfaces at the atomic level. It uses the quantum tunneling phenomenon, which allows particles to pass through a barrier that would normally be insurmountable. Electrons tunnel through the vacuum between two planar metal electrodes, on each of which is an adsorbed atom, providing a tunneling-current density that can be measured. Scanning one atom (taken as the tip) as it moves past the other (the sample) permits plotting of tip displacement versus lateral separation for a constant current. The calculation shows the extent to which scanning-tunneling-microscope images of an individual atom are visible. It confirms that for low bias, the microscope images the space-averaged dimensions of the electron orbitals across closely packed energy levels—the Fermi level local density of states.

An atom can be ionized by removing one of its electrons. The electric charge causes the trajectory of an atom to bend when it passes through a magnetic field. The radius by which the trajectory of a moving ion is turned by the magnetic field is determined by the mass of the atom. The mass spectrometer uses this principle to measure the mass-to-charge ratio of ions. If a sample contains multiple isotopes, the mass spectrometer can determine the proportion of each isotope in the sample by measuring the intensity of the different beams of ions. Techniques to vaporize atoms include inductively coupled plasma atomic emission spectroscopy and inductively coupled plasma mass spectrometry, both of which use a plasma to vaporize samples for analysis.

A more area-selective method is electron energy loss spectroscopy, which measures the energy loss of an electron beam within a transmission electron microscope when it interacts with a portion of a sample. The atom-probe tomograph has sub-nanometer resolution in 3-D and can chemically identify individual atoms using time-of-flight mass spectrometry.

Spectra of excited states can be used to analyze the atomic composition of distant stars. Specific light wavelengths contained in the observed light from stars can be separated out and related to the quantized transitions in free gas atoms. These colors can be replicated using a gas-discharge lamp containing the same element. Helium was discovered in this way in the spectrum of the Sun 23 years before it was found on Earth.

Origin and Current State

Atoms form about 4% of the total energy density of the observable Universe, with an average density of about 0.25 atoms/m^3. Within a galaxy such as the Milky Way, atoms have a much higher concentration, with the density of matter in the interstellar medium (ISM) ranging from 10^5 to 10^9 atoms/m^3. The Sun is believed to be inside the Local Bubble, a region of highly ionized gas, so the density in the solar neighborhood is only about 10^3 atoms/m^3. Stars form from dense clouds in the ISM, and the evolutionary processes of stars result in the steady enrichment of the ISM with elements more massive than hydrogen and helium. Up to 95% of the Milky Way's atoms are concentrated inside stars and the total mass of atoms forms about 10% of the mass of the galaxy. (The remainder of the mass is an unknown dark matter.)

Formation

Electrons are thought to exist in the Universe since early stages of the Big Bang. Atomic nuclei forms in nucleosynthesis reactions. In about three minutes Big Bang nucleosynthesis produced most of the helium, lithium, and deuterium in the Universe, and perhaps some of the beryllium and boron.

Ubiquitousness and stability of atoms relies on their binding energy, which means that an atom has a lower energy than an unbound system of the nucleus and electrons. Where the temperature is much higher than ionization potential, the matter exists in the form of plasma—a gas of positively charged ions (possibly, bare nuclei) and electrons. When the temperature drops below the ionization potential, atoms become statistically favorable. Atoms (complete with bound electrons) became to dominate over charged particles 380,000 years after the Big Bang—an epoch called recombination, when the expanding Universe cooled enough to allow electrons to become attached to nuclei.

Since the Big Bang, which produced no carbon or heavier elements, atomic nuclei have been combined in stars through the process of nuclear fusion to produce more of the element helium, and (via the triple alpha process) the sequence of elements from carbon up to iron; see stellar nucleosynthesis for details.

Isotopes such as lithium-6, as well as some beryllium and boron are generated in space through cosmic ray spallation. This occurs when a high-energy proton strikes an atomic nucleus, causing large numbers of nucleons to be ejected.

Elements heavier than iron were produced in supernovae through the r-process and in AGB stars through the s-process, both of which involve the capture of neutrons by atomic nuclei. Elements such as lead formed largely through the radioactive decay of heavier elements.

Earth

Most of the atoms that make up the Earth and its inhabitants were present in their current form in the nebula that collapsed out of a molecular cloud to form the Solar System. The rest are the result of radioactive decay, and their relative proportion can be used to determine the age of the Earth through radiometric dating. Most of the helium in the crust of the Earth (about 99% of the helium from gas wells, as shown by its lower abundance of helium-3) is a product of alpha decay.

There are a few trace atoms on Earth that were not present at the beginning (i.e., not "primordial"), nor are results of radioactive decay. Carbon-14 is continuously generated by cosmic rays in the atmosphere. Some atoms on Earth have been artificially generated either deliberately or as by-products of nuclear reactors or explosions. Of the transuranic elements—those with atomic numbers greater than 92—only plutonium and neptunium occur naturally on Earth. Transuranic elements have radioactive lifetimes shorter than the current age of the Earth and thus identifiable quantities of these elements have long since decayed, with the exception of traces of plutonium-244 possibly deposited by cosmic dust. Natural deposits of plutonium and neptunium are produced by neutron capture in uranium ore.

The Earth contains approximately 1.33×10^{50} atoms. Although small numbers of independent atoms of noble gases exist, such as argon, neon, and helium, 99% of the atmosphere is bound in the form of molecules, including carbon dioxide and diatomic oxygen and nitrogen. At the surface of the Earth, an overwhelming majority of atoms combine to form various compounds, including water, salt, silicates and oxides. Atoms can also combine to create materials that do not consist of discrete molecules, including crystals and liquid or solid metals. This atomic matter forms networked arrangements that lack the particular type of small-scale interrupted order associated with molecular matter.

Rare and Theoretical Forms

Superheavy Elements

While isotopes with atomic numbers higher than lead (82) are known to be radioactive, an "island of stability" has been proposed for some elements with atomic numbers above 103. These superheavy elements may have a nucleus that is relatively stable against radioactive decay. The most likely candidate for a stable superheavy atom, unbihexium, has 126 protons and 184 neutrons.

Exotic Matter

Each particle of matter has a corresponding antimatter particle with the opposite electrical charge. Thus, the positron is a positively charged antielectron and the antiproton is a negatively charged equivalent of a proton. When a matter and corresponding antimatter particle meet, they annihilate each other. Because of this, along with an imbalance between the number of matter and antimatter particles, the latter are rare in the universe. The first causes of this imbalance are not yet fully understood, although theories of baryogenesis may offer an explanation. As a result, no antimatter atoms have been discovered in nature. However, in 1996 the antimatter counterpart of the hydrogen atom (antihydrogen) was synthesized at the CERN laboratory in Geneva.

Other exotic atoms have been created by replacing one of the protons, neutrons or electrons with other particles that have the same charge. For example, an electron can be replaced by a more massive muon, forming a muonic atom. These types of atoms can be used to test the fundamental predictions of physics.

Atomic Number

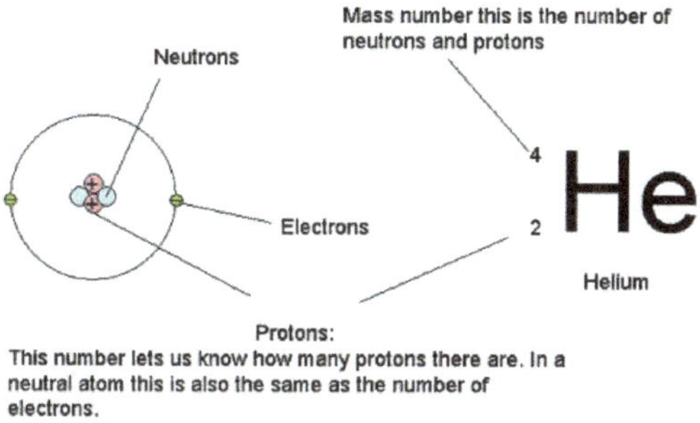

An explanation of the superscripts and subscripts seen in atomic number notation. Atomic number is the number of protons, and therefore also the total positive charge, in the atomic nucleus.

The Rutherford–Bohr model of the hydrogen atom ($Z = 1$) or a hydrogen-like ion ($Z > 1$). In this model it is an essential feature that the photon energy (or frequency) of the electromagnetic radiation emitted (shown) when an electron jumps from one orbital to another, be proportional to the mathematical square of atomic charge (Z^2). Experimental measurement by Henry Moseley of this radiation for many elements (from $Z = 13$ to 92) showed the results as predicted by Bohr. Both the concept of atomic number and the Bohr model were thereby given scientific credence.

In chemistry and physics, the atomic number of a chemical element (also known as its proton number) is the number of protons found in the nucleus of an atom of that element, and therefore identical to the charge number of the nucleus. It is conventionally represented by the symbol ***Z***. The atomic number uniquely identifies a chemical element. In an uncharged atom, the atomic number is also equal to the number of electrons.

The atomic number, Z, should not be confused with the mass number, A, which is the number of nucleons, the total number of protons and neutrons in the nucleus of an atom. The number of neutrons, N, is known as the neutron number of the atom; thus, $A = Z + N$ (these quantities are always whole numbers). Since protons and neutrons have approximately the same mass (and the mass of the electrons is negligible for many purposes) and the mass defect of nucleon binding is always small compared to the nucleon mass, the atomic mass of any atom, when expressed in unified atomic mass units (making a quantity called the "relative isotopic mass"), is roughly (to within 1%) equal to the whole number A.

Atoms with the same atomic number Z but different neutron numbers N, and hence different atomic masses, are known as isotopes. A little more than three-quarters of naturally occurring elements exist as a mixture of isotopes (see monoisotopic elements), and the average isotopic mass of an isotopic mixture for an element (called the relative atomic mass) in a defined environment on Earth, determines the element's standard atomic weight. Historically, it was these atomic weights of elements (in comparison to hydrogen) that were the quantities measurable by chemists in the 19th century.

The conventional symbol *Z* comes from the German word ***Zahl*** meaning number/numeral/figure, which, prior to the modern synthesis of ideas from chemistry and physics, merely denoted an element's numerical place in the periodic table, whose order is approximately, but not completely, consistent with the order of the elements by atomic weights. Only after 1915, with the suggestion and evidence that this *Z* number was also the nuclear charge and a physical characteristic of atoms, did the word *Atomzahl* (and its English equivalent *atomic number*) come into common use in this context.

History

The Periodic Table and a Natural Number for Each Element

Russian chemist Dmitri Mendeleev created a periodic table of the elements that ordered them numerically by atomic weight, yet occasionally used chemical properties in contradiction to weight.

Loosely speaking, the existence or construction of a periodic table of elements creates an ordering of the elements, and so they can be numbered in order.

Dmitri Mendeleev claimed that he arranged his first periodic tables in order of atomic weight ("Atomgewicht"). However, in consideration of the elements' observed chemical properties, he changed the order slightly and placed tellurium (atomic weight 127.6) ahead of iodine (atomic weight 126.9). This placement is consistent with the modern practice of ordering the elements by proton number, *Z*, but that number was not known or suspected at the time.

A simple numbering based on periodic table position was never entirely satisfactory, however. Besides the case of iodine and tellurium, later several other pairs of elements (such as argon and potassium, cobalt and nickel) were known to have nearly identical or reversed atomic weights, thus requiring their placement in the periodic table to be determined by their chemical properties. However the gradual identification of more and more chemically similar lanthanide elements, whose atomic number was not obvious, led to inconsistency and uncertainty in the periodic numbering of elements at least from lutetium (element 71) onwards (hafnium was not known at this time).

Niels Bohr's 1913 Bohr model of the atom required van den Broek's atomic number of nuclear charges, and Bohr believed that Moseley's work contributed greatly to the acceptance of the model.

Henry Moseley helped develop the concept of atomic number by showing experimentally (1913) that Van den Broek's 1911 hypothesis combined with the Bohr model nearly correctly predicted atomic X-ray emissions.

The Rutherford-Bohr Model and Van Den Broek

In 1911, Ernest Rutherford gave a model of the atom in which a central core held most of the atom's mass and a positive charge which, in units of the electron's charge, was to be approximately equal to half of the atom's atomic weight, expressed in numbers of hydrogen atoms. This central charge would thus be approximately half the atomic weight (though it was almost 25% different from the atomic number of gold ($Z = 79, A = 197$), the single element from which Rutherford made his guess). Nevertheless, in spite of Rutherford's estimation that gold had a central charge of about 100 (but was element Z = 79 on the periodic table), a month after Rutherford's paper appeared, Antonius van den Broek first formally suggested that the central charge and number of electrons in an atom was *exactly* equal to its place in the periodic table (also known as element number, atomic number, and symbolized Z). This proved eventually to be the case.

Moseley's 1913 Experiment

The experimental position improved dramatically after research by Henry Moseley in 1913. Moseley, after discussions with Bohr who was at the same lab (and who had used Van den Broek's hypothesis in his Bohr model of the atom), decided to test Van den Broek's and Bohr's hypothesis directly, by seeing if spectral lines emitted from excited atoms fitted the Bohr theory's postulation that the frequency of the spectral lines be proportional to the square of Z.

To do this, Moseley measured the wavelengths of the innermost photon transitions (K and L lines) produced by the elements from aluminum ($Z = 13$) to gold ($Z = 79$) used as a series of movable anodic targets inside an x-ray tube. The square root of the frequency of these photons (x-rays) increased from one target to the next in an arithmetic progression. This led to the conclusion (Moseley's law) that the atomic number does closely correspond (with an offset of one unit for K-lines, in Moseley's work) to the calculated electric charge of the nucleus, i.e. the element number Z. Among other things, Moseley demonstrated that the lanthanide series (from lanthanum to lutetium inclusive) must have 15 members—no fewer and no more—which was far from obvious from the chemistry at that time.

The Proton and the Idea of Nuclear Electrons

In 1915 the reason for nuclear charge being quantized in units of Z, which were now recognized to be the same as the element number, was not understood. An old idea called Prout's hypothesis had postulated that the elements were all made of residues (or "protyles") of the lightest element hydrogen, which in the Bohr-Rutherford model had a single electron and a nuclear charge of one. However, as early as 1907 Rutherford and Thomas Royds had shown that alpha particles, which had a charge of +2, were the nuclei of helium atoms, which had a mass four times that of hydrogen, not two times. If Prout's hypothesis were true, something had to be neutralizing some of the charge of the hydrogen nuclei present in the nuclei of heavier atoms.

In 1917 Rutherford succeeded in generating hydrogen nuclei from a nuclear reaction between alpha particles and nitrogen gas, and believed he had proven Prout's law. He called the new heavy nuclear particles protons in 1920 (alternate names being proutons and protyles). It had been immediately apparent from the work of Moseley that the nuclei of heavy atoms have more than twice as much mass as would be expected from their being made of hydrogen nuclei, and thus there was required a hypothesis for the neutralization of the extra protons presumed present in all heavy nuclei. A helium nucleus was presumed to be composed of four protons plus two "nuclear electrons" (electrons bound inside the nucleus) to cancel two of the charges. At the other end of the periodic table, a nucleus of gold with a mass 197 times that of hydrogen, was thought to contain 118 nuclear electrons in the nucleus to give it a residual charge of + 79, consistent with its atomic number.

The Discovery of the Neutron Makes Z the Proton Number

All consideration of nuclear electrons ended with James Chadwick's discovery of the neutron in 1932. An atom of gold now was seen as containing 118 neutrons rather than 118 nuclear electrons, and its positive charge now was realized to come entirely from a content of 79 protons. After 1932, therefore, an element's atomic number Z was also realized to be identical to the proton number of its nuclei.

The Symbol of Z

The conventional symbol *Z* possibly comes from the German word *Atomzahl* (atomic number). However, prior to 1915, the word *Zahl* (simply *number*) was used for an element's assigned number in the periodic table.

Chemical Properties

Each element has a specific set of chemical properties as a consequence of the number of electrons present in the neutral atom, which is *Z* (the atomic number). The configuration of these electrons follows from the principles of quantum mechanics. The number of electrons in each element's electron shells, particularly the outermost valence shell, is the primary factor in determining its chemical bonding behavior. Hence, it is the atomic number alone that determines the chemical properties of an element; and it is for this reason that an element can be defined as consisting of *any* mixture of atoms with a given atomic number.

New Elements

The quest for new elements is usually described using atomic numbers. As of 2010, elements with atomic numbers 1 to 118 have been observed. Synthesis of new elements is accomplished by bombarding target atoms of heavy elements with ions, such that the sum of the atomic numbers of the target and ion elements equals the atomic number of the element being created. In general, the half-life becomes shorter as atomic number increases, though an "island of stability" may exist for undiscovered isotopes with certain numbers of protons and neutrons.

Subatomic Particle

In the physical sciences, subatomic particles are particles much smaller than atoms.

There are two types of subatomic particles: elementary particles, which according to current theories are not made of other particles; and *composite* particles. Particle physics and nuclear physics study these particles and how they interact.

In particle physics, the concept of a particle is one of several concepts inherited from classical physics. But it also reflects the modern understanding that at the quantum scale matter and energy behave very differently from what much of everyday experience would lead us to expect.

The idea of a particle underwent serious rethinking when experiments showed that light could behave like a stream of particles (called photons) as well as exhibit wave-like properties. This led to the new concept of wave–particle duality to reflect that quantum-scale "particles" behave like both particles and waves (also known as wavicles). Another new concept, the uncertainty principle, states that some of their properties taken together, such as their simultaneous position and momentum, cannot be measured exactly. In more recent times, wave–particle duality has been shown to apply not only to photons but to increasingly massive particles as well.

Interactions of particles in the framework of quantum field theory are understood as creation and annihilation of *quanta* of corresponding fundamental interactions. This blends particle physics with field theory.

Classification

By Statistics

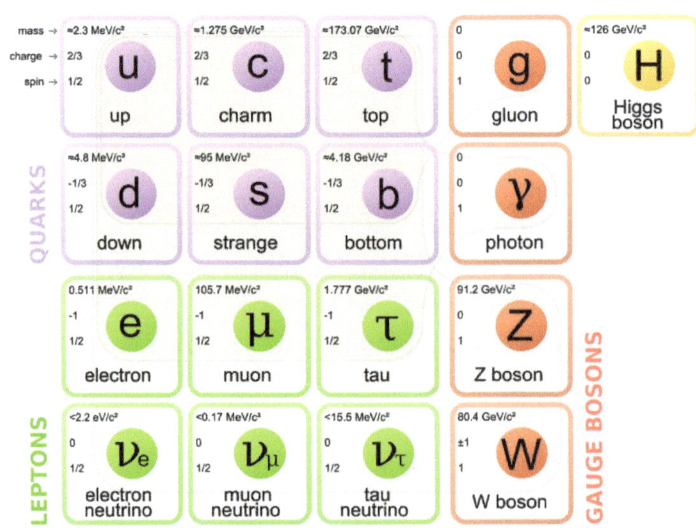

The Standard Model classification of particles

Any subatomic particle, like any particle in the 3-dimensional space that obeys laws of quantum mechanics, can be either a boson (an integer spin) or a fermion (a half-integer spin).

By Composition

The elementary particles of the Standard Model include:

- Six "flavors" of quarks: up, down, bottom, top, strange, and charm;
- Six types of leptons: electron, electron neutrino, muon, muon neutrino, tau, tau neutrino;
- Twelve gauge bosons (force carriers): the photon of electromagnetism, the three W and Z bosons of the weak force, and the eight gluons of the strong force;
- The Higgs boson.

Various extensions of the Standard Model predict the existence of an elementary graviton particle and many other elementary particles.

Composite subatomic particles (such as protons or atomic nuclei) are bound states of two or more elementary particles. For example, a proton is made of two up quarks and one down quark, while the atomic nucleus of helium-4 is composed of two protons and two neutrons. The neutron is made of two down quarks and one up quark. Composite particles include all hadrons: these include baryons (such as protons and neutrons) and mesons (such as pions and kaons).

By mass

In special relativity, the energy of a particle at rest equals its mass times the speed of light squared, $E = mc^2$. That is, mass can be expressed in terms of energy and vice versa. If a particle has a frame of reference where it lies at rest, then it has a positive rest mass and is referred to as *massive*.

All composite particles are massive. Baryons (meaning "heavy") tend to have greater mass than mesons (meaning "intermediate"), which in turn tend to be heavier than leptons (meaning "lightweight"), but the heaviest lepton (the tau particle) is heavier than the two lightest flavours of baryons (nucleons). It is also certain that any particle with an electric charge is massive.

All massless particles (particles whose invariant mass is zero) are elementary. These include the photon and gluon, although the latter cannot be isolated.

Other Properties

Through the work of Albert Einstein, Louis de Broglie, and many others, current scientific theory holds that *all* particles also have a wave nature. This has been verified not only for elementary particles but also for compound particles like atoms and even molecules. In fact, according to traditional formulations of non-relativistic quantum mechanics, wave–particle duality applies to all objects, even macroscopic ones; although the wave properties of macroscopic objects cannot be detected due to their small wavelengths.

Interactions between particles have been scrutinized for many centuries, and a few simple laws underpin how particles behave in collisions and interactions. The most fundamental of these are the laws of conservation of energy and conservation of momentum, which let us make calculations of particle interactions on scales of magnitude that range from stars to quarks. These are the prerequisite basics of Newtonian mechanics, a series of statements and equations in *Philosophiae Naturalis Principia Mathematica*, originally published in 1687.

Dividing an Atom

The negatively charged electron has a mass equal to $^1/_{1837 \text{ or } 1836}$ of that of a hydrogen atom. The remainder of the hydrogen atom's mass comes from the positively charged proton. The atomic number of an element is the number of protons in its nucleus. Neutrons are neutral particles having a mass slightly greater than that of the proton. Different isotopes of the same element contain the same number of protons but differing numbers of neutrons. The mass number of an isotope is the total number of nucleons (neutrons and protons collectively).

Chemistry concerns itself with how electron sharing binds atoms into structures such as crystals and molecules. Nuclear physics deals with how protons and neutrons arrange themselves in nuclei. The study of subatomic particles, atoms and molecules, and their structure and interactions, requires quantum mechanics. Analyzing processes that change the numbers and types of particles requires quantum field theory. The study of subatomic particles *per se* is called particle physics. The term *high-energy physics* is nearly synonymous to "particle physics" since creation of particles requires high energies: it occurs only as a result of cosmic rays, or in particle accelerators. Particle phenomenology systematizes the knowledge about subatomic particles obtained from these experiments. Total known Isotopes of Hydrogen - Deuterium and tritium - 2

History

The term "*subatomic* particle" is largely a retronym of 1960s made to distinguish a big number of baryons and mesons (that comprise hadrons) from particles that are now thought to be truly elementary. Before that hadrons were usually classified as "elementary" because their composition was unknown.

Atomic Orbital

In quantum mechanics, an atomic orbital is a mathematical function that describes the wave-like behavior of either one electron or a pair of electrons in an atom. This function can be used to calculate the probability of finding any electron of an atom in any specific region around the atom's nucleus. The term, atomic orbital, may also refer to the physical region or space where the electron can be calculated to be present, as defined by the particular mathematical form of the orbital.

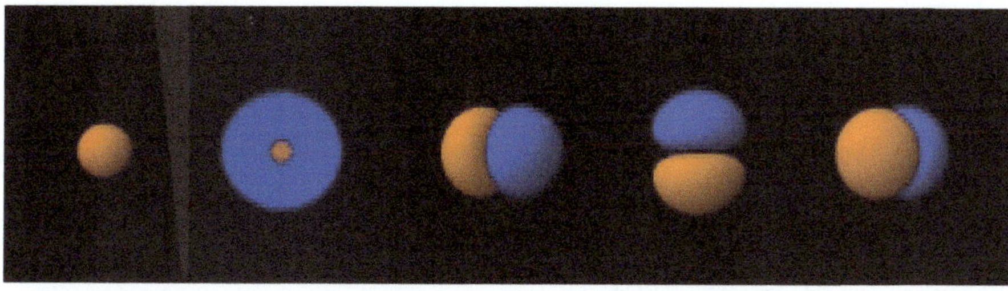

The shapes of the first five atomic orbitals are: 1s, 2s, 2px, 2py, and 2pz. The two colors show the phase or sign of the wave function in each region. These are graphs of $\psi(x, y, z)$ functions which depend on the coordinates of one electron. To see the elongated shape of $\psi(x, y, z)2$ functions that show probability density more directly, see the graphs of d-orbitals below.

Each orbital in an atom is characterized by a unique set of values of the three quantum numbers n, ℓ, and m, which respectively correspond to the electron's energy, angular momentum, and an angular momentum vector component (the magnetic quantum number). Each such orbital can be occupied by a maximum of two electrons, each with its own spin quantum number s. The simple names s orbital, p orbital, d orbital and f orbital refer to orbitals with angular momentum quantum number $\ell = 0, 1, 2$ and 3 respectively. These names, together with the value of n, are used to describe the electron configurations of atoms. They are derived from the description by early spectroscopists of certain series of alkali metal spectroscopic lines as sharp, principal, diffuse, and fundamental. Orbitals for $\ell > 3$ continue alphabetically, omitting j (g, h, i, k, ...) because some languages do not distinguish between the letters "i" and "j":

Atomic orbitals are the basic building blocks of the atomic orbital model (alternatively known as the electron cloud or wave mechanics model), a modern framework for visualizing the submicroscopic behavior of electrons in matter. In this model the electron cloud of a multi-electron atom may be seen as being built up (in approximation) in an electron configuration that is a product of simpler hydrogen-like atomic orbitals. The repeating *periodicity* of the blocks of 2, 6, 10, and 14 elements within sections of the periodic table arises naturally from the total number of electrons that occupy a complete set of s, p, d and f atomic orbitals, respectively, although for higher values of the quantum number n, particularly when the atom in question bears a positive charge, the energies of certain sub-shells become very similar and so the order in which they are said to be populated by electrons (e.g. Cr = [Ar]$4s^1 3d^5$ and Cr^{2+} = [Ar]$3d^4$) can only be rationalized somewhat arbitrarily.

Electron Properties

With the development of quantum mechanics and experimental findings (such as the two slits diffraction of electrons), it was found that the orbiting electrons around a nucleus could not be fully described as particles, but needed to be explained by the wave-particle duality. In this sense, the electrons have the following properties:

Wave-like properties:

1. The electrons do not orbit the nucleus in the manner of a planet orbiting the sun, but instead exist as standing waves. The lowest possible energy an electron can take is therefore analogous to the fundamental frequency of a wave on a string. Higher energy states are then similar to harmonics of the fundamental frequency.

2. The electrons are never in a single point location, although the probability of interacting with the electron at a single point can be found from the wave function of the electron.

Particle-like properties:

1. There is always an integer number of electrons orbiting the nucleus.

2. Electrons jump between orbitals in a particle-like fashion. For example, if a single photon strikes the electrons, only a single electron changes states in response to the photon.

3. The electrons retain particle-like properties such as: each wave state has the same electrical charge as the electron particle. Each wave state has a single discrete spin (spin up or spin down). This can depend upon its superposition.

Thus, despite the popular analogy to planets revolving around the Sun, electrons cannot be described simply as solid particles. In addition, atomic orbitals do not closely resemble a planet's elliptical path in ordinary atoms. A more accurate analogy might be that of a large and often oddly shaped "atmosphere" (the electron), distributed around a relatively tiny planet (the atomic nucleus). Atomic orbitals exactly describe the shape of this "atmosphere" only when a single electron is present in an atom. When more electrons are added to a single atom, the additional electrons tend to more evenly fill in a volume of space around the nucleus so that the resulting collection (sometimes termed the atom's "electron cloud") tends toward a generally spherical zone of probability describing where the atom's electrons will be found. The is due to the uncertainty principle.

Formal Quantum Mechanical Definition

Atomic orbitals may be defined more precisely in formal quantum mechanical language. Specifically, in quantum mechanics, the state of an atom, i.e., an eigenstate of the atomic Hamiltonian, is approximated by an expansion (see configuration interaction expansion and basis set) into linear combinations of anti-symmetrized products (Slater determinants) of one-electron functions. The spatial components of these one-electron functions are called atomic orbitals. (When one considers also their spin component, one speaks of atomic spin orbitals.) A state is actually a function of the coordinates of all the electrons, so that their motion is correlated, but this is often approximated by this independent-particle model of products of single electron wave functions. (The London dispersion force, for example, depends on the correlations of the motion of the electrons.)

In atomic physics, the atomic spectral lines correspond to transitions (quantum leaps) between quantum states of an atom. These states are labeled by a set of quantum numbers summarized in the term symbol and usually associated with particular electron configurations, i.e., by occupation schemes of atomic orbitals (for example, $1s^2\ 2s^2\ 2p^6$ for the ground state of neon—term symbol: $^1S_0$).

This notation means that the corresponding Slater determinants have a clear higher weight in the configuration interaction expansion. The atomic orbital concept is therefore a key concept for visualizing the excitation process associated with a given transition. For example, one can say for a given transition that it corresponds to the excitation of an electron from an occupied orbital to a given unoccupied orbital. Nevertheless, one has to keep in mind that electrons are fermions ruled by the Pauli exclusion principle and cannot be distinguished from the other electrons in the atom. Moreover, it sometimes happens that the configuration interaction expansion converges very slowly and that one cannot speak about simple one-determinant wave function at all. This is the case when electron correlation is large.

Fundamentally, an atomic orbital is a one-electron wave function, even though most electrons do not exist in one-electron atoms, and so the one-electron view is an approximation. When thinking about orbitals, we are often given an orbital vision which (even if it is not spelled out) is heavily influenced by this Hartree–Fock approximation, which is one way to reduce the complexities of molecular orbital theory.

Types of Orbitals

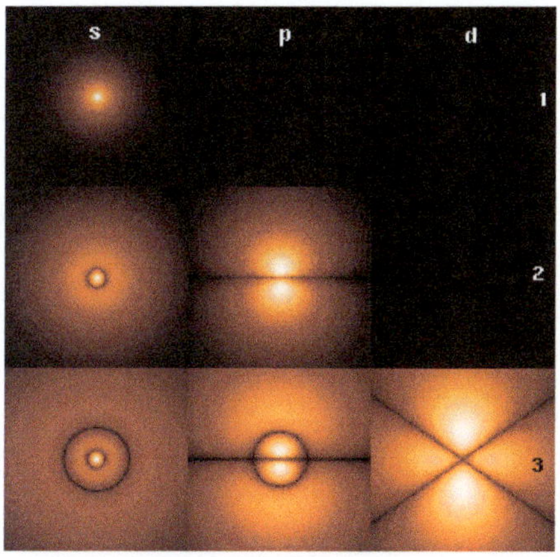

False-color density images of some hydrogen-like atomic orbitals (f orbitals and higher are not shown)

Atomic orbitals can be the hydrogen-like "orbitals" which are exact solutions to the Schrödinger equation for a hydrogen-like "atom" (i.e., an atom with one electron). Alternatively, atomic orbitals refer to functions that depend on the coordinates of one electron (i.e., orbitals) but are used as starting points for approximating wave functions that depend on the simultaneous coordinates of all the electrons in an atom or molecule. The coordinate systems chosen for atomic orbitals are usually spherical coordinates (r, θ, φ) in atoms and cartesians (x, y, z) in polyatomic molecules. The advantage of spherical coordinates (for atoms) is that an orbital wave function is a product of three factors each dependent on a single coordinate: $\psi(r, \theta, \varphi) = R(r)\Theta(\theta)\Phi(\varphi)$. The angular factors of atomic orbitals $\Theta(\theta)\Phi(\varphi)$ generate s, p, d, etc. functions as real combinations of spherical harmonics $Y_{\ell m}(\theta, \varphi)$ (where ℓ and m are quantum numbers). There are typically three mathematical forms for the radial functions $R(r)$ which can be chosen as a starting point for the calculation of the properties of atoms and molecules with many electrons:

1. The *hydrogen-like atomic orbitals* are derived from the exact solution of the Schrödinger Equation for one electron and a nucleus, for a hydrogen-like atom. The part of the function that depends on the distance from the nucleus has nodes (radial nodes) and decays as $e^{-(\text{constant} \times \text{distance})}$.

2. The Slater-type orbital (STO) is a form without radial nodes but decays from the nucleus as does the hydrogen-like orbital.

3. The form of the Gaussian type orbital (Gaussians) has no radial nodes and decays as $e^{(-\text{distance squared})}$.

Although hydrogen-like orbitals are still used as pedagogical tools, the advent of computers has made STOs preferable for atoms and diatomic molecules since combinations of STOs can replace the nodes in hydrogen-like atomic orbital. Gaussians are typically used in molecules with three or more atoms. Although not as accurate by themselves as STOs, combinations of many Gaussians can attain the accuracy of hydrogen-like orbitals.

History

The term "orbital" was coined by Robert Mulliken in 1932 as an abbreviation for *one-electron orbital wave function*. However, the idea that electrons might revolve around a compact nucleus with definite angular momentum was convincingly argued at least 19 years earlier by Niels Bohr, and the Japanese physicist Hantaro Nagaoka published an orbit-based hypothesis for electronic behavior as early as 1904. Explaining the behavior of these electron "orbits" was one of the driving forces behind the development of quantum mechanics.

Early Models

With J.J. Thomson's discovery of the electron in 1897, it became clear that atoms were not the smallest building blocks of nature, but were rather composite particles. The newly discovered structure within atoms tempted many to imagine how the atom's constituent parts might interact with each other. Thomson theorized that multiple electrons revolved in orbit-like rings within a positively charged jelly-like substance, and between the electron's discovery and 1909, this "plum pudding model" was the most widely accepted explanation of atomic structure.

Shortly after Thomson's discovery, Hantaro Nagaoka predicted a different model for electronic structure. Unlike the plum pudding model, the positive charge in Nagaoka's "Saturnian Model" was concentrated into a central core, pulling the electrons into circular orbits reminiscent of Saturn's rings. Few people took notice of Nagaoka's work at the time, and Nagaoka himself recognized a fundamental defect in the theory even at its conception, namely that a classical charged object cannot sustain orbital motion because it is accelerating and therefore loses energy due to electromagnetic radiation. Nevertheless, the Saturnian model turned out to have more in common with modern theory than any of its contemporaries.

Bohr Atom

In 1909, Ernest Rutherford discovered that the bulk of the atomic mass was tightly condensed into a nucleus, which was also found to be positively charged. It became clear from his analysis in 1911 that the plum pudding model could not explain atomic structure. In 1913 as Rutherford's post-doctoral student, Niels Bohr proposed a new model of the atom, wherein electrons orbited the nucleus with classical periods, but were only permitted to have discrete values of angular momentum, quantized in units $h/2\pi$. This constraint automatically permitted only certain values of electron energies. The Bohr model of the atom fixed the problem of energy loss from radiation from a ground state (by declaring that there was no state below this), and more importantly explained the origin of spectral lines.

After Bohr's use of Einstein's explanation of the photoelectric effect to relate energy levels in atoms with the wavelength of emitted light, the connection between the structure of electrons in atoms and the emission and absorption spectra of atoms became an increasingly useful tool in the understanding of electrons in atoms. The most prominent feature of emission and absorption spectra (known experimentally since the middle of the 19th century), was that these atomic spectra contained discrete lines. The significance of the Bohr model was that it related the lines in emission and absorption spectra to the energy differences between the orbits that electrons could take around an atom. This was, however, *not* achieved by Bohr through giving the electrons some kind of wave-like properties, since the idea that electrons could behave as matter waves was not suggested until eleven years later. Still, the Bohr model's use of quantized angular momenta and therefore quantized energy levels was a significant step towards the understanding of electrons in atoms, and also a significant step towards the development of quantum mechanics in suggesting that quantized restraints must account for all discontinuous energy levels and spectra in atoms.

With de Broglie's suggestion of the existence of electron matter waves in 1924, and for a short time before the full 1926 Schrödinger equation treatment of hydrogen-like atom, a Bohr electron "wavelength" could be seen to be a function of its momentum, and thus a Bohr orbiting electron was seen to orbit in a circle at a multiple of its half-wavelength (this physically incorrect Bohr model is still often taught to beginning students). The Bohr model for a short time could be seen as a classical model with an additional constraint provided by the 'wavelength' argument. However, this period was immediately superseded by the full three-dimensional wave mechanics of 1926. In our current understanding of physics, the Bohr model is called a semi-classical model because of its quantization of angular momentum, not primarily because of its relationship with electron wavelength, which appeared in hindsight a dozen years after the Bohr model was proposed.

The Bohr model was able to explain the emission and absorption spectra of hydrogen. The energies of electrons in the $n = 1, 2, 3$, etc. states in the Bohr model match those of current physics. However, this did not explain similarities between different atoms, as expressed by the periodic table, such as the fact that helium (two electrons), neon (10 electrons), and argon (18 electrons) exhibit similar chemical inertness. Modern quantum mechanics explains this in terms of electron shells and subshells which can each hold a number of electrons determined by the Pauli exclusion principle. Thus the $n = 1$ state can hold one or two electrons, while the $n = 2$ state can hold up to eight electrons in 2s and 2p subshells. In helium, all $n = 1$ states are fully occupied; the same for $n = 1$ and $n = 2$ in neon. In argon the 3s and 3p subshells are similarly fully occupied by eight electrons; quantum mechanics also allows a 3d subshell but this is at higher energy than the 3s and 3p in argon (contrary to the situation in the hydrogen atom) and remains empty.

Modern Conceptions and Connections to the Heisenberg Uncertainty Principle

Immediately after Heisenberg discovered his uncertainty principle, it was noted by Bohr that the existence of any sort of wave packet implies uncertainty in the wave frequency and wavelength, since a spread of frequencies is needed to create the packet itself. In quantum mechanics, where all particle momenta are associated with waves, it is the formation of such a wave packet which localizes the wave, and thus the particle, in space. In states where a quantum mechanical particle is bound, it must be localized as a wave packet, and the existence of the packet and its minimum size implies a spread and minimal value in particle wavelength, and thus also momentum and energy. In quantum mechanics, as a particle is localized to a smaller region in space, the associated compressed wave packet requires a larger and larger range of momenta, and thus larger kinetic energy. Thus, the binding energy to contain or trap a particle in a smaller region of space, increases without bound, as the region of space grows smaller. Particles cannot be restricted to a geometric point in space, since this would require an infinite particle momentum.

In chemistry, Schrödinger, Pauling, Mulliken and others noted that the consequence of Heisenberg's relation was that the electron, as a wave packet, could not be considered to have an exact location in its orbital. Max Born suggested that the electron's position needed to be described by a probability distribution which was connected with finding the electron at some point in the wave-function which described its associated wave packet. The new quantum mechanics did not give exact results, but only the probabilities for the occurrence of a variety of possible such results. Heisenberg held that the path of a moving particle has no meaning if we cannot observe it, as we cannot with electrons in an atom.

In the quantum picture of Heisenberg, Schrödinger and others, the Bohr atom number *n* for each orbital became known as an *n-sphere* in a three dimensional atom and was pictured as the mean energy of the probability cloud of the electron's wave packet which surrounded the atom.

Orbital Names

Orbitals are given names in the form:

$$X\,\mathrm{type}^{y}$$

where X is the energy level corresponding to the principal quantum number n, type is a lower-case letter denoting the shape or subshell of the orbital and it corresponds to the angular quantum number ℓ, and y is the number of electrons in that orbital.

For example, the orbital $1s^2$ (pronounced "one ess two") has two electrons and is the lowest energy level ($n = 1$) and has an angular quantum number of $\ell = 0$. In X-ray notation, the principal quantum number is given a letter associated with it. For $n = 1, 2, 3, 4, 5, \ldots$, the letters associated with those numbers are K, L, M, N, O, ... respectively.

Hydrogen-like Orbitals

The simplest atomic orbitals are those that are calculated for systems with a single electron, such as the hydrogen atom. An atom of any other element ionized down to a single electron is very similar to hydrogen, and the orbitals take the same form. In the Schrödinger equation for this system of one negative and one positive particle, the atomic orbitals are the eigenstates of the Hamiltonian operator for the energy. They can be obtained analytically, meaning that the resulting orbitals are products of a polynomial series, and exponential and trigonometric functions. (see hydrogen atom).

For atoms with two or more electrons, the governing equations can only be solved with the use of methods of iterative approximation. Orbitals of multi-electron atoms are *qualitatively* similar to those of hydrogen, and in the simplest models, they are taken to have the same form. For more rigorous and precise analysis, the numerical approximations must be used.

A given (hydrogen-like) atomic orbital is identified by unique values of three quantum numbers: n, ℓ, and m_ℓ. The rules restricting the values of the quantum numbers, and their energies (see below), explain the electron configuration of the atoms and the periodic table.

The stationary states (quantum states) of the hydrogen-like atoms are its atomic orbitals. However, in general, an electron's behavior is not fully described by a single orbital. Electron states are best represented by time-depending "mixtures" (linear combinations) of multiple orbitals. See Linear combination of atomic orbitals molecular orbital method.

The quantum number n first appeared in the Bohr model where it determines the radius of each circular electron orbit. In modern quantum mechanics however, n determines the mean distance of the electron from the nucleus; all electrons with the same value of n lie at the same average distance. For this reason, orbitals with the same value of n are said to comprise a "shell". Orbitals with the same value of n and also the same value of ℓ are even more closely related, and are said to comprise a "subshell".

Quantum Numbers

Because of the quantum mechanical nature of the electrons around a nucleus, atomic orbitals can be uniquely defined by a set of integers known as quantum numbers. These quantum numbers only occur in certain combinations of values, and their physical interpretation changes depending on whether real or complex versions of the atomic orbitals are employed.

Complex Orbitals

In physics, the most common orbital descriptions are based on the solutions to the hydrogen atom, where orbitals are given by the product between a radial function and a pure spherical harmonic. The quantum numbers, together with the rules governing their possible values, are as follows:

The principal quantum number n describes the energy of the electron and is always a positive integer. In fact, it can be any positive integer, but for reasons discussed below, large numbers are seldom encountered. Each atom has, in general, many orbitals associated with each value of n; these orbitals together are sometimes called *electron shells.*

The azimuthal quantum number ℓ describes the orbital angular momentum of each electron and is a non-negative integer. Within a shell where n is some integer n_0, ℓ ranges across all (integer) values satisfying the relation $0 \le \ell \le n_0 - 1$. For instance, the $n = 1$ shell has only orbitals with $\ell = 0$, and the $n = 2$ shell has only orbitals with $\ell = 0$, and $\ell = 1$. The set of orbitals associated with a particular value of ℓ are sometimes collectively called a *subshell.*

The magnetic quantum number, m_ℓ, describes the magnetic moment of an electron in an arbitrary direction, and is also always an integer. Within a subshell where ℓ is some integer ℓ_0, m_ℓ ranges thus: $-\ell_0 \le m_\ell \le \ell_0$.

The above results may be summarized in the following table. Each cell represents a subshell, and lists the values of m_ℓ available in that subshell. Empty cells represent subshells that do not exist.

Subshells are usually identified by their n- and ℓ-values. n is represented by its numerical value, but ℓ is represented by a letter as follows: 0 is represented by 's', 1 by 'p', 2 by 'd', 3 by 'f', and 4 by 'g'. For instance, one may speak of the subshell with $n = 2$ and $\ell = 0$ as a '2s subshell'.

Each electron also has a spin quantum number, s, which describes the spin of each electron (spin up or spin down). The number s can be +1/2 or −1/2.

The Pauli exclusion principle states that no two electrons can occupy the same quantum state: every electron in an atom must have a unique combination of quantum numbers.

The above conventions imply a preferred axis (for example, the *z* direction in Cartesian coordinates), and they also imply a preferred direction along this preferred axis. Otherwise there would be no sense in distinguishing $m = +1$ from $m = -1$. As such, the model is most useful when applied to physical systems that share these symmetries. The Stern–Gerlach experiment — where an atom is exposed to a magnetic field — provides one such example.

Real Orbitals

An atom that is embedded in a crystalline solid feels multiple preferred axes, but no preferred direction. Instead of building atomic orbitals out of the product of radial functions and a single spherical harmonic, linear combinations of spherical harmonics are typically used, designed so that the imaginary part of the spherical harmonics cancel out. These real orbitals are the building blocks most commonly shown in orbital visualizations.

In the real hydrogen-like orbitals, for example, n and ℓ have the same interpretation and significance as their complex counterparts, but m is no longer a good quantum number (though its absolute value is). The orbitals are given new names based on their shape with respect to a standardized Cartesian basis. The real hydrogen-like p orbitals are given by the following

$$p_z = p_0$$

$$p_x = \frac{1}{\sqrt{2}}\left(p_1 + p_{-1}\right)$$

$$p_y = \frac{1}{i\sqrt{2}}(p_1 - p_{-1})$$

where $p_0 = R_{n1}Y_{10}$, $p_1 = R_{n1}Y_{11}$, and $p_{-1} = R_{n1}Y_{1-1}$, are the complex orbitals corresponding to $\ell = 1$.

The equations for the p_x and p_y orbitals depend on the phase convention used for the spherical harmonics. The above equations suppose that the spherical harmonics are defined by $Y_\ell^m(\theta,\varphi) = Ne^{im\varphi}P_\ell^m(\cos\theta)$.. However some quantum physicists include a phase factor $(-1)^m$ in these definitions, which has the effect of relating the p_x orbital to a *difference* of spherical harmonics and the p_y orbital to the corresponding *sum*. (For more detail, see Spherical harmonics#Conventions).

Shapes of Orbitals

Cross-section of computed hydrogen atom orbital ($\psi(r, \theta, \varphi)2$) for the 6s ($n = 6$, $\ell = 0$, $m = 0$) orbital. Note that s orbitals, though spherically symmetrical, have radially placed wave-nodes for $n > 1$. However, only s orbitals invariably have a center anti-node; the other types never do.

Simple pictures showing orbital shapes are intended to describe the angular forms of regions in space where the electrons occupying the orbital are likely to be found. The diagrams cannot, however, show the entire region where an electron can be found, since according to quantum mechanics there is a non-zero probability of finding the electron (almost) anywhere in space. Instead the diagrams are approximate representations of boundary or contour surfaces where the probability density $|\psi(r,\theta,\varphi)|^2$ has a constant value, chosen so that there is a certain probability (for example 90%) of finding the electron within the contour. Although $|\psi|^2$ as the square of an absolute value is everywhere non-negative, the sign of the wave function $\psi(r,\theta,\varphi)$ is often indicated in each subregion of the orbital picture.

Sometimes the ψ function will be graphed to show its phases, rather than the $|\psi(r,\theta,\varphi)|^2$ which shows probability density but has no phases (which have been lost in the process of taking the absolute value, since $\psi(r,\theta,\varphi)$ is a complex number). $|\psi(r,\theta,\varphi)|^2$ orbital graphs tend to have less spherical, thinner lobes than $\psi(r,\theta,\varphi)$ graphs, but have the same number of lobes in the same places, and otherwise are recognizable. This article, in order to show wave function phases, shows mostly $\psi(r,\theta,\varphi)$ graphs.

The lobes can be viewed as standing wave interference patterns between the two counter rotating, ring resonant travelling wave "m" and "$-m$" modes, with the projection of the orbital onto the xy plane having a resonant "m" wavelengths around the circumference. Though rarely depicted the travelling wave solutions can be viewed as rotating banded tori, with the bands representing phase information. For each m there are two standing wave solutions $\langle m\rangle+\langle -m\rangle$ and $\langle m\rangle-\langle -m\rangle$. For the case where $m = 0$ the orbital is vertical, counter rotating information is unknown, and the orbital is z-axis symmetric. For the case where $\ell = 0$ there are no counter rotating modes. There are only radial modes and the shape is spherically symmetric. For any given n, the smaller ℓ is, the more radial nodes there are. Loosely speaking n is energy, ℓ is analogous to eccentricity, and m is orientation. For the record, in the classical case, a ring resonant travelling wave, for example in a circular transmission line, unless actively forced, will spontaneously decay into a ring resonant standing wave because reflections will build up over time at even the smallest imperfection or discontinuity.

Generally speaking, the number n determines the size and energy of the orbital for a given nucleus: as n increases, the size of the orbital increases. However, in comparing different elements, the higher nuclear charge Z of heavier elements causes their orbitals to contract by comparison to lighter ones, so that the overall size of the whole atom remains very roughly constant, even as the number of electrons in heavier elements (higher Z) increases.

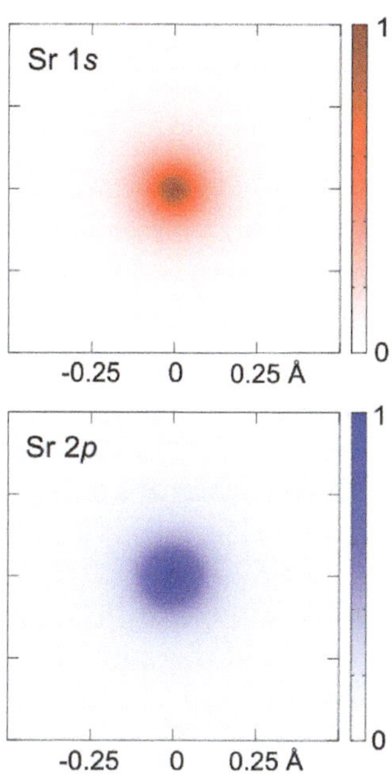

Experimentally imaged 1s and 2p core-electron orbitals of Sr, including the effects of atomic thermal vibrations and excitation broadening, retrieved from energy dispersive x-ray spectroscopy (EDX) in scanning transmission electron microscopy (STEM).

Also in general terms, ℓ determines an orbital's shape, and m_ℓ its orientation. However, since some orbitals are described by equations in complex numbers, the shape sometimes depends on m_ℓ also. Together, the whole set of orbitals for a given ℓ and n fill space as symmetrically as possible, though with increasingly complex sets of lobes and nodes.

The single s-orbitals ($\ell = 0$) are shaped like spheres. For $n = 1$ it is roughly a solid ball (it is most dense at the center and fades exponentially outwardly), but for $n = 2$ or more, each single s-orbital is composed of spherically symmetric surfaces which are nested shells (i.e., the "wave-structure" is radial, following a sinusoidal radial component as well). See illustration of a cross-section of these nested shells, at right. The s-orbitals for all n numbers are the only orbitals with an anti-node (a region of high wave function density) at the center of the nucleus. All other orbitals (p, d, f, etc.) have angular momentum, and thus avoid the nucleus (having a wave node *at* the nucleus). Recently, there has been an effort to experimentally image the 1*s* and 2*p* orbitials in a $SrTiO_3$ crystal using scanning transmission electron microscopy with energy dispersive x-ray spectroscopy. Because the imaging was conducted using an electron beam, Coulombic beam-orbital interaction that is often termed as the impact parameter effect is included in the final outcome (see the figure at right).

The shapes of p, d and f-orbitals are described verbally here and shown graphically in the *Orbitals table* below. The three p-orbitals for $n = 2$ have the form of two ellipsoids with a point of tangency at the nucleus (the two-lobed shape is sometimes referred to as a "dumbbell"—there are two lobes pointing in opposite directions from each other). The three p-orbitals in each shell are oriented at right angles to each other, as determined by their respective linear combination of values of m_ℓ. The overall result is a lobe pointing along each direction of the primary axes.

Four of the five d-orbitals for $n = 3$ look similar, each with four pear-shaped lobes, each lobe tangent at right angles to two others, and the centers of all four lying in one plane. Three of these planes are the xy-, xz-, and yz-planes—the lobes are between the pairs of primary axes—and the fourth has the centres along the x and y axes themselves. The fifth and final d-orbital consists of three regions of high probability density: a torus with two pear-shaped regions placed symmetrically on its z axis. The overall total of 18 directional lobes point in every primary axis direction and between every pair.

There are seven f-orbitals, each with shapes more complex than those of the d-orbitals.

Additionally, as is the case with the s orbitals, individual p, d, f and g orbitals with n values higher than the lowest possible value, exhibit an additional radial node structure which is reminiscent of harmonic waves of the same type, as compared with the lowest (or fundamental) mode of the wave. As with s orbitals, this phenomenon provides p, d, f, and g orbitals at the next higher possible value of n (for example, 3p orbitals vs. the fundamental 2p), an additional node in each lobe. Still higher values of n further increase the number of radial nodes, for each type of orbital.

The shapes of atomic orbitals in one-electron atom are related to 3-dimensional spherical harmonics. These shapes are not unique, and any linear combination is valid, like a transformation to cubic harmonics, in fact it is possible to generate sets where all the d's are the same shape, just like the p_x, p_y, and p_z are the same shape.

Orbitals Table

This table shows all orbital configurations for the real hydrogen-like wave functions up to 7s, and therefore covers the simple electronic configuration for all elements in the periodic table up to radium. "ψ" graphs are shown with – and + wave function phases shown in two different colors (arbitrarily red and blue). The p_z orbital is the same as the p_0 orbital, but the p_x and p_y are formed by taking linear combinations of the p_{+1} and p_{-1} orbitals (which is why they are listed under the $m = \pm 1$ label). Also, the p_{+1} and p_{-1} are not the same shape as the p_0, since they are pure spherical harmonics.

Qualitative Understanding of Shapes

The shapes of atomic orbitals can be understood qualitatively by considering the analogous case of standing waves on a circular drum. To see the analogy, the mean vibrational displacement of each bit of drum membrane from the equilibrium point over many cycles (a measure of average drum membrane velocity and momentum at that point) must be considered relative to that point's distance from the center of the drum head. If this displacement is taken as being analogous to the probability of finding an electron at a given distance from the nucleus, then it will be seen that the many modes of the vibrating disk form patterns that trace the various shapes of atomic orbitals. The basic reason for this correspondence lies in the fact that the distribution of kinetic energy and momentum in a matter-wave is predictive of where the particle associated with the wave will be. That is, the probability of finding an electron at a given place is also a function of the electron's average momentum at that point, since high electron momentum at a given position tends to "localize" the electron in that position, via the properties of electron wave-packets (see the Heisenberg uncertainty principle for details of the mechanism).

This relationship means that certain key features can be observed in both drum membrane modes and atomic orbitals. For example, in all of the modes analogous to s orbitals (the top row in the animated illustration below), it can be seen that the very center of the drum membrane vibrates most strongly, corresponding to the antinode in all s orbitals in an atom. This antinode means the electron is most likely to be at the physical position of the nucleus (which it passes straight through without scattering or striking it), since it is moving (on average) most rapidly at that point, giving it maximal momentum.

A mental "planetary orbit" picture closest to the behavior of electrons in s orbitals, all of which have no angular momentum, might perhaps be that of a Keplerian orbit with the orbital eccentricity of 1 but a finite major axis, not physically possible (because particles were to collide), but can be imagined as a limit of orbits with equal major axes but increasing eccentricity.

Below, a number of drum membrane vibration modes and the respective wave functions of the hydrogen atom are shown. A correspondence can be considered where the wave functions of a vibrating drum head are for a two-coordinate system $\psi(r, \theta)$ and the wave functions for a vibrating sphere are three-coordinate $\psi(r, \theta, \varphi)$.

s-type drum modes and wave functions		
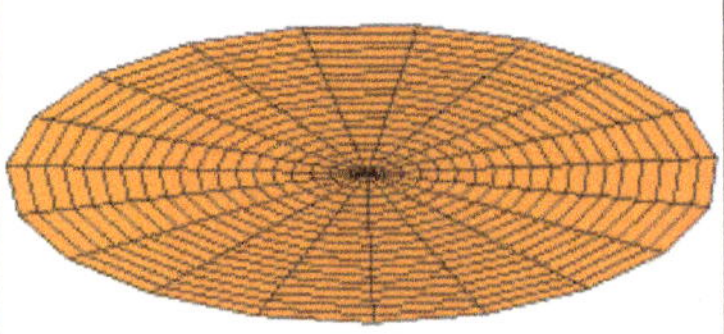 Drum mode	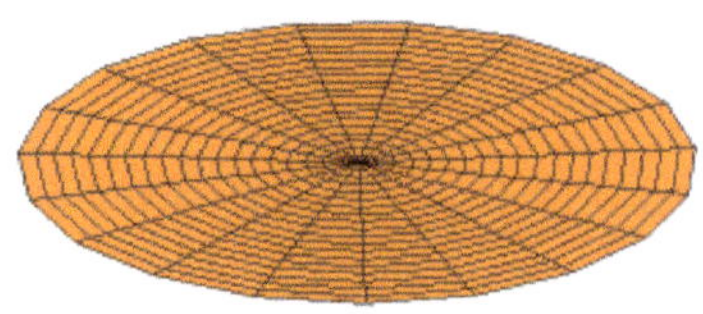 Drum mode	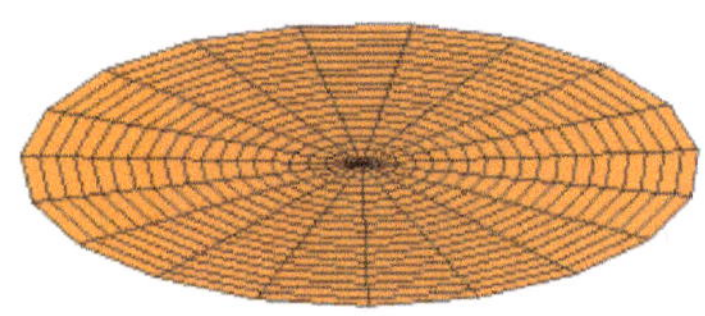 Drum mode
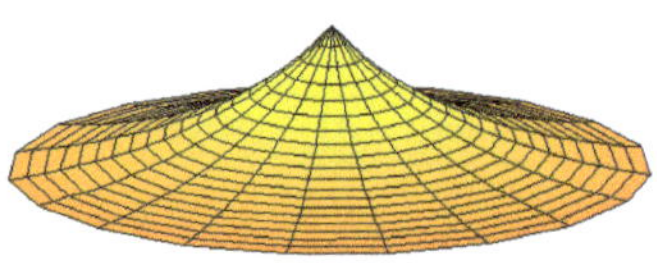 Wave function of 1s orbital (real part, 2D-cut, $r_{max} = 2a_0$))	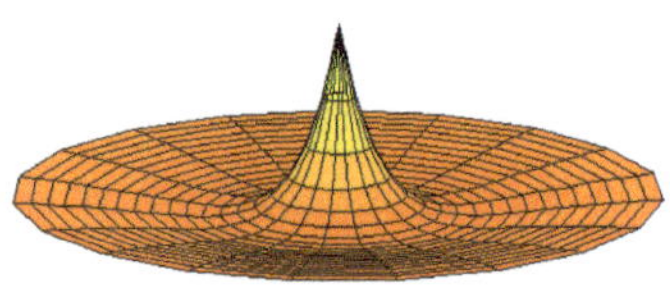 Wave function of 2s orbital (real part, 2D-cut, $r_{max} = 10a_0$))	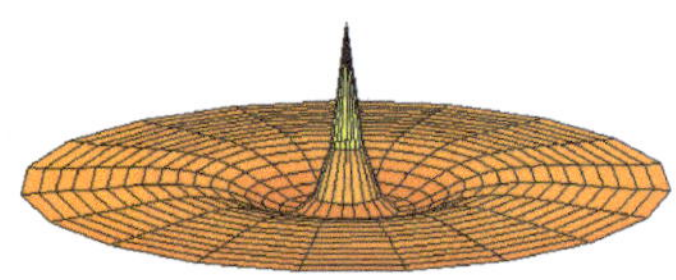 Wave function of 3s orbital (real part, 2D-cut, $r_{max} = 20a_0$))

None of the other sets of modes in a drum membrane have a central antinode, and in all of them the center of the drum does not move. These correspond to a node at the nucleus for all non-s orbitals in an atom. These orbitals all have some angular momentum, and in the planetary model, they correspond to particles in orbit with eccentricity less than 1.0, so that they do not pass straight through the center of the primary body, but keep somewhat away from it.

In addition, the drum modes analogous to p and d modes in an atom show spatial irregularity along the different radial directions from the center of the drum, whereas all of the modes analogous to s modes are perfectly symmetrical in radial direction. The non radial-symmetry properties of non-s orbitals are necessary to localize a particle with angular momentum and a wave nature in an orbital where it must tend to stay away from the central attraction force, since any particle localized at the point of central attraction could have no angular momentum. For these modes, waves in the drum head tend to avoid the central point. Such features again emphasize that the shapes of atomic orbitals are a direct consequence of the wave nature of electrons.

p-type drum modes and wave functions		
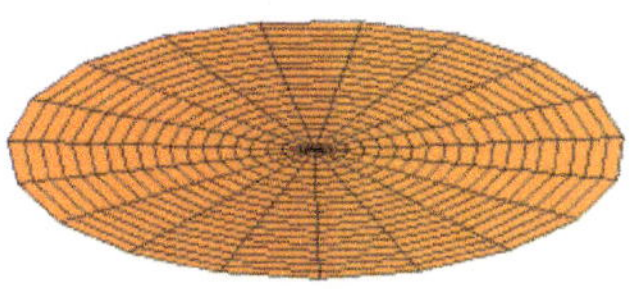 Drum mode u_{11}	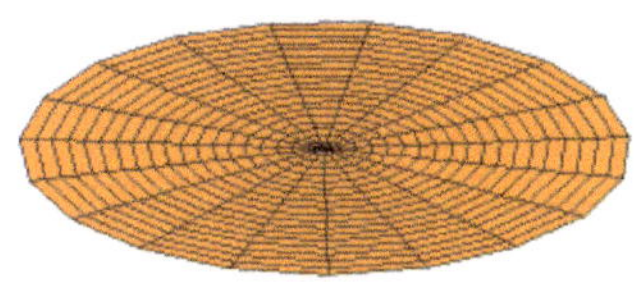 Drum mode u_{12}	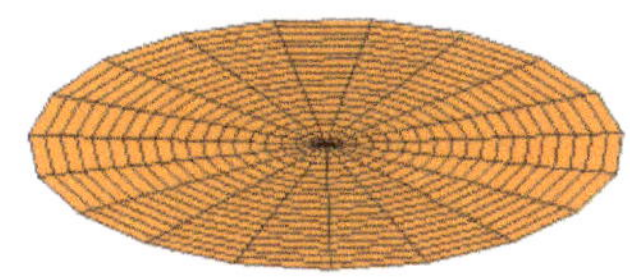 Drum mode u_{13}
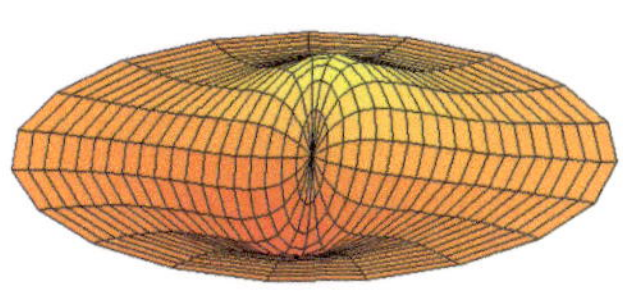 Wave function of 2p orbital (real part, 2D-cut, $r_{max} = 10a_0$))	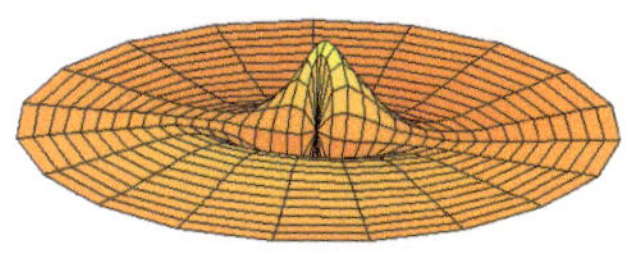 Wave function of 3p orbital (real part, 2D-cut, $r_{max} = 20a_0$)	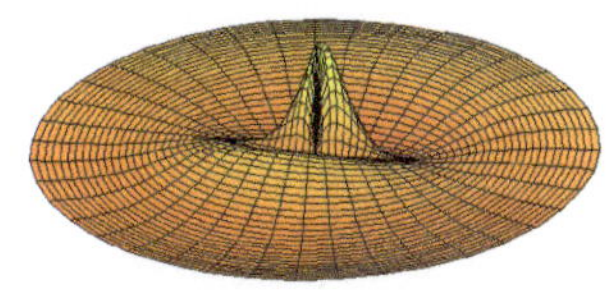 Wave function of 4p orbital (real part, 2D-cut, $r_{max} = 25a_0$))

d-type drum Modes		
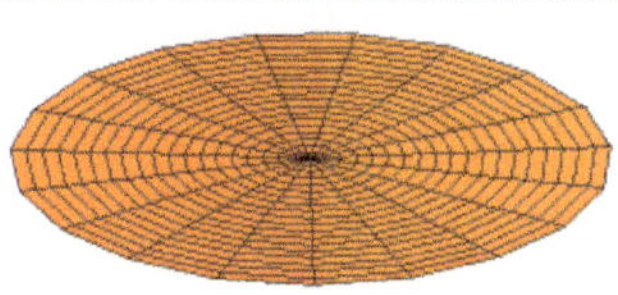 Mode u_{21} (3d orbital)	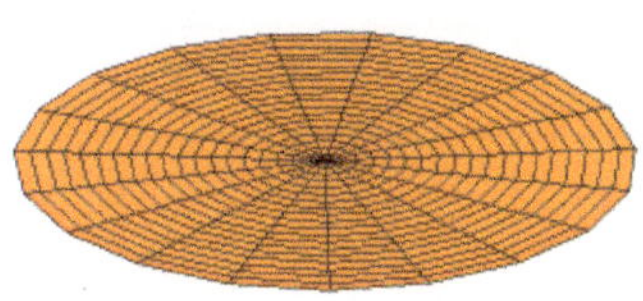 Mode u_{22} (4d orbital)	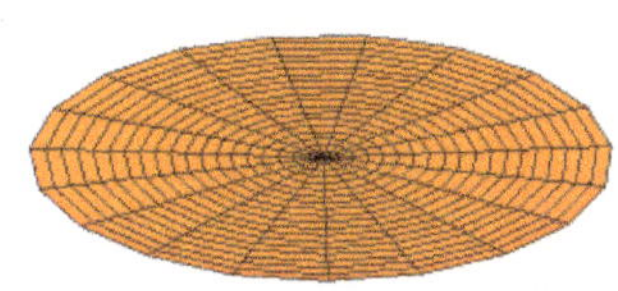 Mode u_{23} (5d orbital)

Orbital Energy

In atoms with a single electron (hydrogen-like atoms), the energy of an orbital (and, consequently, of any electrons in the orbital) is determined exclusively by n. The $n = 1$ orbital has the lowest possible energy in the atom. Each successively higher value of n has a higher level of energy, but the difference decreases as n increases. For high n, the level of energy becomes so high that the electron can easily escape from the atom. In single electron atoms, all levels with different ℓ within a given n are (to a good approximation) degenerate, and have the same energy. This approximation is broken to a slight extent by the effect of the magnetic field of the nucleus, and by quantum electrodynamics effects. The latter induce tiny binding energy differences especially for s electrons that go nearer the nucleus, since these feel a very slightly different nuclear charge, even in one-electron atoms; see Lamb shift.

In atoms with multiple electrons, the energy of an electron depends not only on the intrinsic properties of its orbital, but also on its interactions with the other electrons. These interactions depend on the detail of its spatial probability distribution, and so the energy levels of orbitals depend not only on n but also on ℓ. Higher values of ℓ are associated with higher values of energy; for instance, the 2p state is higher than the 2s state. When $\ell = 2$, the increase in energy of the orbital becomes so large as to push the energy of orbital above the energy of the s-orbital in the next higher shell; when $\ell = 3$ the energy is pushed into the shell two steps higher. The filling of the 3d orbitals does not occur until the 4s orbitals have been filled.

The increase in energy for subshells of increasing angular momentum in larger atoms is due to electron–electron interaction effects, and it is specifically related to the ability of low angular momentum electrons to penetrate more effectively toward the nucleus, where they are subject to less screening from the charge of intervening electrons. Thus, in atoms of higher atomic number, the ℓ of electrons becomes more and more of a determining factor in their energy, and the principal quantum numbers n of electrons becomes less and less important in their energy placement.

The energy sequence of the first 35 subshells (e.g., 1s, 2p, 3d, etc.) is given in the following table. Each cell represents a subshell with n and ℓ given by its row and column indices, respectively. The number in the cell is the subshell's position in the sequence. For a linear listing of the subshells in terms of increasing energies in multielectron atoms, see the section below.

	s	p	d	f	g	h
1	1					
2	2	3				
3	4	5	7			
4	6	8	10	13		
5	9	11	14	17	*21*	
6	12	15	18	*22*	*26*	*31*
7	16	19	*23*	*27*	*32*	*37*
8	*20*	*24*	*28*	*33*	*38*	*44*
9	*25*	*29*	*34*	*39*	*45*	*51*
10	*30*	*35*	*40*	*46*	*52*	*59*

Note: empty cells indicate non-existent sublevels, while numbers in italics indicate sublevels that could (potentially) exist, but which do not hold electrons in any element currently known.

Electron Placement and the Periodic Table

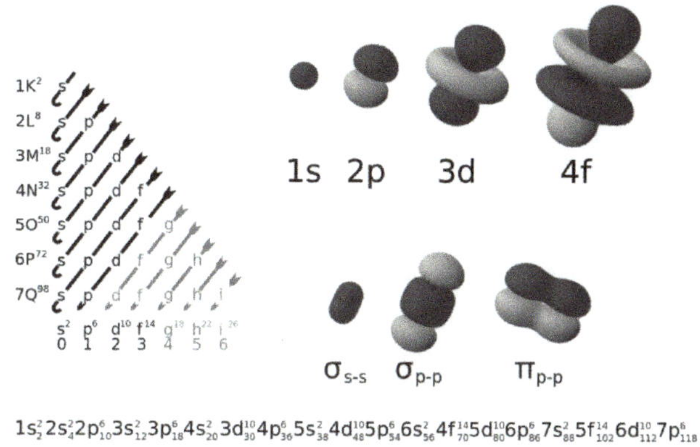

Electron atomic and molecular orbitals. The chart of orbitals (left) is arranged by increasing energy (see Madelung rule). *Note that atomic orbits are functions of three variables (two angles, and the distance r from the nucleus). These images are faithful to the angular component of the orbital, but not entirely representative of the orbital as a whole.*

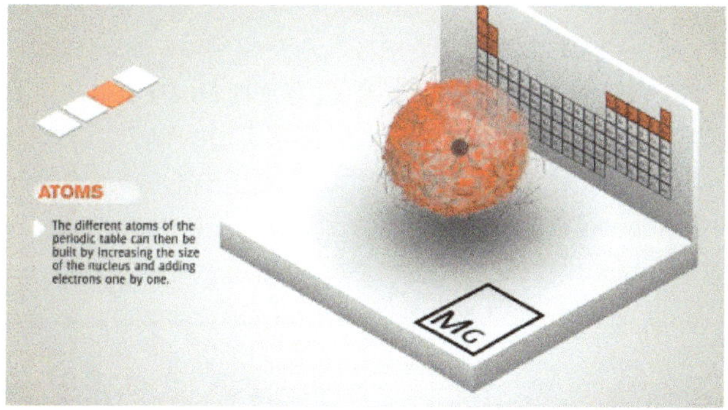

Atomic orbitals and periodic table construction

Several rules govern the placement of electrons in orbitals (*electron configuration*). The first dictates that no two electrons in an atom may have the same set of values of quantum numbers (this is the Pauli exclusion principle). These quantum numbers include the three that define orbitals, as well as *s*, or spin quantum number. Thus, two electrons may occupy a single orbital, so long as they have different values of *s*. However, *only* two electrons, because of their spin, can be associated with each orbital.

Additionally, an electron always tends to fall to the lowest possible energy state. It is possible for it to occupy any orbital so long as it does not violate the Pauli exclusion principle, but if lower-energy orbitals are available, this condition is unstable. The electron will eventually lose energy (by releasing a photon) and drop into the lower orbital. Thus, electrons fill orbitals in the order specified by the energy sequence given above.

This behavior is responsible for the structure of the periodic table. The table may be divided into several rows (called 'periods'), numbered starting with 1 at the top. The presently known elements occupy seven periods. If a certain period has number *i*, it consists of elements whose outermost electrons fall in the *i*th shell. Niels Bohr was the first to propose (1923) that the periodicity in the properties of the elements might be explained by the periodic filling of the electron energy levels, resulting in the electronic structure of the atom.

The periodic table may also be divided into several numbered rectangular 'blocks'. The elements belonging to a given block have this common feature: their highest-energy electrons all belong to the same ℓ-state (but the n associated with that ℓ-state depends upon the period). For instance, the leftmost two columns constitute the 's-block'. The outermost electrons of Li and Be respectively belong to the 2s subshell, and those of Na and Mg to the 3s subshell.

The following is the order for filling the "subshell" orbitals, which also gives the order of the "blocks" in the periodic table:

1s, 2s, 2p, 3s, 3p, 4s, 3d, 4p, 5s, 4d, 5p, 6s, 4f, 5d, 6p, 7s, 5f, 6d, 7p

The "periodic" nature of the filling of orbitals, as well as emergence of the s, p, d and f "blocks", is more obvious if this order of filling is given in matrix form, with increasing principal quantum numbers starting the new rows ("periods") in the matrix. Then, each subshell (composed of the first two quantum numbers) is repeated as many times as required for each pair of electrons it may contain. The result is a compressed periodic table, with each entry representing two successive elements:

1s															
2s													2p	2p	2p
3s													3p	3p	3p
4s								3d	3d	3d	3d	3d	4p	4p	4p
5s								4d	4d	4d	4d	4d	5p	5p	5p
6s	4f	4f	4f	4f	4f	4f	4f	5d	5d	5d	5d	5d	6p	6p	6p
7s	5f	5f	5f	5f	5f	5f	5f	6d	6d	6d	6d	6d	7p	7p	7p

Although this is the general order of orbital filling according to the Madelung rule, there are exceptions, and the actual electronic energies of each element are also dependent upon additional details of the atoms (see Electron configuration#Atoms: Aufbau principle and Madelung rule).

The number of electrons in an electrically neutral atom increases with the atomic number. The electrons in the outermost shell, or *valence electrons*, tend to be responsible for an element's chemical behavior. Elements that contain the same number of valence electrons can be grouped together and display similar chemical properties.

Relativistic Effects

For elements with high atomic number Z, the effects of relativity become more pronounced, and especially so for s electrons, which move at relativistic velocities as they penetrate the screening electrons near the core of high-Z atoms. This relativistic increase in momentum for high speed electrons causes a corresponding decrease in wavelength and contraction of 6s orbitals relative to 5d orbitals (by comparison to corresponding s and d electrons in lighter elements in the same column of the periodic table); this results in 6s valence electrons becoming lowered in energy.

Examples of significant physical outcomes of this effect include the lowered melting temperature of mercury (which results from 6s electrons not being available for metal bonding) and the golden color of gold and caesium (which results from narrowing of 6s to 5d transition energy to the point that visible light begins to be absorbed).

In the Bohr Model, an $n = 1$ electron has a velocity given by $v = Z\alpha c$, where Z is the atomic number, α is the fine-structure constant, and c is the speed of light. In non-relativistic quantum mechanics, therefore, any atom with an atomic number greater than 137 would require its 1s electrons to be traveling faster than the speed of light. Even in the Dirac equation, which accounts for relativistic effects, the wave function of the electron for atoms with $\boldsymbol{Z} > 137$ is oscillatory and unbounded. The significance of element 137, also known as untriseptium, was first pointed out by the physicist Richard Feynman. Element 137 is sometimes informally called feynmanium (symbol Fy). However, Feynman's approximation fails to predict the exact critical value of Z due to the non-point-charge nature of the nucleus and very small orbital radius of inner electrons, resulting in a potential seen by inner electrons which is effectively less than Z. The critical Z value which makes the atom unstable with regard to high-field breakdown of the vacuum and production of electron-positron pairs, does not occur until Z is about 173. These conditions are not seen except transiently in collisions of very heavy nuclei such as lead or uranium in accelerators, where such electron-positron production from these effects has been claimed to be observed. See Extension of the periodic table beyond the seventh period.

There are no nodes in relativistic orbital densities, although individual components of the wave function will have nodes.

Transitions Between Orbitals

Bound quantum states have discrete energy levels. When applied to atomic orbitals, this means that the energy differences between states are also discrete. A transition between these states (i.e., an electron absorbing or emitting a photon) can thus only happen if the photon has an energy corresponding with the exact energy difference between said states.

Consider two states of the hydrogen atom:

State 1) $n = 1, \ell = 0, m_\ell = 0$ and $s = +1/2$

State 2) $n = 2, \ell = 0, m_\ell = 0$ and $s = +1/2$

By quantum theory, state 1 has a fixed energy of E_1, and state 2 has a fixed energy of E_2. Now, what would happen if an electron in state 1 were to move to state 2? For this to happen, the electron would need to gain an energy of exactly $E_2 - E_1$. If the electron receives energy that is less than or greater than this value, it cannot jump from state 1 to state 2. Now, suppose we irradiate the atom with a broad-spectrum of light. Photons that reach the atom that have an energy of exactly $E_2 - E_1$ will be absorbed by the electron in state 1, and that electron will jump to state 2. However, photons that are greater or lower in energy cannot be absorbed by the electron, because the electron can only jump to one of the orbitals, it cannot jump to a state between orbitals. The result is that only photons of a specific frequency will be absorbed by the atom. This creates a line in the spectrum, known as an absorption line, which corresponds to the energy difference between states 1 and 2.

The atomic orbital model thus predicts line spectra, which are observed experimentally. This is one of the main validations of the atomic orbital model.

The atomic orbital model is nevertheless an approximation to the full quantum theory, which only recognizes many electron states. The predictions of line spectra are qualitatively useful but are not quantitatively accurate for atoms and ions other than those containing only one electron.

Atomic Mass

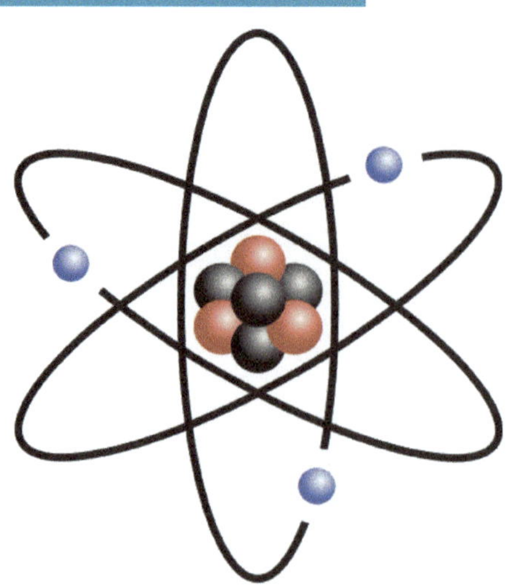

Stylized lithium-7 atom: 3 protons, 4 neutrons, & 3 electrons (total electrons are ~1/4300th of the mass of the nucleus). It has a mass of 7.016 u. Rare lithium-6 (mass of 6.015 u) has only 3 neutrons, reducing the atomic weight (average) of lithium to 6.941.

The atomic mass (m_a) is the mass of an atomic particle, sub-atomic particle, or molecule. It is commonly expressed in unified atomic mass units (u) where by international agreement, 1 unified atomic mass unit is defined as 1/12 of the mass of a single carbon-12 atom (at rest). For atoms, the protons and neutrons of the nucleus account for almost all of the mass, and the atomic mass measured in u has nearly the same value as the mass number.

When divided by unified atomic mass units or daltons to form a pure number ratio, the atomic mass of an atom becomes a dimensionless number called the relative isotopic mass (see section below). Thus, the atomic mass of a carbon-12 atom is 12 u or 12 daltons (Da), but the relative isotopic mass of a carbon-12 atom is simply 12.

The atomic mass or relative isotopic mass refers to the mass of a single particle, and is fundamentally different from the quantities elemental atomic weight (also called "relative *atomic* mass") and standard atomic weight, both of which refer to averages (mathematical means) of naturally-occurring atomic mass values for samples of elements. Most elements have more than one stable nuclide; for those elements, such an average depends on the mix of nuclides present, which may vary to some limited extent depending on the source of the sample, as each nuclide has a different mass. (However, a typical value can be established, which is called the standard atomic weight.) By contrast, atomic mass figures refer to an individual particle species: as atoms of the same species are identical, atomic mass values are expected to have no intrinsic variance at all. Atomic mass figures are thus commonly reported to many more significant figures than atomic weights. Standard atomic weight is related to atomic mass by the abundance ranking of isotopes for each element. It is usually *about* the same value as the atomic mass of the most abundant isotope, other than what looks like (but is not actually) a rounding difference.

The atomic mass of atoms, ions, or atomic nuclei is slightly less than the sum of the masses of their constituent protons, neutrons, and electrons, due to binding energy mass loss (as per $E=mc^2$).

Relative Isotopic Mass: the Same Quantity as Atomic Mass, But With Different Units

Relative isotopic mass (a property of a single atom) is not to be confused with the averaged quantity "relative *atomic* mass," which is the same as atomic weight (see above), and is an average of values for many atoms in a given sample of a chemical element.

Relative isotopic mass is similar to atomic mass and has exactly the same numerical value as atomic mass, whenever atomic mass is expressed in unified atomic mass units. The only difference in that case, is that relative isotopic mass is a pure number with no units. This loss of units results from the use of a scaling ratio with respect to a carbon-12 standard, and the word "relative" in the term "relative isotopic mass" refers to this scaling *relative* to carbon-12.

The relative isotopic mass, then, is the mass of a given isotope (specifically, any single nuclide), when this value is scaled by the mass of carbon-12, when the latter is set equal to 12. Equivalently, the relative isotopic mass of an isotope or nuclide is the mass of the isotope relative to 1/12 of the mass of a carbon-12 atom.

For example, the relative isotopic mass of a carbon-12 atom is exactly 12. For comparison, the atomic mass of a carbon-12 atom is exactly 12 daltons or 12 unified atomic mass units. Alternately, the atomic mass of a carbon-12 atom may be expressed in any other mass units: for example, the atomic mass of a carbon-12 atom is about $1.998467052 \times 10^{-26}$ kilogram.

As in the case of atomic mass, no nuclides other than carbon-12 have exactly whole-number values of relative isotopic mass. As is the case for the related *atomic mass* when expressed in unified atomic mass units or daltons, the relative isotopic mass numbers of nuclides other than carbon-12 are not whole numbers, but are always close to whole numbers. This is discussed more fully below.

Similar Terms for Different Quantities

The atomic mass and relative isotopic mass are sometimes confused, or incorrectly used, as synonyms of relative atomic mass (also known as atomic weight) and the standard atomic weight (a particular variety of atomic weight, in the sense that is a standardized atomic weight). However, as noted in the introduction, atomic weight and standard atomic weight represent terms for (abundance-weighted) averages of atomic masses in elemental samples, not for single nuclides. As such, atomic weight and standard atomic weight often differ numerically from relative isotopic mass and atomic mass, and they can also have different units than atomic mass when this quantity is not expressed in unified atomic mass units.

The atomic mass (relative isotopic mass) is defined as the mass of a single atom, which can only be one isotope (nuclide) at a time, and is not an abundance-weighted average, as in the case of relative atomic mass/atomic weight. The atomic mass or relative isotopic mass of each isotope and nuclide of a chemical element is therefore a number that can in principle be measured to a very great precision, since every specimen of such a nuclide is expected to be exactly identical to every other specimen, as all atoms of a given type in the same energy state, and every specimen of a particular nuclide, are expected to be exactly identical in mass to every other specimen of that nuclide. For example, every atom of oxygen-16 is expected to have exactly the same atomic mass (relative isotopic mass) as every other atom of oxygen-16.

In the case of many elements that have one naturally occurring isotope (mononuclidic elements) or one dominant isotope, the actual numerical similarity/difference between the atomic mass of the most common isotope, and the (standard) relative atomic mass or (standard) atomic weight can be small or even nil, and does affect most bulk calculations. However, such an error can exist and even be important when considering individual atoms for elements that are not mononuclidic.

For non-mononuclidic elements that have more than one common isotope, the numerical difference in relative atomic mass (atomic weight) from even the most common relative isotopic mass, can be half a mass unit or more (e.g. see the case of chlorine where atomic weight and standard atomic weight are about 35.45). The atomic mass (relative isotopic mass) of an uncommon isotope can differ from the relative atomic mass, atomic weight, or standard atomic weight, by several mass units.

Atomic masses expressed in unified atomic mass units (i.e. relative isotopic masses) are always close to whole-number values, but never (except in the case of carbon-12) exactly a whole number, for two reasons:

- protons and neutrons have different masses, and different nuclides have different ratios of protons and neutrons.
- atomic masses are reduced, to different extents, by their binding energies.

The ratio of atomic mass to mass number (number of nucleons) varies from about 0.99884 for ^{56}Fe to 1.00782505 for ^{1}H.

Any mass defect due to nuclear binding energy is experimentally a small fraction (less than 1%) of the mass of equal number of free nucleons. When compared to the average mass per nucleon in carbon-12, which is moderately strongly-bound compared with other atoms, the mass defect of binding for most atoms is an even smaller fraction of a dalton (unified atomic mass unit, based on carbon-12). Since free protons and neutrons differ from each other in mass by a small fraction of a dalton (about 0.0014 u), rounding the relative isotopic mass, or the atomic mass of any given nuclide given in daltons to the nearest whole number always gives the nucleon count, or mass number. Additionally, the neutron count (neutron number) may then be derived by subtracting the number of protons (atomic number) from the mass number (nucleon count).

Mass Defects in Atomic Masses

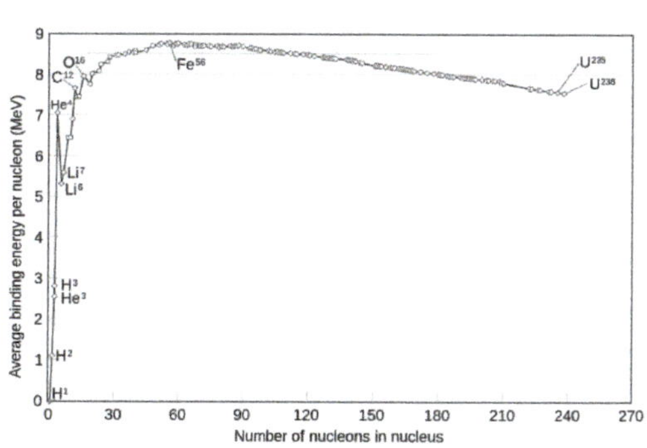

Binding energy per nucleon of common isotopes. A graph of the ratio of mass number to atomic mass would be similar.

The amount that the ratio of atomic masses to mass number deviates from 1 is as follows: the deviation starts positive at hydrogen-1, then decreases until it reaches a local minimum at helium-4. Isotopes of lithium, beryllium, and boron are less strongly bound than helium, as shown by their increasing mass-to-mass number ratios.

At carbon, the ratio of mass (in daltons) to mass number is defined as 1, and after carbon it becomes less than one until a minimum is reached at iron-56 (with only slightly higher values for iron-58 and nickel-62), then increases to positive values in the heavy isotopes, with increasing atomic number. This corresponds to the fact that nuclear fission in an element heavier than zirconium produces energy, and fission in any element lighter than niobium requires energy. On the other hand, nuclear fusion of two atoms of an element lighter than scandium (except for helium) produces energy, whereas fusion in elements heavier than calcium requires energy. The fusion of two atoms of He-4 to give beryllium-8 would require energy, and the beryllium would quickly fall apart again. He-4 can fuse with tritium (H-3) or with He-3, and these processes occurred during Big Bang nucleosynthesis. The formation of elements with more than seven nucleons requires the fusion of three atoms of He-4 in the so-called triple alpha process, skipping over lithium, beryllium, and boron to produce carbon.

Here are some values of the ratio of atomic mass to mass number:

Nuclide	Ratio of atomic mass to mass number
^{1}H	1.00782505
^{2}H	1.0070508885
^{3}H	1.0053497592
^{3}He	1.0053431064
^{4}He	1.0006508135
^{6}Li	1.0025204658
^{12}C	1
^{14}N	1.0002195718
^{16}O	0.9996821637
^{56}Fe	0.9988381696
^{210}Po	0.9999184462
^{232}Th	1.0001640315
^{238}U	1.0002133958

Measurement of Atomic Masses

Direct comparison and measurement of the masses of atoms is achieved with mass spectrometry.

Conversion Factor Between Atomic Mass Units and Grams

The standard scientific unit used to quantify the amount of a substance in macroscopic quantities is the mole (symbol: mol), which is defined arbitrarily as the amount of a substance which has as many atoms or molecules as there are atoms in 12 grams of the carbon isotope C-12. The number of atoms in a mole is called Avogadro's number, the value of which is approximately 6.022×10^{23}.

One mole of a substance always contains almost exactly the *relative atomic mass* or *molar mass* of that substance; however, this may or may not be true for the *atomic mass,* depending on whether or not the element exists naturally in more than one isotope. For example, the relative atomic mass of iron is 55.847 g/mol, and therefore one mole of iron as commonly found on earth has a mass of 55.847 grams. The *atomic mass* of the ^{56}Fe isotope is 55.935 u and one mole of ^{56}Fe atoms would then in theory have a mass of 55.935 g, but such amounts of pure ^{56}Fe have never been found (or separated out) on Earth. However, there are 22 mononuclidic elements of which essentially only a single isotope is found in nature (common examples are fluorine, sodium, aluminum and phosphorus) and for these elements the relative atomic mass and atomic mass are the same. Samples of these elements therefore may serve as reference standards for certain atomic mass values.

The formula for conversion between atomic mass units and SI mass in grams for a single atom is:

$$1\ \mathrm{u} = \frac{M_\mathrm{u}}{N_\mathrm{A}} = \frac{1\ \mathrm{g/mol}}{N_\mathrm{A}}$$

where M_u is the Molar mass constant and N_A is the Avogadro constant.

Relationship Between Atomic and Molecular Masses

Similar definitions apply to molecules. One can compute the molecular mass of a compound by adding the atomic masses of its constituent atoms (nuclides). One can compute the molar mass of a compound by adding the relative atomic masses of the elements given in the chemical formula. In both cases the multiplicity of the atoms (the number of times it occurs) must be taken into account, usually by multiplication of each unique mass by its multiplicity.

History

The first scientists to determine relative atomic masses were John Dalton and Thomas Thomson between 1803 and 1805 and Jöns Jakob Berzelius between 1808 and 1826. Relative atomic mass (*Atomic weight*) was originally defined relative to that of the lightest element, hydrogen, which was taken as 1.00, and in the 1820s Prout's hypothesis stated that atomic masses of all elements would prove to be exact multiples of that of hydrogen. Berzelius, however, soon proved that this was not even approximately true, and for some elements, such as chlorine, relative atomic mass, at about 35.5, falls almost exactly halfway between two integral multiples of that of hydrogen. Still later, this was shown to be largely due to a mix of isotopes, and that the atomic masses of pure isotopes, or nuclides, are multiples of the hydrogen mass, to within about 1%.

In the 1860s Stanislao Cannizzaro refined relative atomic masses by applying Avogadro's law (notably at the Karlsruhe Congress of 1860). He formulated a law to determine relative atomic masses of elements: *the different quantities of the same element contained in different molecules are all whole multiples of the atomic weight* and determined relative atomic masses and molecular masses by comparing the vapor density of a collection of gases with molecules containing one or more of the chemical element in question.

In the 20th century, until the 1960s chemists and physicists used two different atomic-mass scales. The chemists used a "atomic mass unit" (amu) scale such that the natural mixture of oxygen isotopes had an atomic mass 16, while the physicists assigned the same number 16 to only the atomic mass of the most common oxygen isotope (O-16, containing eight protons and eight neutrons). However, because oxygen-17 and oxygen-18 are also present in natural oxygen this led to two different tables of atomic mass. The unified scale based on carbon-12, ^{12}C, met the physicists' need to base the scale on a pure isotope, while being numerically close to the chemists' scale.

The term *atomic weight* is being phased out slowly and being replaced by *relative atomic mass*, in most current usage. This shift in nomenclature reaches back to the 1960s and has been the source of much debate in the scientific community, which was triggered by the adoption of the unified atomic mass unit and the realization that weight was in some ways an inappropriate term. The argument for keeping the term "atomic weight" was primarily that it was a well understood term to those in the field, that the term "atomic mass" was already in use (as it is currently defined) and that the term "relative atomic mass" might be easily confused with *relative isotopic mass* (the mass of a single atom of a given nuclide, expressed dimensionlessly relative to 1/12 of the mass of carbon-12; see section above).

In 1979, as a compromise, the term "relative atomic mass" was introduced as a secondary synonym for atomic weight. Twenty years later the primacy of these synonyms was reversed, and the term "relative atomic mass" is now the preferred term.

However, the term "*standard* atomic weights" (referring to the standardized expectation atomic weights of differing samples) have maintained the same name. In the case of this latter term, simple replacement of the "atomic weight" term with "relative atomic mass" would have resulted in the term "standard relative atomic mass."

Atomic Radius

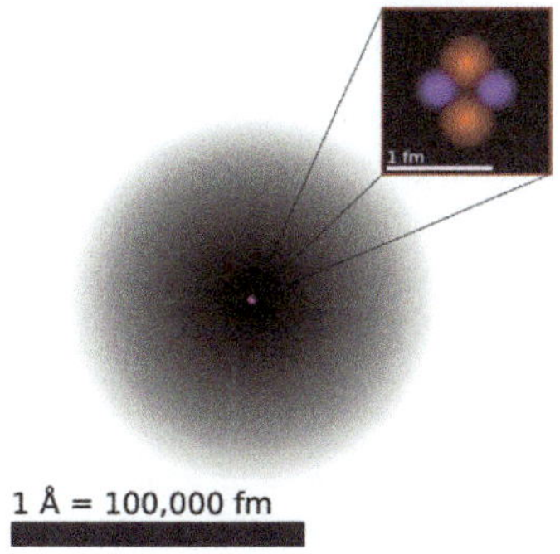

Diagram of a helium atom, showing the electron probability density as shades of gray.

The atomic radius of a chemical element is a measure of the size of its atoms, usually the mean or typical distance from the center of the nucleus to the boundary of the surrounding cloud of electrons. Since the boundary is not a well-defined physical entity, there are various non-equivalent definitions of atomic radius. Three widely used definitions of atomic radius are: Van der Waals radius, ionic radius, and covalent radius.

Depending on the definition, the term may apply only to isolated atoms, or also to atoms in condensed matter, covalently bound in molecules, or in ionized and excited states; and its value may be obtained through experimental measurements, or computed from theoretical models. The value of the radius may depend on the atom's state and context.

Electrons do not have definite orbits, or sharply defined ranges. Rather, their positions must be described as probability distributions that taper off gradually as one moves away from the nucleus, without a sharp cutoff. Moreover, in condensed matter and molecules, the electron clouds of the atoms usually overlap to some extent, and some of the electrons may roam over a large region encompassing two or more atoms.

Under most definitions the radii of isolated neutral atoms range between 30 and 300 pm (trillionths of a meter), or between 0.3 and 3 ångströms. Therefore, the radius of an atom is more than 10,000 times the radius of its nucleus (1–10 fm), and less than 1/1000 of the wavelength of visible light (400–700 nm).

The approximate shape of a molecule of ethanol, CH3CH2OH. Each atom is modeled by a sphere with the element's Van der Waals radius.

For many purposes, atoms can be modeled as spheres. This is only a crude approximation, but it can provide quantitative explanations and predictions for many phenomena, such as the density of liquids and solids, the diffusion of fluids through molecular sieves, the arrangement of atoms and ions in crystals, and the size and shape of molecules.

Atomic radii vary in a predictable and explicable manner across the periodic table. For instance, the radii generally decrease along each period (row) of the table, from the alkali metals to the noble gases; and increase down each group (column). The radius increases sharply between the noble gas at the end of each period and the alkali metal at the beginning of the next period. These trends of the atomic radii (and of various other chemical and physical properties of the elements) can be explained by the electron shell theory of the atom; they provided important evidence for the development and confirmation of quantum theory. The atomic radii decrease across the Periodic Table because as the atomic number increases, the number of protons increases across the period, but the extra electrons are only added to the same quantum shell. Therefore, the effective nuclear charge towards the outermost electrons increases, drawing the outermost electrons closer. As a result, the electron cloud contracts and the atomic radius decreases.

History

In 1920, shortly after it had become possible to determine the sizes of atoms using X-ray crystallography, it was suggested that all atoms of the same element have the same radii. However, in 1923, when more crystal data had become available, it was found that the approximation of an atom as a sphere does not necessarily hold when comparing the same atom in different crystal structures.

Definitions

Widely used definitions of atomic radius include:

- Van der Waals radius: in principle, half the minimum distance between the nuclei of two atoms of the element that are not bound to the same molecule.
- Ionic radius: the nominal radius of the ions of an element in a specific ionization state, deduced from the spacing of atomic nuclei in crystalline salts that include that ion. In principle, the spacing between two adjacent oppositely charged ions (the length of the ionic bond between them) should equal the sum of their ionic radii.
- Covalent radius: the nominal radius of the atoms of an element when covalently bound to other atoms, as deduced from the separation between the atomic nuclei in molecules. In principle, the distance between two atoms that are bound to each other in a molecule (the length of that covalent bond) should equal the sum of their covalent radii.

- Metallic radius: the nominal radius of atoms of an element when joined to other atoms by metallic bonds.
- Bohr radius: the radius of the lowest-energy electron orbit predicted by Bohr model of the atom (1913). It is only applicable to atoms and ions with a single electron, such as hydrogen, singly ionized helium, and positronium. Although the model itself is now obsolete, the Bohr radius for the hydrogen atom is still regarded as an important physical constant.

Empirically Measured Atomic Radius

The following table shows empirically measured covalent radii for the elements, as published by J. C. Slater in 1964. The values are in picometers (pm or 1×10^{-12} m), with an accuracy of about 5 pm. The shade of the box ranges from red to yellow as the radius increases; gray indicates lack of data.

Group (column)	1	2	3	4	5	6	7	8	9	10	11	12	13	14	15	16	17	18
Period (row)																		
1	H 25																	He
2	Li 145	Be 105											B 85	C 70	N 65	O 60	F 50	Ne
3	Na 180	Mg 150											Al 125	Si 110	P 100	S 100	Cl 100	Ar
4	K 220	Ca 180	Sc 160	Ti 140	V 135	Cr 140	Mn 140	Fe 140	Co 135	Ni 135	Cu 135	Zn 135	Ga 130	Ge 125	As 115	Se 115	Br 115	Kr
5	Rb 235	Sr 200	Y 180	Zr 155	Nb 145	Mo 145	Tc 135	Ru 130	Rh 135	Pd 140	Ag 160	Cd 155	In 155	Sn 145	Sb 145	Te 140	I 140	Xe
6	Cs 260	Ba 215	*	Hf 155	Ta 145	W 135	Re 135	Os 130	Ir 135	Pt 135	Au 135	Hg 150	Tl 190	Pb 180	Bi 160	Po 190	At	Rn
7	Fr	Ra 215	**	Rf	Db	Sg	Bh	Hs	Mt	Ds	Rg	Cn	Uut	Fl	Uup	Lv	Uus	Uuo
Lantha-nides	*	La 195	Ce 185	Pr 185	Nd 185	Pm 185	Sm 185	Eu 185	Gd 180	Tb 175	Dy 175	Ho 175	Er 175	Tm 175	Yb 175	Lu 175		
Actinides	**	Ac 195	Th 180	Pa 180	U 175	Np 175	Pu 175	Am 175	Cm	Bk	Cf	Es	Fm	Md	No	Lr		

Explanation of the General Trends

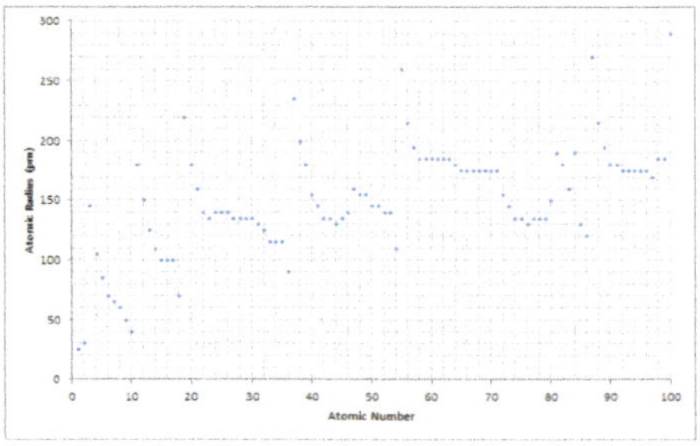

A graph comparing the atomic radius of elements with atomic numbers 1–100. Accuracy of ±5 pm.

The way the atomic radius varies with increasing atomic number can be explained by the arrangement of electrons in shells of fixed capacity. The shells are generally filled in order of increasing radius, since the negatively charged electrons are attracted by the positively charged protons in the nucleus. As the atomic number increases along each row of the periodic table, the additional electrons go into the same outermost shell; whose radius gradually contracts, due to the increasing nuclear charge. In a noble gas, the outermost shell is completely filled; therefore, the additional electron of next alkali metal will go into the next outer shell, accounting for the sudden increase in the atomic radius.

The increasing nuclear charge is partly counterbalanced by the increasing number of electrons, a phenomenon that is known as shielding; which explains why the size of atoms usually increases down each column. However, there is one notable exception, known as the lanthanide contraction: the 5d block of elements are much smaller than one would expect, due to the shielding caused by the 4f electrons.

The following table summarizes the main phenomena that influence the atomic radius of an element:

factor	**principle**	**increase with...**	***tend to***	**effect on radius**
electron shells	quantum mechanics	principal and azimuthal quantum numbers	increase atomic radius	increases down each column
nuclear charge	attractive force acting on electrons by protons in nucleus	atomic number	decrease atomic radius	decreases along each period
shielding	repulsive force acting on outermost shell electrons by inner electrons	number of electrons in inner shells	increase atomic radius	reduces the effect of the 2nd factor

Lanthanide Contraction

The electrons in the 4f-subshell, which is progressively filled from cerium ($Z = 58$) to lutetium ($Z = 71$), are not particularly effective at shielding the increasing nuclear charge from the sub-shells further out. The elements immediately following the lanthanides have atomic radii which are smaller than would be expected and which are almost identical to the atomic radii of the elements immediately above them. Hence hafnium has virtually the same atomic radius (and chemistry) as zirconium, and tantalum has an atomic radius similar to niobium, and so forth. The effect of the lanthanide contraction is noticeable up to platinum ($Z = 78$), after which it is masked by a relativistic effect known as the inert pair effect.

Due to lanthanide contraction, the 5 following observations can be drawn:

1. The size of Ln^{3+} ions regularly decreases with atomic number. According to Fajans' rules, decrease in size of Ln^{3+} ions increases the covalent character and decreases the basic character between Ln^{3+} and OH^- ions in $Ln(OH)_3$, to the point that $Yb(OH)_3$ and $Lu(OH)_3$ can dissolve with difficulty in hot concentrated NaOH. Hence the order of size of Ln^{3+} is given: $La^{3+} > Ce^{3+} > \ldots, \ldots > Lu^{3+}$.

2. There is a regular decrease in their ionic radii.

3. There is a regular decrease in their tendency to act as a reducing agent, with increase in atomic number.

4. The second and third rows of d-block transition elements are quite close in properties.

5. Consequently, these elements occur together in natural minerals and are difficult to separate.

D-Block Contraction

The d-block contraction is less pronounced than the lanthanide contraction but arises from a similar cause. In this case, it is the poor shielding capacity of the 3d-electrons which affects the atomic radii and chemistries of the elements immediately following the first row of the transition metals, from gallium ($Z = 31$) to bromine ($Z = 35$).

Calculated Atomic Radii

The following table shows atomic radii computed from theoretical models, as published by Enrico Clementi and others in 1967. The values are in picometres (pm).

References

- Mazo, Robert M. (2002). Brownian Motion: Fluctuations, Dynamics, and Applications. Oxford University Press. pp. 1–7. ISBN 0-19-851567-7. OCLC 48753074.
- Pais, Abraham (1986). Inward Bound: Of Matter and Forces in the Physical World. New York: Oxford University Press. pp. 228–230. ISBN 0-19-851971-0.
- Scerri, Eric R. (2007). The periodic table: its story and its significance. Oxford University Press US. pp. 205–226. ISBN 0-19-530573-6.
- Bowden, Mary Ellen (1997). "Otto Hahn, Lise Meitner, and Fritz Strassmann". Chemical achievers : the human face of the chemical sciences. Philadelphia, PA: Chemical Heritage Foundation. pp. 76–80, 125. ISBN 9780941901123.
- Demtröder, Wolfgang (2002). Atoms, Molecules and Photons: An Introduction to Atomic-Molecular- and Quantum Physics (1st ed.). Springer. pp. 39–42. ISBN 3-540-20631-0. OCLC 181435713.
- Woan, Graham (2000). The Cambridge Handbook of Physics. Cambridge University Press. p. 8. ISBN 0-521-57507-9. OCLC 224032426.
- MacGregor, Malcolm H. (1992). The Enigmatic Electron. Oxford University Press. pp. 33–37. ISBN 0-19-521833-7. OCLC 223372888.
- Pfeffer, Jeremy I.; Nir, Shlomo (2000). Modern Physics: An Introductory Text. Imperial College Press. pp. 330–336. ISBN 1-86094-250-4. OCLC 45900880.
- Shultis, J. Kenneth; Faw, Richard E. (2002). Fundamentals of Nuclear Science and Engineering. CRC Press. pp. 10–17. ISBN 0-8247-0834-2. OCLC 123346507.
- a b Sills, Alan D. (2003). Earth Science the Easy Way. Barron's Educational Series. pp. 131–134. ISBN 0-7641-2146-4. OCLC 51543743.
- Zumdahl, Steven S. (2002). Introductory Chemistry: A Foundation (5th ed.). Houghton Mifflin. ISBN 0-618-34342-3. OCLC 173081482. Archived from the original on 4 March 2008. Retrieved 5 February 2008.
- L'Annunziata, Michael F. (2003). Handbook of Radioactivity Analysis. Academic Press. pp. 3–56. ISBN 0-12-436603-1. OCLC 16212955.

3

Atomic Theory and Models

As early as the 19th century, it was recognized that all matter consisted of smaller units called atoms. Experiments in electromagnetism and radioactivity paved way for a better understanding of the atomic structure and the previously thought indivisible units were found to be composed of even smaller components like quarks, bosons, alpha and beta particles. This chapter chronologically illustrates atomic models like plum pudding model, Rutherford model, Bohr model and theories like atomic theory and atomism. A section of the chapter outlines atomic orbital and its classification.

Atomic Theory

In chemistry and physics, atomic theory is a scientific theory of the nature of matter, which states that matter is composed of discrete units called atoms. It began as a philosophical concept in ancient Greece and entered the scientific mainstream in the early 19th century when discoveries in the field of chemistry showed that matter did indeed behave as if it were made up of atoms.

The word *atom* comes from the Ancient Greek adjective *atomos*, meaning "indivisible". 19th century chemists began using the term in connection with the growing number of irreducible chemical elements. While seemingly apropos, around the turn of the 20th century, through various experiments with electromagnetism and radioactivity, physicists discovered that the so-called "uncuttable atom" was actually a conglomerate of various subatomic particles (chiefly, electrons, protons and neutrons) which can exist separately from each other. In fact, in certain extreme environments, such as neutron stars, extreme temperature and pressure prevents atoms from existing at all. Since atoms were found to be divisible, physicists later invented the term "elementary particles" to describe the "uncuttable", though not indestructible, parts of an atom. The field of science which studies subatomic particles is particle physics, and it is in this field that physicists hope to discover the true fundamental nature of matter.

History

Philosophical Atomism

The idea that matter is made up of discrete units is a very old one, appearing in many ancient cultures such as Greece and India. However, these ideas were founded in philosophical and theological reasoning rather than evidence and experimentation. Because of this, they could not convince everybody, so atomism was but one of a number of competing theories on the nature of matter. It was not until the 19th century that the idea was embraced and refined by scientists, as the blossoming science of chemistry produced discoveries that could easily be explained using the concept of atoms.

John Dalton

Near the end of the 18th century, two laws about chemical reactions emerged without referring to the notion of an atomic theory. The first was the law of conservation of mass, formulated by Antoine Lavoisier in 1789, which states that the total mass in a chemical reaction remains constant (that is, the reactants have the same mass as the products). The second was the law of definite proportions. First proven by the French chemist Joseph Louis Proust in 1799, this law states that if a compound is broken down into its constituent elements, then the masses of the constituents will always have the same proportions, regardless of the quantity or source of the original substance.

John Dalton studied and expanded upon this previous work and developed the law of multiple proportions: if two elements can be combined to form a number of possible compounds, then the ratios of the masses of the second element which combine with a fixed mass of the first element will be ratios of small whole numbers. For example: Proust had studied tin oxides and found that their masses were either 88.1% tin and 11.9% oxygen or 78.7% tin and 21.3% oxygen (these were tin(II) oxide and tin dioxide respectively). Dalton noted from these percentages that 100g of tin will combine either with 13.5g or 27g of oxygen; 13.5 and 27 form a ratio of 1:2. Dalton found that an atomic theory of matter could elegantly explain this common pattern in chemistry. In the case of Proust's tin oxides, one tin atom will combine with either one or two oxygen atoms.

Dalton believed atomic theory could explain why water absorbed different gases in different proportions - for example, he found that water absorbed carbon dioxide far better than it absorbed nitrogen. Dalton hypothesized this was due to the differences in mass and complexity of the gases' respective particles. Indeed, carbon dioxide molecules (CO_2) are heavier and larger than nitrogen molecules (N_2).

Dalton proposed that each chemical element is composed of atoms of a single, unique type, and though they cannot be altered or destroyed by chemical means, they can combine to form more complex structures (chemical compounds). This marked the first truly scientific theory of the atom, since Dalton reached his conclusions by experimentation and examination of the results in an empirical fashion.

In 1803 Dalton orally presented his first list of relative atomic weights for a number of substances. This paper was published in 1805, but he did not discuss there exactly how he obtained these figures. The method was first revealed in 1807 by his acquaintance Thomas Thomson, in the third edition of Thomson's textbook, *A System of Chemistry*. Finally, Dalton published a full account in his own textbook, *A New System of Chemical Philosophy*, 1808 and 1810.

Dalton estimated the atomic weights according to the mass ratios in which they combined, with the hydrogen atom taken as unity. However, Dalton did not conceive that with some elements atoms exist in molecules—e.g. pure oxygen exists as O_2. He also mistakenly believed that the simplest compound between any two elements is always one atom of each (so he thought water was HO, not H_2O). This, in addition to the crudity of his equipment, flawed his results. For instance, in 1803 he believed that oxygen atoms were 5.5 times heavier than hydrogen atoms, because in water he measured 5.5 grams of oxygen for every 1 gram of hydrogen and believed the formula for water was HO. Adopting better data, in 1806 he concluded that the atomic weight of oxygen must actually be 7 rather than 5.5, and he retained this weight for the rest of his life. Others at this time had already concluded that the oxygen atom must weigh 8 relative to hydrogen equals 1, if one assumes Dalton's formula for the water molecule (HO), or 16 if one assumes the modern water formula (H_2O).

Avogadro

The flaw in Dalton's theory was corrected in principle in 1811 by Amedeo Avogadro. Avogadro had proposed that equal volumes of any two gases, at equal temperature and pressure, contain equal numbers of molecules (in other words, the mass of a gas's particles does not affect the volume that it occupies). Avogadro's law allowed him to deduce the diatomic nature of numerous gases by studying the volumes at which they reacted. For instance: since two liters of hydrogen will react with just one liter of oxygen to produce two liters of water vapor (at constant pressure and temperature), it meant a single oxygen molecule splits in two in order to form two particles of water. Thus, Avogadro was able to offer more accurate estimates of the atomic mass of oxygen and various other elements, and made a clear distinction between molecules and atoms.

Brownian Motion

In 1827, the British botanist Robert Brown observed that dust particles inside pollen grains floating in water constantly jiggled about for no apparent reason. In 1905, Albert Einstein theorized that this Brownian motion was caused by the water molecules continuously knocking the grains about, and developed a hypothetical mathematical model to describe it. This model was validated experimentally in 1908 by French physicist Jean Perrin, thus providing additional validation for particle theory (and by extension atomic theory).

Discovery of Subatomic Particles

Atoms were thought to be the smallest possible division of matter until 1897 when J.J. Thomson discovered the electron through his work on cathode rays.

A Crookes tube is a sealed glass container in which two electrodes are separated by a vacuum. When a voltage is applied across the electrodes, cathode rays are generated, creating a glowing patch where they strike the glass at the opposite end of the tube.

Through experimentation, Thomson discovered that the rays could be deflected by an electric field (in addition to magnetic fields, which was already known). He concluded that these rays, rather than being a form of light, were composed of very light negatively charged particles he called "corpuscles" (they would later be renamed electrons by other scientists). He measured the mass-to-charge ratio and discovered it was 1800 times smaller than that of hydrogen, the smallest atom. These corpuscles were a particle unlike any other previously known.

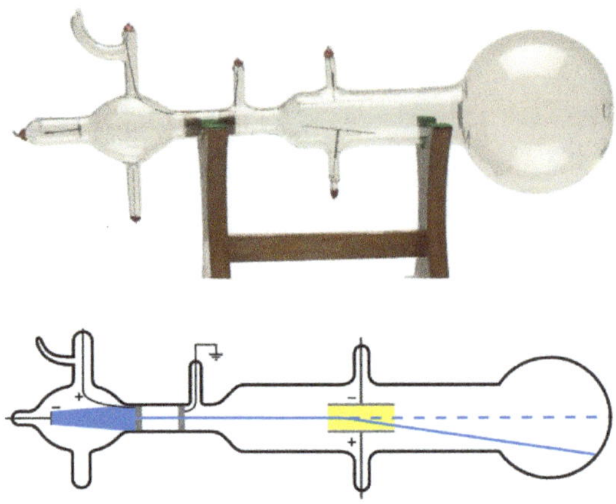

The cathode rays (blue) were emitted from the cathode, sharpened to a beam by the slits, then deflected as they passed between the two electrified plates.

Thomson suggested that atoms were divisible, and that the corpuscles were their building blocks. To explain the overall neutral charge of the atom, he proposed that the corpuscles were distributed in a uniform sea of positive charge; this was the plum pudding model as the electrons were embedded in the positive charge like plums in a plum pudding (although in Thomson's model they were not stationary).

Discovery of the Nucleus

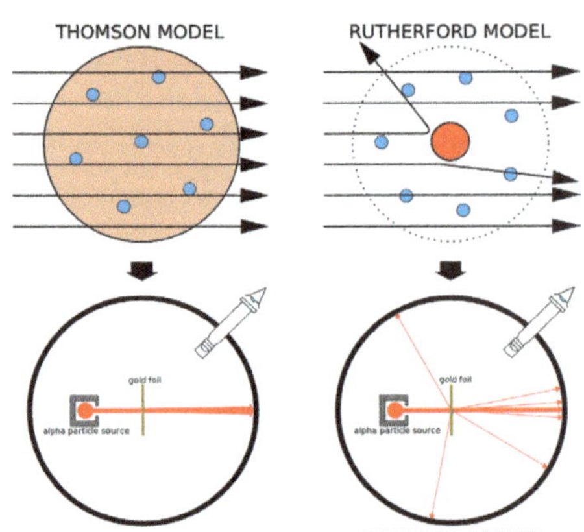

The Geiger-Marsden experiment *Left:* Expected results: alpha particles passing through the plum pudding model of the atom with negligible deflection. *Right:* Observed results: a small portion of the particles were deflected by the concentrated positive charge of the nucleus.

Thomson's plum pudding model was disproved in 1909 by one of his former students, Ernest Rutherford, who discovered that most of the mass and positive charge of an atom is concentrated in a very small fraction of its volume, which he assumed to be at the very center.

In the Geiger–Marsden experiment, Hans Geiger and Ernest Marsden (colleagues of Rutherford working at his behest) shot alpha particles at thin sheets of metal and measured their deflection through the use of a fluorescent screen. Given the very small mass of the electrons, the high momentum of the alpha particles, and the low concentration of the positive charge of the plum pudding model, the experimenters expected all the alpha particles to pass through the metal foil without significant deflection. To their astonishment, a small fraction of the alpha particles experienced heavy deflection. Rutherford concluded that the positive charge of the atom must be concentrated in a very tiny volume to produce an electric field sufficiently intense to deflect the alpha particles so strongly.

This led Rutherford to propose a planetary model in which a cloud of electrons surrounded a small, compact nucleus of positive charge. Only such a concentration of charge could produce the electric field strong enough to cause the heavy deflection.

First Steps Toward a Quantum Physical Model of the Atom

The planetary model of the atom had two significant shortcomings. The first is that, unlike planets orbiting a sun, electrons are charged particles. An accelerating electric charge is known to emit electromagnetic waves according to the Larmor formula in classical electromagnetism. An orbiting charge should steadily lose energy and spiral toward the nucleus, colliding with it in a small fraction of a second. The second problem was that the planetary model could not explain the highly peaked emission and absorption spectra of atoms that were observed.

Quantum theory revolutionized physics at the beginning of the 20th century, when Max Planck and Albert Einstein postulated that light energy is emitted or absorbed in discrete amounts known as quanta (singular, *quantum*). In 1913, Niels Bohr incorporated this idea into his Bohr model of the atom, in which an electron could only orbit the nucleus in particular circular orbits with fixed angular momentum and energy, its distance from the nucleus (i.e., their radii) being proportional to its energy. Under this model an electron could not spiral into the nucleus because it could not lose energy in a continuous manner; instead, it could only make instantaneous "quantum leaps" between the fixed energy levels. When this occurred, light was emitted or absorbed at a frequency proportional to the change in energy (hence the absorption and emission of light in discrete spectra).

Bohr's model was not perfect. It could only predict the spectral lines of hydrogen; it couldn't predict those of multielectron atoms. Worse still, as spectrographic technology improved, additional spectral lines in hydrogen were observed which Bohr's model couldn't explain. In 1916, Arnold Sommerfeld added elliptical orbits to the Bohr model to explain the extra emission lines, but this made the model very difficult to use, and it still couldn't explain more complex atoms.

Discovery of Isotopes

While experimenting with the products of radioactive decay, in 1913 radiochemist Frederick Soddy discovered that there appeared to be more than one element at each position on the periodic table. The term isotope was coined by Margaret Todd as a suitable name for these elements.

That same year, J.J. Thomson conducted an experiment in which he channeled a stream of neon ions through magnetic and electric fields, striking a photographic plate at the other end. He observed two glowing patches on the plate, which suggested two different deflection trajectories. Thomson concluded this was because some of the neon ions had a different mass. The nature of this differing mass would later be explained by the discovery of neutrons in 1932.

Discovery of Nuclear Particles

In 1917 Rutherford bombarded nitrogen gas with alpha particles and observed hydrogen nuclei being emitted from the gas (Rutherford recognized these, because he had previously obtained them bombarding hydrogen with alpha particles, and observing hydrogen nuclei in the products). Rutherford concluded that the hydrogen nuclei emerged from the nuclei of the nitrogen atoms themselves (in effect, he had split a nitrogen).

From his own work and the work of his students Bohr and Henry Moseley, Rutherford knew that the positive charge of any atom could always be equated to that of an integer number of hydrogen nuclei. This, coupled with the atomic mass of many elements being roughly equivalent to an integer number of hydrogen atoms - then assumed to be the lightest particles - led him to conclude that hydrogen nuclei were singular particles and a basic constituent of all atomic nuclei. He named such particles protons. Further experimentation by Rutherford found that the nuclear mass of most atoms exceeded that of the protons it possessed; he speculated that this surplus mass was composed of hitherto unknown neutrally charged particles, which were tentatively dubbed "neutrons".

In 1928, Walter Bothe observed that beryllium emitted a highly penetrating, electrically neutral radiation when bombarded with alpha particles. It was later discovered that this radiation could knock hydrogen atoms out of paraffin wax. Initially it was thought to be high-energy gamma radiation, since gamma radiation had a similar effect on electrons in metals, but James Chadwick found that the ionization effect was too strong for it to be due to electromagnetic radiation, so long as energy and momentum were conserved in the interaction. In 1932, Chadwick exposed various elements, such as hydrogen and nitrogen, to the mysterious "beryllium radiation", and by measuring the energies of the recoiling charged particles, he deduced that the radiation was actually composed of electrically neutral particles which could not be massless like the gamma ray, but instead were required to have a mass similar to that of a proton. Chadwick now claimed these particles as Rutherford's neutrons. For his discovery of the neutron, Chadwick received the Nobel Prize in 1935.

Quantum Physical Models of the Atom

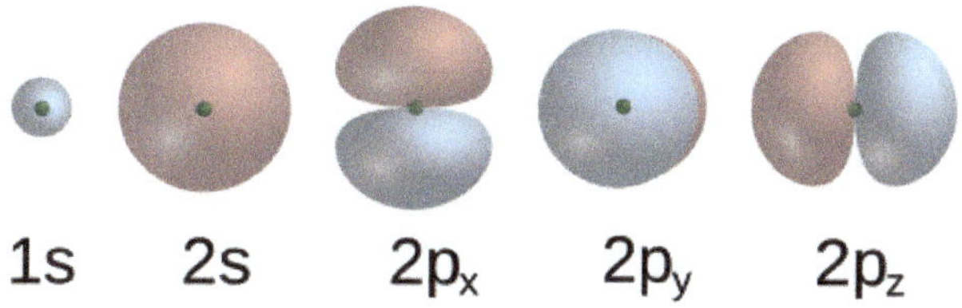

The five filled atomic orbitals of a neon atom separated and arranged in order of increasing energy from left to right, with the last three orbitals being equal in energy. Each orbital holds up to two electrons, which most probably exist in the zones represented by the colored bubbles. Each electron is equally present in both orbital zones, shown here by color only to highlight the different wave phase.

In 1924, Louis de Broglie proposed that all moving particles—particularly subatomic particles such as electrons—exhibit a degree of wave-like behavior. Erwin Schrödinger, fascinated by this idea, explored whether or not the movement of an electron in an atom could be better explained as a wave rather than as a particle. Schrödinger's equation, published in 1926, describes an electron as a wavefunction instead of as a point particle. This approach elegantly predicted many of the spectral phenomena that Bohr's model failed to explain. Although this concept was mathematically convenient, it was difficult to visualize, and faced opposition. One of its critics, Max Born, proposed instead that Schrödinger's wavefunction described not the electron but rather all its possible states, and thus could be used to calculate the probability of finding an electron at any given location around the nucleus. This reconciled the two opposing theories of particle versus wave electrons and the idea of wave–particle duality was introduced. This theory stated that the electron may exhibit the properties of both a wave and a particle. For example, it can be refracted like a wave, and has mass like a particle.

A consequence of describing electrons as waveforms is that it is mathematically impossible to simultaneously derive the position and momentum of an electron. This became known as the Heisenberg uncertainty principle after the theoretical physicist Werner Heisenberg, who first described it and published it in 1927. This invalidated Bohr's model, with its neat, clearly defined circular orbits. The modern model of the atom describes the positions of electrons in an atom in terms of probabilities. An electron can potentially be found at any distance from the nucleus, but, depending on its energy level, exists more frequently in certain regions around the nucleus than others; this pattern is referred to as its atomic orbital. The orbitals come in a variety of shapes-sphere, dumbbell, torus, etc.-with the nucleus in the middle.

Atomism

Atomism (from Greek ἄτομον, *atomon*, i.e. "uncuttable", "indivisible") is a natural philosophy that developed in several ancient traditions. The atomists theorized that nature consists of two fundamental principles: *atom* and *void*. Unlike their modern scientific namesake in atomic theory, philosophical atoms come in an infinite variety of shapes and sizes, each indestructible, immutable and surrounded by a void where they collide with the others or hook together forming a cluster. Clusters of different shapes, arrangements, and positions give rise to the various macroscopic substances in the world.

References to the concept of atomism and its atoms are found in ancient India and ancient Greece. In the West, atomism emerged in the 5th century BCE with Leucippus and Democritus. In India the Jain, Ajivika and Carvaka schools of atomism may date back to the 4th century BCE. The Nyaya and Vaisheshika schools later developed theories on how atoms combined into more complex objects. Whether Indian culture influenced Greek or vice versa or whether both evolved independently is a matter of dispute.

The particles of chemical matter for which chemists and other natural philosophers of the early 19th century found experimental evidence were thought to be indivisible, and therefore were given the name "atom", long used by the atomist philosophy.

However, in the 20th century, the "atoms" of the chemists were found to be composed of even smaller entities: electrons, neutrons and protons, and further experiments showed that protons and neutrons are made of quarks. Although the connection to historical atomism is at best tenuous, elementary particles have thus become a modern analog of philosophical atoms, despite the misnomer in chemistry.

Reductionism

Philosophical atomism is a reductive argument: not only that everything is composed of atoms and void, but that nothing they compose really exists: the only things that really exist are atoms ricocheting off each other mechanistically in an otherwise empty void. Atomism stands in contrast to a substance theory wherein a prime material continuum remains qualitatively invariant under division (for example, the ratio of the four classical elements would be the same in any portion of a homogeneous material).

Indian Buddhists, such as Dharmakirti and others, also developed distinctive theories of atomism, for example, involving momentary (instantaneous) atoms, that flash in and out of existence (Kalapas).

Greek Atomism

In the 5th century B.C.E., Leucippus and his pupil Democritus proposed that all matter was composed of small indivisible particles called atoms, in order to reconcile two conflicting schools of thought on the nature of reality. On one side was Heraclitus, who believed that the nature of all existence is change. On the other side was Parmenides, who believed instead that all change is illusion.

Parmenides denied the existence of motion, change and void. He believed all existence to be a single, all-encompassing and unchanging mass (a concept known as monism), and that change and motion were mere illusions. This conclusion, as well as the reasoning that led to it, may indeed seem baffling to the modern empirical mind, but Parmenides explicitly rejected sensory experience as the path to an understanding of the universe, and instead used purely abstract reasoning. Firstly, he believed there is no such thing as void, equating it with non-being (i.e. "if the void ***is***, then it is not nothing; therefore it is not the void"). This in turn meant that motion is impossible, because there is no void to move into. He also wrote all that *is* must be an indivisible unity, for if it were manifold, then there would have to be a void that could divide it (and he did not believe the void exists). Finally, he stated that the all encompassing Unity is unchanging, for the Unity already encompasses all that is and can be.

Democritus accepted most of Parmenides' arguments, except for the idea that change is an illusion. He believed change was real, and if it was not then at least the illusion had to be explained. He thus supported the concept of void, and stated that the universe is made up of many Parmenidean entities that move around in the void. The void is infinite and provides the space in which the atoms can pack or scatter differently. The different possible packings and scatterings within the void make up the shifting outlines and bulk of the objects that organisms feel, see, eat, hear, smell, and taste. While organisms may feel hot or cold, hot and cold actually have no real existence. They are simply sensations produced in organisms by the different packings and scatterings of the atoms in the void that compose the object that organisms sense as being "hot" or "cold".

The work of Democritus only survives in secondhand reports, some of which are unreliable or conflicting. Much of the best evidence of Democritus' theory of atomism is reported by Aristotle in his discussions of Democritus' and Plato's contrasting views on the types of indivisibles composing the natural world.

Geometry and Atoms

Element	Polyhedron		Number of Faces	Number of Triangles
Fire	Tetrahedron (Animation)	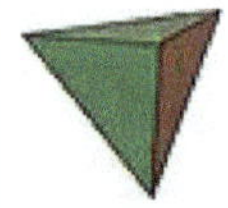	4	24
Air	Octahedron (Animation)	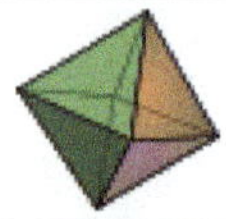	8	48
Water	Icosahedron (Animation)	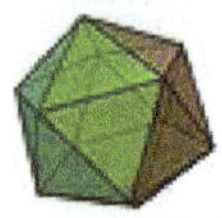	20	120
Earth	Cube (Animation)	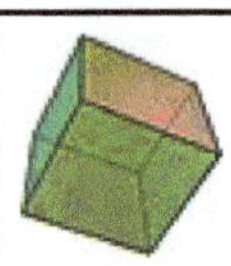	6	24

Geometrical Simple Bodies According to Plato

Plato (c. 427 — c. 347 B.C.E.), were he familiar with the atomism of Democritus, would have objected to its mechanistic materialism. He argued that atoms just crashing into other atoms could never produce the beauty and form of the world. In Plato's *Timaeus*, (28B – 29A) the character of Timeaus insisted that the cosmos was not eternal but was created, although its creator framed it after an eternal, unchanging model.

One part of that creation were the four simple bodies of fire, air, water, and earth. But Plato did not consider these corpuscles to be the most basic level of reality, for in his view they were made up of an unchanging level of reality, which was mathematical. These simple bodies were geometric solids, the faces of which were, in turn, made up of triangles. The square faces of the cube were each made up of four isosceles right-angled triangles and the triangular faces of the tetrahedron, octahedron, and icosahedron were each made up of six right-angled triangles.

He postulated the geometric structure of the simple bodies of the four elements as summarized in the table to the right. The cube, with its flat base and stability, was assigned to earth; the tetrahedron was assigned to fire because its penetrating points and sharp edges made it mobile. The points and edges of the octahedron and icosahedron were blunter and so these less mobile bodies were assigned to air and water. Since the simple bodies could be decomposed into triangles, and the triangles reassembled into atoms of different elements, Plato's model offered a plausible account of changes among the primary substances.

The Rejection of Atoms

Sometime before 330 B.C.E. Aristotle asserted that the elements of fire, air, earth, and water were not made of atoms, but were continuous. Aristotle considered the existence of a void, which was required by atomic theories, to violate physical principles. Change took place not by the rearrangement of atoms to make new structures, but by transformation of matter from what it was in potential to a new actuality. A piece of wet clay, when acted upon by a potter, takes on its potential to be an actual drinking mug. Aristotle has often been criticized for rejecting atomism, but in ancient Greece the atomic theories of Democritus remained "pure speculations, incapable of being put to any experimental test. Granted that atomism was, in the long run, to prove far more fruitful than any qualitative theory of matter, in the short run the theory that Aristotle proposed must have seemed in some respects more promising".

Later Ancient Atomism

Epicurus (341–270) studied atomism with Nausiphanes who had been a student of Democritus. Although Epicurus was certain of the existence of atoms and the void, he was less sure we could adequately explain specific natural phenomena such as earthquakes, lightning, comets, or the phases of the Moon (Lloyd 1973, 25–6). Few of Epicurus's writings survive and those that do reflect his interest in applying Democritus's theories to assist people in taking responsibility for themselves and for their own happiness—since he held there are no gods around that can help them. He understood gods' role as moral ideals.

His ideas are also represented in the works of his follower Lucretius, who wrote *On the Nature of Things*. This scientific work in poetic form illustrates several segments of Epicurean theory on how the universe came into its current stage and it shows that the phenomena we perceive are actually composite forms. The atoms and the void are eternal and in constant motion. Atomic collisions create objects, which are still composed of the same eternal atoms whose motion for a while is incorporated into the created entity. Human sensations and meteorological phenomena are also explained by Lucretius in terms of atomic motion.

Atomism and Ethics

Some later philosophers attributed the idea, that man created gods and the gods did not create man, to Democritus. For example, Sextus Empiricus noted:

> Some people think that we arrived at the idea of gods from the remarkable things that happen in the world. Democritus ... says that the people of ancient times were frightened by happenings in the heavens such as thunder, lightning, ..., and thought that they were caused by gods.

Three hundred years after Epicurus, Lucretius in his epic poem *On the Nature of Things* would depict him as the hero who crushed the monster Religion through educating the people in what was possible in the atoms and what was *not* possible in the atoms. However, Epicurus expressed a non-aggressive attitude characterized by his statement: “The man who best knows how to meet external threats makes into one family all the creatures he can; and those he can not, he at any rate does not treat as aliens; and where he finds even this impossible, he avoids all dealings, and, so far as is advantageous, excludes them from his life.”

Indian Atomism

The Indian atomistic position, like many movements in Indian Philosophy and Mathematics, starts with an argument from Linguistics. The Vedic etymologist and grammarian Yaska (c. 7th century BC) in his Nirukta, in dealing with models for how linguistic structures get to have their meanings, takes the atomistic position that words are the “primary” carrier of meaning – i.e. words have a preferred ontological status in defining meaning. This position was to be the subject of a fierce debate in the Indian tradition from the early Christian era till the 18th century, involving different philosophers from the Nyaya, Mimamsa and Buddhist schools.

In the pratishakhya text (c. 2nd century BCE), the gist of the controversy was stated cryptically in the sutra form as “saMhitA pada-prakr^tiH”. According to the atomist view, the words (*pada*) would be the primary elements (*prakrti*) out of which the sentence is constructed, while the holistic view considers the sentence as the primary entity, originally “given” in its context of utterance, and the words are arrived at only through analysis and abstraction.

These two positions came to be called *a-kShaNDa-pakSha* (indivisibility or sentence-holism), a position developed later by Bhartrihari (c. 500 AD), vs. *kShaNDa-pakSha* (atomism), a position adopted by the Mimamsa and Nyaya schools (Note: *kShanDa* = fragmented; “a-kShanDa” = whole).

Between the 5th and 3rd centuries BC, the atom (anu or aṇor) is mentioned in the Bhagavad Gita (Chapter 8, Verse 9):

kaviṁ purāṇam anuśāsitāram ***aṇor*** *aṇīyāṁsam anusmared yaḥ sarvasya dhātāram acintya-rūpam āditya-varṇaṁ tamasaḥ parastāt*One meditates on the omniscient, primordial, the controller, smaller than the atom, yet the maintainer of everything; whose form is inconceivable, resplendent like the sun and totally transcendental to material nature

The ancient *shāshvata-vāda* doctrine of eternalism, which held that elements are eternal, is also suggestive of a possible starting point for atomism (Gangopadhyaya 1981).

The atomist position had transcended language into epistemology by the time that

Nyaya–Vaisesika, Buddhist and Jaina theology were developing mature philosophical positions.

Will Durant wrote in *Our Oriental Heritage*:

"Two systems of Indian thought propound physical theories suggestively similar to those of Greece. Kanada, founder of the Vaisheshika philosophy, held that the world was composed of atoms as many in kind as the various elements. The Jains more nearly approximated to Democritus by teaching that all atoms were of the same kind, producing different effects by diverse modes of combinations. Kanada believed light and heat to be varieties of the same substance; Udayana taught that all heat comes from the sun; and Vachaspati, like Newton, interpreted light as composed of minute particles emitted by substances and striking the eye."

Indian atomism in the Middle Ages was still mostly philosophical and/or religious in intent, though it was also scientific. Because the "infallible Vedas", the oldest Hindu texts, do not mention atoms (though they do mention elements), atomism was not orthodox in many schools of Hindu philosophy, although accommodationist interpretations or assumptions of lost text justified the use of atomism for non-orthodox schools of Hindu thought. The Buddhist and Jaina schools, however, were more willing to accept the ideas of atomism.

Nyaya–Vaisesika School

The Nyaya–Vaisesika school developed one of the earliest forms of atomism; scholars date the Nyaya and Vaisesika texts from the 6th to 1st centuries BC. Like the Buddhist atomists, the Vaisesika had a pseudo-Aristotelian theory of atomism. They posited the four elemental atom types, but in Vaisesika physics atoms had 24 different possible qualities, divided between general extensive properties and specific (intensive) properties. Like the Jaina school, the Nyaya–Vaisesika atomists had elaborate theories of how atoms combine. In both Jaina and Vaisesika atomism, atoms first combine in pairs (dyads), and then group into trios of pairs (triads), which are the smallest visible units of matter.

Buddhist School

The Buddhist atomists had very qualitative, Aristotelian-style atomic theory. According to ancient Buddhist atomism, which probably began developing before the 4th century BC, there are four kinds of atoms, corresponding to the standard elements. Each of these elements has a specific property, such as solidity or motion, and performs a specific function in mixtures, such as providing support or causing growth. Like the Hindu Jains, the Buddhists were able to integrate a theory of atomism with their theological presuppositions. Later Indian Buddhist philosophers, such as Dharmakirti and Dignāga, considered atoms to be point-sized, durationless, and made of energy.

Jaina School

The most elaborate and well-preserved Indian theory of atomism comes from the philosophy of the Jaina school, dating back to at least the 6th century BC. Some of the Jain texts that refer to matter and atoms are Pancastikayasara, Kalpasutra, Tattvarthasutra and Pannavana Suttam. The Jains envisioned the world as consisting wholly of atoms, except for souls. Paramāṇus or atoms were considered as the basic building blocks of all matter. Their concept of atoms was very similar to classical atomism, differing primarily in the specific properties of atoms. Each atom, according to Jain philosophy, has one kind of taste, one smell, one color, and two kinds of touch, though it is unclear what was meant by "kind of touch". Atoms can exist in one of two states: subtle, in which case they can fit in infinitesimally small spaces, and gross, in which case they have extension and occupy a finite space. Certain characteristics of Paramāṇu correspond with that sub-atomic particles. For example, Paramāṇu is characterized by continuous motion either in a straight line or in case of attractions from other Paramāṇus, it follows a curved path. This corresponds with the description of orbit of electrons across the Nucleus. Ultimate particles are also described as particles with positive (Snigdha i.e. smooth charge) and negative (Rūksa – rough) charges that provide them the binding force. Although atoms are made of the same basic substance, they can combine based on their eternal properties to produce any of six "aggregates", which seem to correspond with the Greek concept of "elements": earth, water, shadow, sense objects, karmic matter, and unfit matter. To the Jains, karma was real, but was a naturalistic, mechanistic phenomenon caused by buildups of subtle karmic matter within the soul. They also had detailed theories of how atoms could combine, react, vibrate, move, and perform other actions, all of which were thoroughly deterministic.

Time as Explained in Srimad-Bhagavatam

"The material manifestation's ultimate particle, which is indivisible and not formed into a body, is called the atom (*parama-aṇuh*). It exists always as an invisible identity, even after the dissolution of all forms. The material body is but a combination of such atoms, but it is misunderstood by the common man.

Atoms are the ultimate state of the manifest universe. When they stay in their own forms without forming different bodies, they are called the unlimited oneness. There are certainly different bodies in physical forms, but the atoms themselves form the complete manifestation.

One can estimate time by measuring the movement of the atomic combination of bodies. Time is the potency of the almighty Personality of Godhead, Hari, who controls all physical movement although He is not visible in the physical world.

The division of gross time is calculated as follows: two atoms make one double atom, and three double atoms make one hexatom. This hexatom is visible in the sunshine which enters through the holes of a window screen. One can clearly see that the hexatom goes up towards the sky.

The atom is described as an invisible particle, but when six such atoms combine together, they are called a *trasareṇu*, and this is visible in the sunshine pouring through the holes of a window screen."

– Excerpted from *Srimad-Bhagavatam* (*SB* 3.11.1–5) by A. C. Bhaktivedanta Swami Prabhupada, courtesy of the Bhaktivedanta Book Trust International (prabhupadabooks.com)

Islamic Atomism

Atomistic philosophies are found very early in Islamic philosophy and was influenced by earlier Greek and to some extent Indian philosophy. Like both the Greek and Indian versions, Islamic atomism was a charged topic that had the potential for conflict with the prevalent religious orthodoxy, but it was instead more often favoured by orthodox Islamic theologians. It was such a fertile and flexible idea that, as in Greece and India, it flourished in some leading schools of Islamic thought.

Asharite Atomism

The most successful form of Islamic atomism was in the Asharite school of Islamic theology, most notably in the work of the theologian al-Ghazali (1058–1111). In Asharite atomism, atoms are the only perpetual, material things in existence, and all else in the world is "accidental" meaning something that lasts for only an instant. Nothing accidental can be the cause of anything else, except perception, as it exists for a moment. Contingent events are not subject to natural physical causes, but are the direct result of God's constant intervention, without which nothing could happen. Thus nature is completely dependent on God, which meshes with other Asharite Islamic ideas on causation, or the lack thereof (Gardet 2001). Al-Ghazali also used the theory to support his theory of occasionalism. In a sense, the Asharite theory of atomism has far more in common with Indian atomism than it does with Greek atomism.

Averroism

Other traditions in Islam rejected the atomism of the Asharites and expounded on many Greek texts, especially those of Aristotle. An active school of philosophers in Spain, including the noted commentator Averroes (AD 1126–1198) explicitly rejected the thought of al-Ghazali and turned to an extensive evaluation of the thought of Aristotle. Averroes commented in detail on most of the works of Aristotle and his commentaries did much to guide the interpretation of Aristotle in later Jewish and Christian scholastic thought.

Medieval European speculations

While Aristotelian philosophy eclipsed the importance of the atomists in late Roman and medieval Europe, their work was still preserved and exposited through commentaries on the works of Aristotle. In the 2nd century, Galen (AD 129–216) presented extensive discussions of the Greek atomists, especially Epicurus, in his Aristotle commentaries. According to historian of atomism Joshua Gregory, there was no serious work done with atomism from the time of Galen until Gassendi and Descartes resurrected it in the 17th century; "the gap between these two 'modern naturalists' and the ancient Atomists marked "the exile of the atom" and "it is universally admitted that the Middle Ages had abandoned Atomism, and virtually lost it."

However, although the ancient atomists' works were unavailable, Scholastic thinkers still had Aristotle's critiques of atomism. In the medieval universities there were expressions of atomism. For example, in the 14th century Nicholas of Autrecourt considered that matter, space, and time were all made up of indivisible atoms, points, and instants and that all generation and corruption took place by the rearrangement of material atoms. The similarities of his ideas with those of al-Ghazali suggest that Nicholas may have been familiar with Ghazali's work, perhaps through Averroes' refutation of it (Marmara, 1973–74).

Scholastic Minima Naturalia

Although the atomism of Epicurus had fallen out of favor in the centuries of Scholasticism, a related Aristotelian concept, that of *minima naturalia* (natural minima) received extensive consideration. *Minima naturalia* were theorized by Aristotle as the smallest parts into which a homogeneous natural substance (e.g., flesh, bone, or wood) could be divided and still retain its essential character. Speculation on *minima naturalia* provided philosophical background for the mechanistic philosophy of early modern thinkers like Descartes, and for the alchemical works of Geber and Daniel Sennert, who in turn influenced the corpuscularian alchemist Robert Boyle, one of the founders of modern chemistry.

Unlike the atomism of Leucippus, Democritus, and Epicurus, and also unlike the later atomic theory of John Dalton, the Aristotelian natural minimum was not conceptualized as physically indivisible--"atomic" in the contemporary sense. Instead, the concept was rooted in Aristotle's hylomorphic worldview, which held that every physical thing is a compound of matter (Greek *hyle*) and an immaterial substantial form (Greek *morphe*) that imparts its essential nature and structure. For instance, a rubber ball for a hylomorphist like Aristotle would be rubber (matter) structured by spherical shape (form).

Aristotle's intuition was that there is some smallest size beyond which matter could no longer be structured as flesh, or bone, or wood, or some other such organic substance that for Aristotle, living before the microscope, could be considered homogeneous. For instance, if flesh were divided beyond its natural minimum, what would be left might be a large amount of the element water, and smaller amounts of the other elements. But whatever water or other elements were left, they would no longer have the "nature" of flesh: in hylomorphic terms, they would no longer be matter structured by the form of flesh; instead the remaining water, e.g., would be matter structured by the form of water, not the form of flesh. This is suggestive of modern chemistry, in which, e.g., a bar of gold can be continually divided until one has a single atom of gold, but further division yields only subatomic particles (electrons, quarks, etc.) which are no longer "gold." However, the parallel is not exact: *minima naturalia* are not a direct anticipation of modern chemical and physical concepts.

A chief theme in late Roman and Scholastic commentary on this concept is reconciling *minima naturalia* with the general Aristotelian principle of infinite divisibility. Commentators like John Philoponus and Thomas Aquinas reconciled these aspects of Aristotle's thought by distinguishing between mathematical and "natural" divisibility. With few exceptions, much of the curriculum in the universities of Europe was based on such Aristotelianism for most of the Middle Ages (Kargon 1966). Scholasticism was standard science in the time of Isaac Newton, but in the 17th century, a renewed interest in Epicurean atomism and corpuscularianism as a hybrid or an alternative to Aristotelian physics had begun to mount outside the classroom.

Atomic Renaissance

The main figures in the rebirth of atomism were René Descartes, Pierre Gassendi, and Robert Boyle, as well as other notable figures.

One of the first groups of atomists in England was a cadre of amateur scientists known as the Northumberland circle, led by Henry Percy (1585–1632), the 9th Earl of Northumberland. Although they published little of account, they helped to disseminate atomistic ideas among the burgeoning scientific culture of England, and may have been particularly influential to Francis Bacon, who became an atomist around 1605, though he later rejected some of the claims of atomism. Though they revived the classical form of atomism, this group was among the scientific avant-garde: the Northumberland circle contained nearly half of the confirmed Copernicans prior to 1610 (the year of Galileo's The Starry Messenger). Other influential atomists of late 16th and early 17th centuries include Giordano Bruno, Thomas Hobbes (who also changed his stance on atomism late in his career), and Thomas Hariot. A number of different atomistic theories were blossoming in France at this time, as well (Clericuzio 2000).

Galileo Galilei (1564–1642) was an advocate of atomism in his 1612, *Discourse on Floating Bodies* (Redondi 1969). In The Assayer, Galileo offered a more complete physical system based on a corpuscular theory of matter, in which all phenomena—with the exception of sound—are produced by "matter in motion". Galileo identified some basic problems with Aristotelian physics through his experiments. He utilized a theory of atomism as a partial replacement, but he was never unequivocally committed to it. For example, his experiments with falling bodies and inclined planes led him to the concepts of circular inertial motion and accelerating free-fall. The current Aristotelian theories of impetus and terrestrial motion were inadequate to explain these. While atomism did not explain the law of fall either, it was a more promising framework in which to develop an explanation because motion was conserved in ancient atomism (unlike Aristotelian physics).

René Descartes' (1596–1650) "mechanical" philosophy of corpuscularism had much in common with atomism, and is considered, in some senses, to be a different version of it. Descartes thought everything physical in the universe to be made of tiny *vortices* of matter. Like the ancient atomists, Descartes claimed that sensations, such as taste or temperature, are caused by the shape and size of tiny pieces of matter. The main difference between atomism and Descartes' concept was the existence of the void. For him, there could be no vacuum, and all matter was constantly swirling to prevent a void as corpuscles moved through other matter. Another key distinction between Descartes' view and classical atomism is the mind/body duality of Descartes, which allowed for an independent realm of existence for thought, soul, and most importantly, God. Gassendi's concept was closer to classical atomism, but with no atheistic overtone.

Pierre Gassendi (1592–1655) was a Catholic priest from France who was also an avid natural philosopher. He was particularly intrigued by the Greek atomists, so he set out to "purify" atomism from its heretical and atheistic philosophical conclusions (Dijksterhius 1969). Gassendi formulated his atomistic conception of mechanical philosophy partly in response to Descartes; he particularly opposed Descartes' reductionist view that only purely mechanical explanations of physics are valid, as well as the application of geometry to the whole of physics (Clericuzio 2000).

Corpuscularianism

Corpuscularianism is similar to atomism, except that where atoms were supposed to be indivisible, corpuscles could in principle be divided. In this manner, for example, it was theorized that mercury could penetrate into metals and modify their inner structure, a step on the way towards transmutative production of gold. Corpuscularianism was associated by its leading proponents with the idea that some of the properties that objects appear to have are artifacts of the perceiving mind: 'secondary' qualities as distinguished from 'primary' qualities. Not all corpuscularianism made use of the primary-secondary quality distinction, however. An influential tradition in medieval and early modern alchemy argued that chemical analysis revealed the existence of robust corpuscles that retained their identity in chemical compounds (to use the modern term). William R. Newman has dubbed this approach to matter theory "chymical atomism," and has argued for its significance to both the mechanical philosophy and to the chemical atomism that emerged in the early 19th century. Corpuscularianism stayed a dominant theory over the next several hundred years and retained its links with alchemy in the work of scientists such as Robert Boyle and Isaac Newton in the 17th century. It was used by Newton, for instance, in his development of the corpuscular theory of light. The form that came to be accepted by most English scientists after Robert Boyle (1627–1692) was an amalgam of the systems of Descartes and Gassendi. In The Sceptical Chymist (1661), Boyle demonstrates problems that arise from chemistry, and offers up atomism as a possible explanation. The unifying principle that would eventually lead to the acceptance of a hybrid corpuscular–atomism was mechanical philosophy, which became widely accepted by physical sciences.

Atomic Theory

By the late 18th century, the useful practices of engineering and technology began to influence philosophical explanations for the composition of matter. Those who speculated on the ultimate nature of matter began to verify their "thought experiments" with some repeatable demonstrations, when they could.

Roger Boscovich provided the first general mathematical theory of atomism, based on the ideas of Newton and Leibniz but transforming them so as to provide a programme for atomic physics.

In 1808, John Dalton assimilated the known experimental work of many people to summarize the empirical evidence on the composition of matter. He noticed that distilled water everywhere analyzed to the same elements, hydrogen and oxygen. Similarly, other purified substances decomposed to the same elements in the same proportions by weight.

> Therefore we may conclude that the ultimate particles of all homogeneous bodies are perfectly alike in weight, figure, etc. In other words, every particle of water is like every other particle of water; every particle of hydrogen is like every other particle of hydrogen, etc.

Furthermore, he concluded that there was a unique atom for each element, using Lavoisier's definition of an element as a substance that could not be analyzed into something simpler. Thus, Dalton concluded the following.

> Chemical analysis and synthesis go no farther than to the separation of particles one from another, and to their reunion. No new creation or destruction of matter is within the reach of chemical agency. We might as well attempt to introduce a new planet into the solar system, or to annihilate one already in existence, as to create or destroy a particle of hydrogen. All the changes we can produce, consist in separating particles that are in a state of cohesion or combination, and joining those that were previously at a distance.

And then he proceeded to give a list of relative weights in the compositions of several common compounds, summarizing:

> 1st. That water is a binary compound of hydrogen and oxygen, and the relative weights of the two elementary atoms are as 1:7, nearly;
>
> 2nd. That ammonia is a binary compound of hydrogen and azote nitrogen, and the relative weights of the two atoms are as 1:5, nearly...

Dalton concluded that the fixed proportions of elements by weight suggested that the atoms of one element combined with only a limited number of atoms of the other elements to form the substances that he listed.

Atomic Theory Controversy

Dalton's atomic theory remained controversial throughout the 19th century. Whilst the Law of definite proportion was accepted, the hypothesis that this was due to atoms was not so widely accepted. For example, in 1826 when Sir Humphry Davy presented Dalton the Royal Medal from the Royal Society, Davy said that the theory only became useful when the atomic conjecture was ignored. Sir Benjamin Collins Brodie in 1866 published the first part of his Calculus of Chemical Operations as a non-atomic alternative to the Atomic Theory. He described atomic theory as a 'Thoroughly materialistic bit of joiners work'. Alexander Williamson used his Presidential Address to the London Chemical Society in 1869 to defend the Atomic Theory against its critics and doubters. This in turn led to further meetings at which the positivists again attacked the supposition that there were atoms. The matter was finally resolved in Dalton's favour in the early 20th century with the rise of atomic physics.

Plum Pudding Model

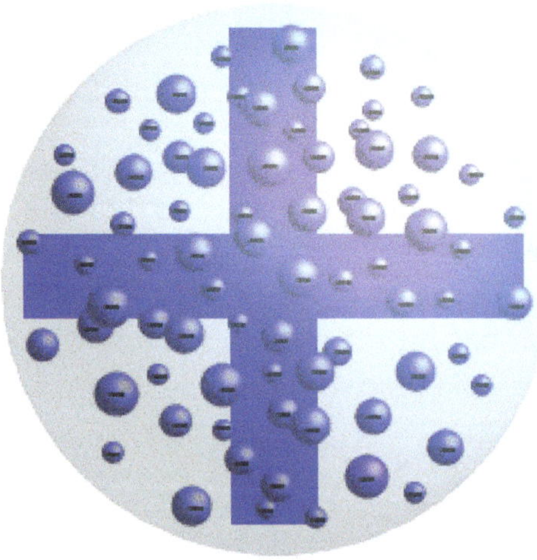

The plum pudding model of the atom.

A Christmas pudding

The plum pudding model is one of several scientific models of the atom. First proposed by J. J. Thomson in 1904 soon after the discovery of the electron, but before the discovery of the atomic nucleus, the model represented an attempt to consolidate the known properties of atoms at the time: 1) electrons are negatively-charged particles and 2) atoms are neutrally-charged.

Overview

In this model, atoms were known to consist of negatively charged electrons. Though Thomson called them "corpuscles," they were more commonly called "electrons" as G. J. Stoney proposed in 1894. At the time, atoms were known to be neutrally charged. To account for this, Thomson knew atoms must also have a source of positive charge to balance the negative charge of the electrons. He considered three plausible models that would satisfy the known properties of atoms at the time:

1. Each negatively-charged electron was paired with a positively-charged particle that followed it everywhere within the atom.
2. Negatively-charged electrons orbited a central region of positive charge having the same magnitude as all the electrons.
3. The negative electrons occupied a region of space that itself was a uniform positive charge (often considered as a kind of "soup" or "cloud" of positive charge).

Thomson chose the third possibility as the most likely structure of atoms. Thomson published his proposed model in the March 1904 edition of the *Philosophical Magazine*, the leading British science journal of the day. In Thomson's view:

... the atoms of the elements consist of a number of negatively electrified corpuscles enclosed in a sphere of uniform positive electrification, ...

With this model, Thomson abandoned his earlier "nebular atom" hypothesis in which atoms were composed of immaterial vortices. Being an astute and practical scientist, Thomson based his atomic model on known experimental evidence of the day. His proposal of a positive volume charge reflects the nature of his scientific approach to discovery which was to propose ideas to guide future experiments.

The orbits of electrons within the model were stabilized by the fact that when an electron moved away from the centre of the positively-charged sphere, it was subjected to a greater net positive inward force, because there was more positive charge inside its orbit (see Gauss's law). Electrons were free to rotate in rings which were further stabilized by interactions among the electrons, and spectroscopic measurements were meant to account for energy differences associated with different electron rings. Thomson attempted unsuccessfully to reshape his model to account for some of the major spectral lines experimentally known for several elements.

The plum pudding model usefully guided his student, Ernest Rutherford, to devise experiments to further explore the composition of atoms. As well, Thomson's model (along with a similar Saturnian ring model for atomic electrons, also put forward in 1904 by Nagaoka after James Clerk Maxwell's model of Saturn's rings), were useful predecessors of the more correct solar-system-like Bohr model of the atom.

The colloquial nickname "plum pudding" was soon attributed to Thomson's model as the distribution of electrons within its positively-charged region of space reminded many scientists of "plums" in the common English dessert, plum pudding.

In 1909 Hans Geiger and Ernest Marsden conducted experiments with thin sheets of gold. Their professor, Ernest Rutherford, expected to find results consistent with Thomson's atomic model. It wasn't until 1911 that Rutherford correctly interpreted the experiment's results which implied the presence of a very small nucleus of positive charge at the center of gold atoms. This led to the development of the Rutherford model of the atom. Immediately after Rutherford published his results, Antonius Van den Broek made the intuitive proposal that the atomic number of an atom is the total number of units of charge present in its nucleus. Henry Moseley's 1913 experiments (see Moseley's law) provided the necessary evidence to support Van den Broek's proposal. The effective nuclear charge was found to be consistent with the atomic number (Moseley found only one unit of charge difference). This work culminated in the solar-system-like (but quantum-limited) Bohr model of the atom in the same year, in which a nucleus containing an atomic number of positive charges is surrounded by an equal number of electrons in orbital shells. As Thomson's model guided Rutherford's experiments, Bohr's model guided Moseley's research.

Related Scientific Problems

The plum pudding model with a single electron was used in part by the physicist Arthur

Erich Haas in 1910 to estimate the numerical value of Planck's constant and the Bohr radius of hydrogen atoms. Haas' work estimated these values to within an order of magnitude and preceded the work of Niels Bohr by three years. Of note, the Bohr model itself only provides substantially-reasonable predictions for atomic and ionic systems having a single effective electron.

A particularly useful mathematics problem related to the plum pudding model is the optimal distribution of equal point charges on a unit sphere called the Thomson problem. The Thomson problem is a natural consequence of the plum pudding model in the absence of its uniform positive background charge.

The classical electrostatic treatment of electrons confined to spherical quantum dots is also similar to their treatment in the plum pudding model. In this classical problem, the quantum dot is modeled as a simple dielectric sphere (in place of a uniform, positively-charged sphere as in the plum pudding model) in which free, or excess, electrons reside. The electrostatic N-electron configurations are found to be exceptionally close to solutions found in the Thomson problem with electrons residing at the same radius within the dielectric sphere. Notably, the plotted distribution of geometry-dependent energetics has been shown to bear a remarkable resemblance to the distribution of anticipated electron orbitals in natural atoms as arranged on the periodic table of elements. Of great interest, solutions of the Thomson problem exhibit this corresponding energy distribution by comparing the energy of each N-electron solution with the energy of its neighbouring (N-1)-electron solution with one charge at the origin. However, when treated within a dielectric sphere model, the features of the distribution are much more pronounced and provide greater fidelity with respect to electron orbital arrangements in real atoms.

Geiger–Marsden Experiment

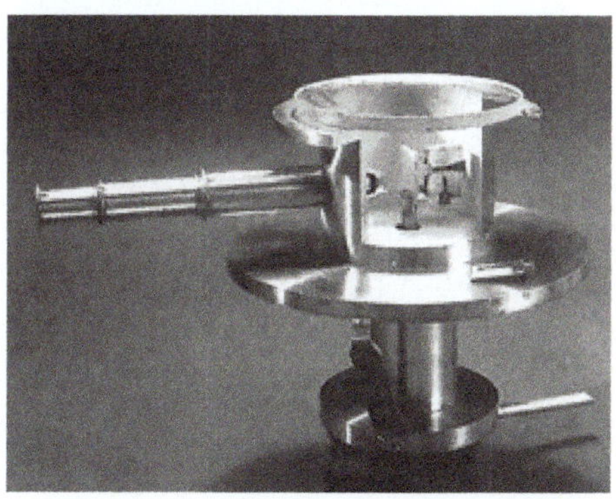

A replica of one of Geiger and Marsden's apparatus

The Geiger–Marsden experiment(s) (also called the Rutherford gold foil experiment) were a landmark series of experiments by which scientists discovered that every atom contains a nucleus where its positive charge and most of its mass are concentrated. They deduced this by measuring how an alpha particle beam is scattered when it strikes a thin metal foil. The experiments were performed between 1908 and 1913 by Hans Geiger and Ernest Marsden under the direction of Ernest Rutherford at the Physical Laboratories of the University of Manchester.

Summary

Contemporary Theories of Atomic Structure

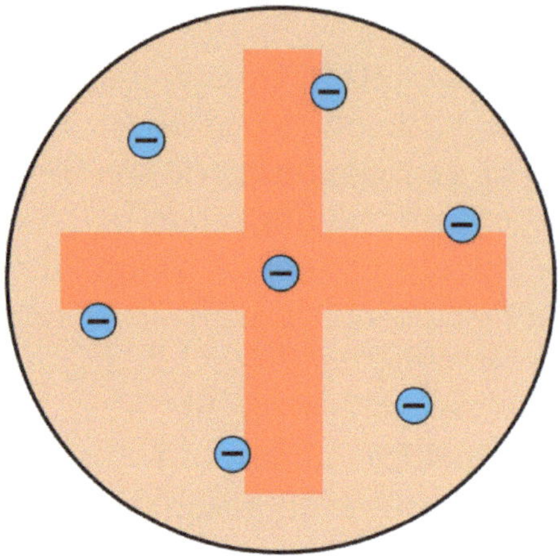

The popular theory of atomic structure at the time of Rutherford's experiment was the "plum pudding model". This model was devised by Lord Kelvin and further developed by J. J. Thomson. Thomson was the scientist who discovered the electron, and that it was a component of every atom. Thomson believed the atom was a sphere of positive charge throughout which the electrons were distributed, a bit like plums in a Christmas pudding. The existence of protons and neutrons was unknown at this time. They knew atoms were very tiny (Rutherford assumed they were in the order of 10^{-8} m in radius). This model was based entirely on classical (Newtonian) physics; the current accepted model uses quantum mechanics.

Thomson's model was not universally accepted even before Rutherford's experiments. Thomson himself was never able to develop a complete and stable model of his concept. A Japanese scientist named Hantaro Nagaoka rejected Thomson's model on the grounds that opposing charges cannot penetrate each other. He proposed instead that electrons orbit the positive charge like the rings around Saturn.

Implications of the Plum Pudding Model

An alpha particle is a sub-microscopic, positively charged particle of matter. According to Thomson's model, if an alpha particle were to collide with an atom, it would just fly straight through, its path being deflected by at most a fraction of a degree. At the atomic scale, the concept of "solid matter" is meaningless, so the alpha particle would not bounce off the atom like a marble; it would be affected only by the atom's electric fields, and Thomson's model predicted that the electric fields in an atom are just too weak to affect a passing alpha particle much (alpha particles tend to move very fast). Both the negative and positive charges within the Thomson atom are spread out over the atom's entire volume. According to Coulomb's Law, the less concentrated a sphere of electric charge is, the weaker its electric field at its surface will be.

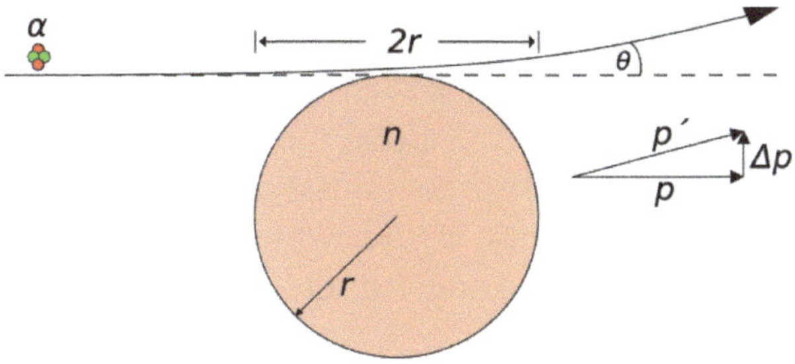

As a worked example, consider an alpha particle passing tangentially to a Thomson gold atom, where it will experience the electric field at its strongest and thus experience the maximum deflection θ. Since the electrons are very light compared to the alpha particle, their influence can be neglected and the atom can be seen as a heavy sphere of positive charge.

Q_n = positive charge of gold atom = 79 e = 1.266×10^{-17} C

Q_α = charge of alpha particle = 2 e = 3.204×10^{-19} C

r = radius of a gold atom = 1.44×10^{-10} m

v_α = velocity of alpha particle = 1.53×10^{7} m/s

m_α = mass of alpha particle = 6.645×10^{-27} kg

k = Coulomb's constant = 8.998×10^{9} N·m²/C²

Using classical physics, the alpha particle's lateral change in momentum Δp can be approximated using the impulse of force relationship and the Coulomb force expression:

$$\Delta p = F\Delta t = k\cdot\frac{Q_\alpha Q_n}{r^2}\cdot\frac{2r}{v_\alpha}$$

$$\theta \approx \frac{\Delta p}{p} < k\cdot\frac{2Q_\alpha Q_n}{m_\alpha r v_\alpha^2} = 8.998\cdot10^9\times\frac{2\times3.204\cdot10^{-19}\times1.266\cdot10^{-17}}{6.645\cdot10^{-27}\times1.44\cdot10^{-10}\times(1.53\cdot10^7)^2}$$

$$\theta < 0.000326\,\text{rad (or } 0.0186^\circ)$$

The above calculation is but an approximation of what happens when an alpha particle comes near a Thomson atom, but it is clear that the deflection at most will be in the order of a small fraction of a degree. If the alpha particle were to pass through a gold foil some 400 atoms thick and experience maximal deflection in the same direction (unlikely), it would still be a small deflection.

The Outcome of the Experiments

At Rutherford's behest, Geiger and Marsden performed a series of experiments where they pointed a beam of alpha particles at a thin foil of metal and measured the scattering pattern by using a fluorescent screen. They spotted alpha particles bouncing off the metal foil in all directions, some right back at the source. This should have been impossible according to Thomson's model; the alpha particles should have all gone straight through. Obviously, those particles had encountered an electrostatic force far greater than Thomson's model suggested they would, which in turn implied that the atom's positive charge was concentrated in a much tinier volume than Thomson imagined.

When Geiger and Marsden shot alpha particles at their metal foils, they noticed only a tiny fraction of the alpha particles were deflected by more than 90°. Most just flew straight through the foil. This suggested that those tiny spheres of intense positive charge were separated by vast gulfs of empty space. Imagine you are standing on the edge of a copse of trees with a large bag full of tennis balls. If you were to blindly throw tennis balls at the trees, you would notice that most of the balls would fly through hitting nothing, while a few would strike tree trunks and bounce off in all directions. This analogy illustrates what Rutherford saw in the scattering pattern of the alpha particles. Most particles went straight through the metal foil because its matter was mostly empty space, but a few had "struck" some small but strong obstacle: the nuclei of the atoms.

Rutherford thus rejected Thomson's model of the atom, and instead proposed a model where the atom consisted of mostly empty space, with all its positive charge concentrated in its center in a very tiny volume, surrounded by a cloud of electrons.

Timeline

Background

Ernest Rutherford

Hans Geiger

Ernest Marsden

Ernest Rutherford was a physics professor at the University of Manchester. He had already received numerous honors for his studies of radiation. He had discovered the existence of alpha rays, beta rays, and gamma rays, and had proved that these were the consequence of the disintegration of atoms. In 1906, he received a visit from a German physicist named Hans Geiger, and was so impressed that he asked Geiger to stay and help him with his research. Ernest Marsden was a physics undergraduate student studying under Geiger.

Alpha particles are tiny, positively charged particles that are spontaneously emitted by certain substances such as uranium and radium. Rutherford himself had discovered them in 1899. In 1908 he was trying to precisely measure their charge-to-mass ratio. To do this, he first needed to know just how many alpha particles his sample of radium was giving off (after which he would measure their total charge and divide one by the other). Alpha particles are too tiny to be seen even with a microscope, but Rutherford knew that alpha particles ionize air molecules, and if the air is within an electric field, the ions will produce an electric current. On this principle, Rutherford and Geiger designed a simple counting device which consisted of two electrodes in a glass tube. Every alpha particle that passed through the tube would create a pulse of electricity that could be counted. It was an early version of the Geiger counter.

The counter that Geiger and Rutherford built proved unreliable because the alpha particles were being too strongly deflected by their collisions with the molecules of air within the detection chamber. The highly variable trajectories of the alpha particles meant that they did not all generate the same number of ions as they passed through the gas, thus producing erratic readings. This puzzled Rutherford because he had thought that alpha particles were just too heavy to be deflected so strongly. Rutherford asked Geiger to investigate just how much matter could scatter alpha rays.

The experiments they designed involved bombarding a metal foil with alpha particles to observe how the foil scattered them in relation to their thickness and material. They used a fluorescent screen to measure the trajectories of the particles. Each impact of an alpha particle on the screen produced a tiny flash of light. Geiger worked in a darkened lab for hours on end, counting these tiny scintillations using a microscope. Rutherford lacked the endurance for this work, which is why he left it to his younger colleagues. For the metal foil, they tested a variety of metals, but they preferred gold because they could make the foil very thin, as gold is very malleable. As a source of alpha particles, Rutherford's substance of choice was radium, a substance several million times more radioactive than uranium.

The 1908 Experiment

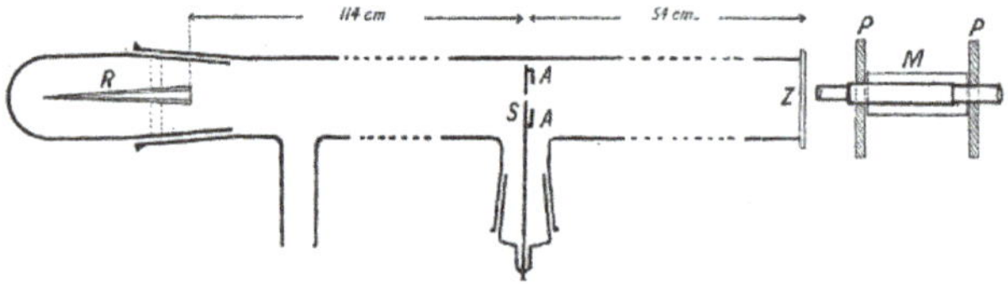

Fig. 1.

This apparatus was described in a 1908 paper by Hans Geiger. It could only measure deflections of a few degrees.

A 1908 paper by Geiger, *On the Scattering of α-Particles by Matter*, describes the following experiment. He constructed a long glass tube, nearly two meters in length. At one end of the tube was a quantity of "radium emanation" (R) that served as a source of alpha particles. The opposite end of the tube was covered with a phosphorescent screen (Z). In the middle of the tube was a 0.9 mm-wide slit. The alpha particles from R passed through the slit and created a glowing patch of light on the screen. A microscope (M) was used to count the scintillations on the screen and measure their spread. Geiger pumped all the air out of the tube so that the alpha particles would be unobstructed, and they left a neat and tight image on the screen that corresponded to the shape of the slit. Geiger then allowed some air in the tube, and the glowing patch became more diffuse. Geiger then pumped out the air and placed some gold foil over the slit at AA. This too caused the patch of light on the screen to become more spread out.

This experiment demonstrated that both air and solid matter could markedly scatter alpha particles. The apparatus, however, could only observe small angles of deflection. Rutherford wanted to know if the alpha particles were being scattered by even larger angles—perhaps larger than 90°.

The 1909 Experiment

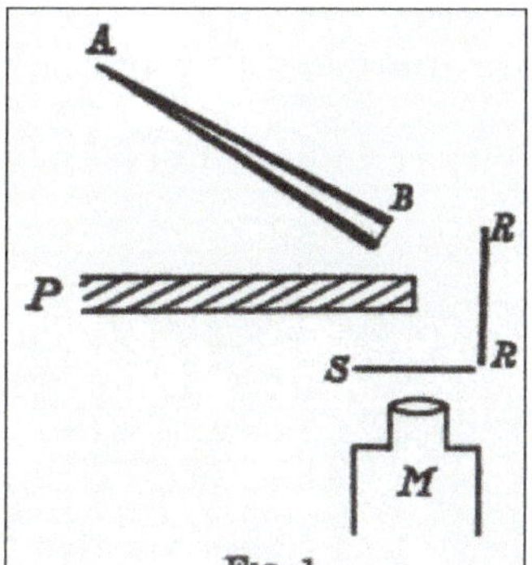

Fig. 1.

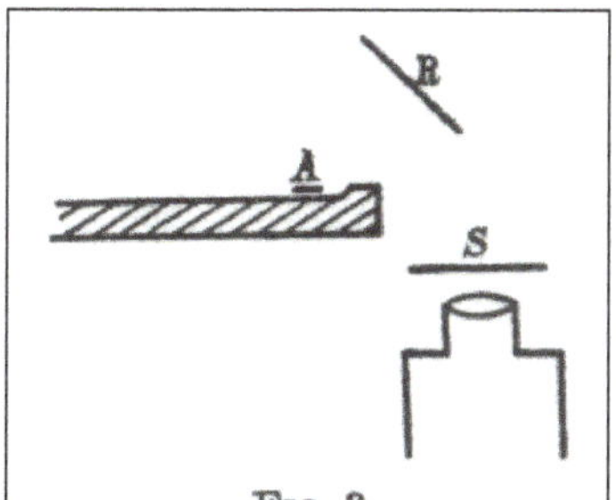

Fig. 3.

In these experiments, alpha particles emitted by a radioactive source (A) were observed bouncing off a metal reflector (R) and onto a fluorescent screen (S) on the other side of a lead plate (P).

In a 1909 paper, *On a Diffuse Reflection of the α-Particles",* Geiger and Marsden described the experiment by which they proved that alpha particles can indeed be scattered by more than 90°. In their experiment, they prepared a small conical glass tube (AB) containing "radium emanation" (radon), "radium A" (actual radium), and "radium C" (bismuth-214); its open end sealed with mica. This was their alpha particle emitter. They then set up a lead plate (P), beneath which they placed a fluorescent screen (S). The tube was held above the plate, such that the alpha particles it emitted could not directly strike the screen. They noticed a few scintillations on the screen—this was because some alpha particles could circumvent the lead plate by bouncing off air molecules (the experiment was not done in a vacuum). They then placed a metal foil (R) to the side of the lead plate. They pointed the tube at the foil to see if the alpha particles would bounce off it and strike the screen on the other side of the plate, and observed an increase in the number of scintillations on the screen. Counting the scintillations, they noticed that metals with higher atomic mass, such as gold, reflected more alpha particles than lighter ones such as aluminium.

Geiger and Marsden then wanted to estimate the total number of alpha particles that were being reflected. The previous setup was unsuitable for doing this because the tube contained several radioactive substances (radium plus its decay products) and thus the alpha particles emitted had varying ranges, and because it was difficult for them to ascertain at what rate the tube was emitting alpha particles. This time, they placed a small quantity of radium C (bismuth-214) on the lead plate, which bounced off a platinum reflector (R) and onto the screen. They found that only a tiny fraction of the alpha particles that struck the reflector bounced onto the screen (in this case, 1 in 8000).

The 1910 Experiment

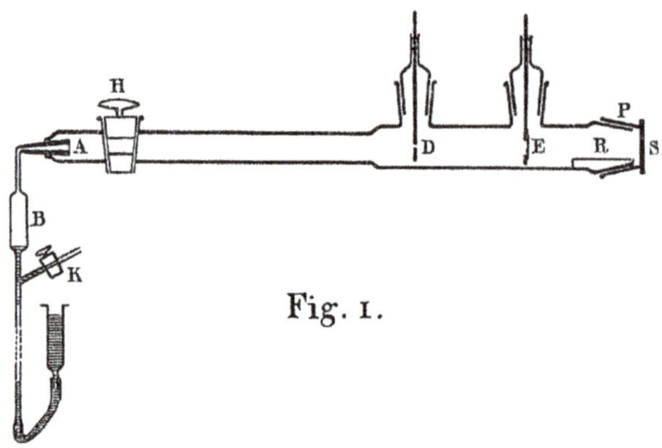

Fig. 1.

This apparatus was described in 1910 paper by Geiger. It was designed to precisely measure how the scattering varied according to the substance and thickness of the foil.

A 1910 paper by Geiger, *The Scattering of the a-Particles by Matter*, describes an experiment by which he sought to measure how the most probable angle through which an a-particle is deflected varies with the material it passes through, the thickness of said material, and the velocity of the alpha particles. He constructed an airtight glass tube from which the air was pumped out. At one end was a bulb (B) containing "radium emanation" (radon-222). By means of mercury, the radon in B was pumped up the narrow glass pipe whose end at A was plugged with mica. At the other end of the tube was a fluorescent zinc sulfide screen (S). The microscope which he used to count the scintillations on the screen was affixed to a vertical millimeter scale with a vernier, which allowed Geiger to precisely measure where the flashes of light appeared on the screen and thus calculate the particles' angles of deflection. The alpha particles emitted from A was narrowed to a beam by a small circular hole at D. Geiger placed a metal foil in the path of the rays at D and E to observe how the zone of flashes changed. He could also vary the velocity of the alpha particles by placing extra sheets of mica or aluminium at A.

From the measurements he took, Geiger came to the following conclusions:

- the most probable angle of deflection increases with the thickness of the material

- the most probable angle of deflection is proportional to the atomic mass of the substance
- the most probable angle of deflection decreases with the velocity of the alpha particles
- the probability that a particle will be deflected by more than 90° is vanishingly small

Rutherford Mathematically Models the Scattering Pattern

Considering the results of the above experiments, Rutherford published a landmark paper in 1911 titled "The Scattering of α and β Particles by Matter and the Structure of the Atom" wherein he proposed that the atom contains at its center a volume of electric charge that is very small and intense (in fact, Rutherford treats it as a point charge in his calculations). For the purpose of his mathematical calculations he assumed this central charge was positive, but he admitted he could not prove this and that he had to wait for other experiments to develop his theory.

Rutherford developed a mathematical equation that modeled how the foil should scatter the alpha particles if all the positive charge and most of the atomic mass was concentrated in a single point at the center of an atom.

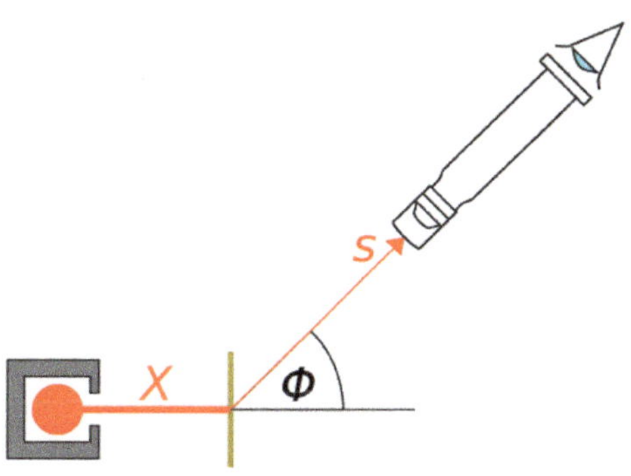

$$s = \frac{Xnt\csc^4\frac{\phi}{2}}{16r^2}\cdot\left(\frac{2Q_nQ_\alpha}{mv^2}\right)^2$$

s = the number of alpha particles falling on unit area at an angle of deflection Φ

r = distance from point of incidence of α rays on scattering material

X = total number of particles falling on the scattering material

n = number of atoms in a unit volume of the material

t = thickness of the foil

Q_n = positive charge of the atomic nucleus

Q_α = positive charge of the alpha particles

m = mass of an alpha particle

v = velocity of the alpha particle

The 1913 Experiment

In a 1913 paper, *The Laws of Deflexion of a Particles through Large Angles*, Geiger and Marsden describe a series of experiments by which they sought to experimentally verify the above equation that Rutherford developed. Rutherford's equation predicted that the number of scintillations per minute *s* that will be observed at a given angle Φ should be proportional to:

1. $csc^4\Phi/2$
2. thickness of foil t
3. magnitude of central charge Q_n
4. $1/(mv^2)^2$

Their 1913 paper describes four experiments by which they proved each of these four relationships.

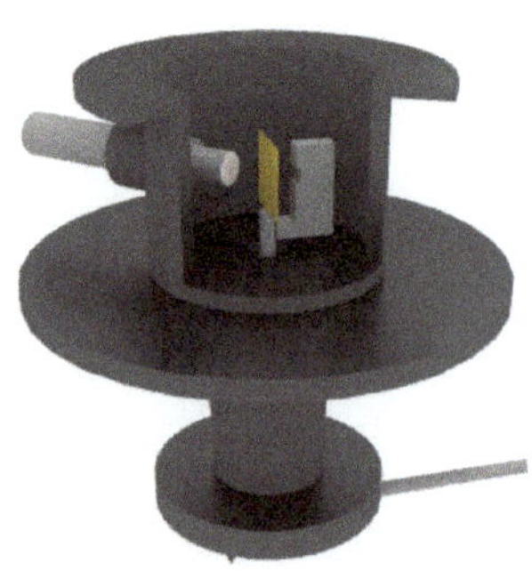

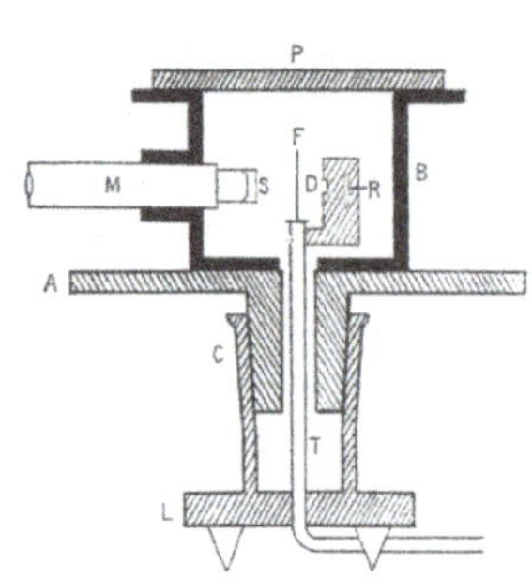

This apparatus was described in a 1913 paper by Geiger and Marsden. It was designed to accurately measure the scattering pattern of the alpha particles produced by the metal foil (F). The microscope (M) and screen (S) were affixed to a rotating cylinder and could be moved a full circle around the foil so that they could count scintillations from every angle.

To test how the scattering varied with the angle of deflection (i.e. if $s \propto csc^4\Phi/2$) Geiger and Marsden built an apparatus that consisted of a hollow metal cylinder mounted on a turntable. Inside the cylinder was a metal foil (F) and a radiation source containing radon (R), mounted on a detached column (T) which allowed the cylinder to rotate independently. The column was also a tube by which air was pumped out of the cylinder. A microscope (M) with its objective lens covered by a fluorescent zinc sulfide screen (S) penetrated the wall of the cylinder and pointed at the metal foil. By turning the table, the microscope could be moved a full circle around the foil, allowing Geiger to observe and count alpha particles deflected by up to 150°. Correcting for experimental error, Geiger and Marsden found that the number of alpha particles that are deflected by a given angle Φ is indeed proportional to $csc^4\Phi/2$.

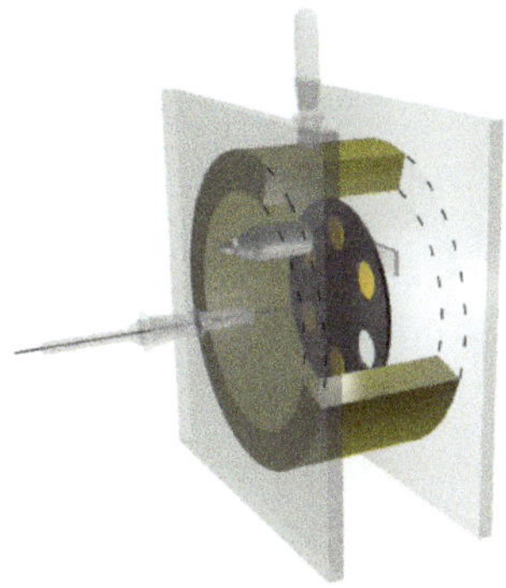

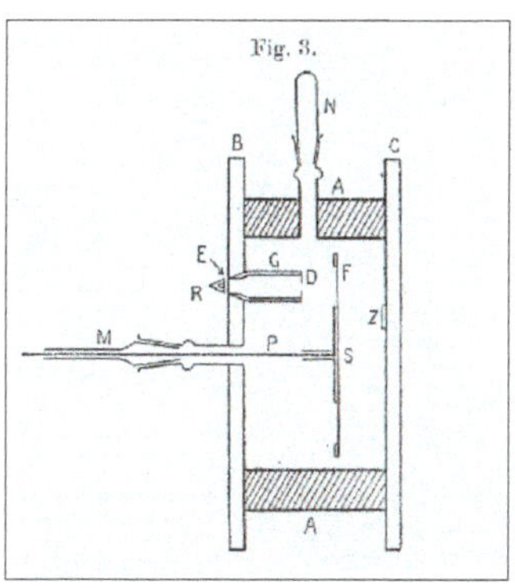

This apparatus was used to measure how the alpha particle scattering pattern varied in relation to the thickness of the foil, the atomic weight of the material, and the velocity of the alpha particles. The rotating disc in the center had six holes which could be covered with foil.

Geiger and Marsden then tested how the scattering varied with the thickness of the foil (i.e. if $s \propto t$). They constructed a disc (S) with six holes drilled in it. The holes were covered with metal foil (F) of varying thickness, or none for control. This disc was then sealed in a brass ring (A) between two glass plates (B and C). The disc could be rotated by means of a rod (P) to bring each window in front of the alpha particle source (R). On the rear glass pane was a zinc sulfide screen (Z). Geiger and Marsden found that the number of scintillations that appeared on the zinc sulfide screen was indeed proportional to the thickness as long as said thickness was small.

Geiger and Marsden reused the above apparatus to measure how the scattering pattern varied with the square of the nuclear charge (i.e. if $s \propto Q_n^2$). Geiger and Marsden didn't know what the positive charge of the nucleus of their metals were (they had only just discovered the nucleus existed at all), but they assumed it was proportional to the atomic weight, so they tested whether the scattering was proportional to the atomic weight squared. Geiger and Marsden covered the holes of the disc with foils of gold, tin, silver, copper, and aluminum. They measured each foil's stopping power by equating it to an equivalent thickness of air. They counted the number of scintillations per minute that each foil produced on the screen. They divided the number of scintillations per minute by the respective foil's air equivalent, then divided again by the square root of the atomic weight (Geiger and Marsden knew that for foils of equal stopping power, the number of atoms per unit area is proportional to the square root of the atomic weight). Thus, for each metal, Geiger and Marsden obtained the number of scintillations that a fixed number of atoms produce. For each metal, they then divided this number by the square of the atomic weight, and found that the ratios were more or less the same. Thus they proved that $s \propto Q_n^2$.

Finally, Geiger and Marsden tested how the scattering varied with the velocity of the alpha particles (i.e. if $s \propto 1/v^4$). Using the same apparatus again, they slowed the alpha particles by placing extra sheets of mica in front of the alpha particle source. They found that, within the range of experimental error, that the number of scinitillations was indeed proportional to $1/v^4$.

Rutherford Determines the Nucleus is Positively Charged

In his 1911 paper (see above), Rutherford assumed that the central charge of the atom was positively charged, but he acknowledged he couldn't say for sure, since either a negative or a positive charge would have fitted his scattering model. The results of other experiments confirmed his hypothesis. In a 1913 paper, Rutherford declared that the "nucleus" (as he now called it) was positively charged, based on the result of experiments exploring the scattering of alpha particles in various gases.

In 1917, Rutherford and his assistant William Kay began exploring the passage of alpha particles through gases such as hydrogen and nitrogen. In an experiment where they shot a beam of alpha particles through hydrogen, the alpha particles knocked the hydrogen nuclei forwards in the direction of the beam, not backwards. In an experiment where they shot alpha particles through nitrogen, he discovered that the alpha particles knocked hydrogen nuclei (i.e. protons) out of the nitrogen nuclei.

Legacy

When Geiger reported to Rutherford that he had spotted alpha particles being strongly deflected, Rutherford was astounded. In a lecture Rutherford delivered at Cambridge University, he said:

It was quite the most incredible event that has ever happened to me in my life. It was almost as incredible as if you fired a 15-inch shell at a piece of tissue paper and it came back and hit you. On consideration, I realized that this scattering backward must be the result of a single collision, and when I made calculations I saw that it was impossible to get anything of that order of magnitude unless you took a system in which the greater part of the mass of the atom was concentrated in a minute nucleus. It was then that I had the idea of an atom with a minute massive centre, carrying a charge.

—Ernest Rutherford

Accolades soon flooded in. Hantaro Nagaoka, who had once proposed a Saturnian model of the atom, wrote to Rutherford from Tokyo in 1911: "Congratulations on the simpleness of the apparatus you employ and the brilliant results you obtained". The astronomer Arthur Eddington called Rutherford's discovery the most important scientific achievement since Democritus proposed the atom ages earlier.

The conclusions of the Geiger–Marsden experiments represent one of the most pivotal scientific discoveries in human history. It deepened mankind's understanding of how all matter in the Universe is structured at the fundamental level and consequently affected every scientific and engineering discipline.

Like most scientific models, Rutherford's atomic model was neither perfect nor complete. According to classical Newtonian physics, it was in fact impossible. Accelerating charged particles radiate electromagnetic waves, so an electron orbiting an atomic nucleus in theory would spiral into the nucleus as it loses energy. To fix this problem, scientists had to incorporate quantum mechanics into Rutherford's model.

Rutherford Model

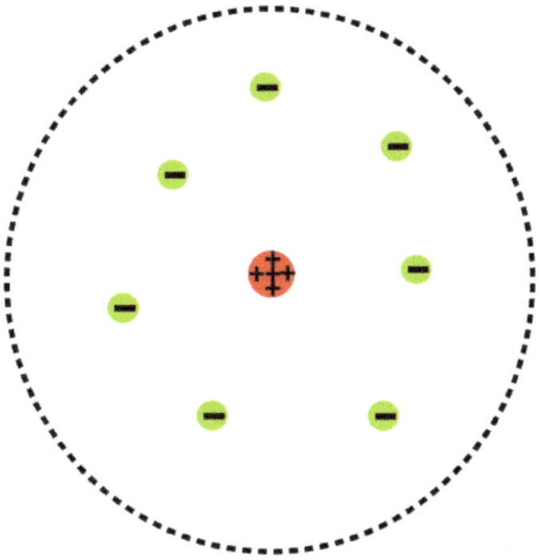

Basic diagram of the atomic nuclear model: electrons in green and nucleus in red

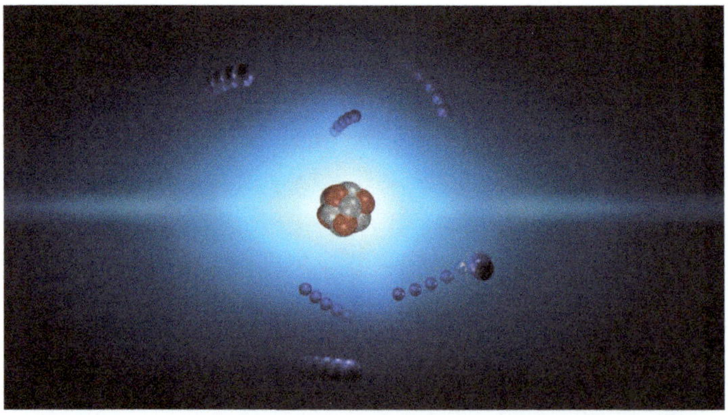

3D animation of an atom incorporating the Rutherford model

The Rutherford model is a model of the atom devised by Ernest Rutherford. Rutherford directed the famous Geiger–Marsden experiment in 1909 which suggested, upon Rutherford's 1911 analysis, that J. J. Thomson's plum pudding model of the atom was incorrect. Rutherford's new model for the atom, based on the experimental results, contained the new features of a relatively high central charge concentrated into a very small volume in comparison to the rest of the atom and with this central volume also containing the bulk of the atomic mass of the atom. This region would be known as the "nucleus" of the atom.

Experimental Basis for the Model

Rutherford overturned Thomson's model in 1911 with his well-known gold foil experiment in which he demonstrated that the atom has a tiny, heavy nucleus. Rutherford designed an experiment to use the alpha particles emitted by a radioactive element as probes to the unseen world of atomic structure. If Thomson was correct, the beam would go straight through the gold foil. Most of the beams went through the foil, but a few were deflected.

Rutherford presented his own physical model for subatomic structure, as an interpretation for the unexpected experimental results. In it, the atom is made up of a central charge (this is the modern atomic nucleus, though Rutherford did not use the term "nucleus" in his paper) surrounded by a cloud of (presumably) orbiting electrons. In this May 1911 paper, Rutherford only commits himself to a small central region of very high positive or negative charge in the atom.

For concreteness, consider the passage of a high speed α particle through an atom having a positive central charge $N e$, and surrounded by a compensating charge of N electrons.

From purely energetic considerations of how far particles of known speed would be able to penetrate toward a central charge of 100 e, Rutherford was able to calculate that the radius of his gold central charge would need to be less (how much less could not be told) than 3.4×10^{-14} meters. This was in a gold atom known to be 10^{-10} meters or so in radius—a very surprising finding, as it implied a strong central charge less than 1/3000th of the diameter of the atom.

The Rutherford model served to concentrate a great deal of the atom's charge and mass to a very small core, but didn't attribute any structure to the remaining electrons and remaining atomic mass. It did mention the atomic model of Hantaro Nagaoka, in which the electrons are arranged in one or more rings, with the specific metaphorical structure of the stable rings of Saturn. The plum pudding model of J. J. Thomson also had rings of orbiting electrons. Jean Baptiste Perrin claimed in his Nobel lecture that he was the first one to suggest the model in his paper dated 1901.

The Rutherford paper suggested that the central charge of an atom might be "proportional" to its atomic mass in hydrogen mass units u (roughly 1/2 of it, in Rutherford's model). For gold, this mass number is 197 (not then known to great accuracy) and was therefore modeled by Rutherford to be possibly 196 u. However, Rutherford did not attempt to make the direct connection of central charge to atomic number, since gold's "atomic number" (at *that* time merely its place number in the periodic table) was 79, and Rutherford had modeled the charge to be about +100 units (he had actually suggested 98 units of positive charge, to make half of 196). Thus, Rutherford did not formally suggest the two numbers (periodic table place, 79, and nuclear charge, 98 or 100) might be exactly the same.

A month after Rutherford's paper appeared, the proposal regarding the exact identity of atomic number and nuclear charge *was* made by Antonius van den Broek, and later confirmed experimentally within two years, by Henry Moseley.

These are the key indicators-

- The atom's electron cloud does not influence alpha particle scattering.
- Much of an atom's positive charge is concentrated in a relatively tiny volume at the center of the atom, known today as the nucleus. The magnitude of this charge is proportional to (up to a charge number that can be approximately half of) the atom's atomic mass—the remaining mass is now known to be mostly attributed to neutrons. This concentrated central mass and charge is responsible for deflecting both alpha and beta particles.
- The mass of heavy atoms such as gold is mostly concentrated in the central charge region, since calculations show it is not deflected or moved by the high speed alpha particles, which have very high momentum in comparison to electrons, but not with regard to a heavy atom as a whole.
- The atom itself is about 100,000 (10^5) times the diameter of the nucleus. This could be related to putting a grain of sand in the middle of a football field.

Contribution to Modern Science

After Rutherford's discovery, scientists started to realize that the atom is not ultimately a single particle, but is made up of far smaller subatomic particles. Subsequent research determined the exact atomic structure which led to Rutherford's gold foil experiment. Scientists eventually discovered that atoms have a positively charged nucleus (with an exact atomic number of charges) in the center, with a radius of about 1.2×10^{-15} meters $\times$ [Atomic Mass Number]$^{1/3}$. Electrons were found to be even smaller.

Later, scientists found the expected number of electrons (the same as the atomic number) in an atom by using X-rays. When an X-ray passes through an atom, some of it is scattered while the rest passes through the atom. Since the X-ray loses its intensity primarily due to scattering at electrons, by noting the rate of decrease in X-ray intensity, the number of electrons contained in an atom can be accurately estimated.

Symbolism

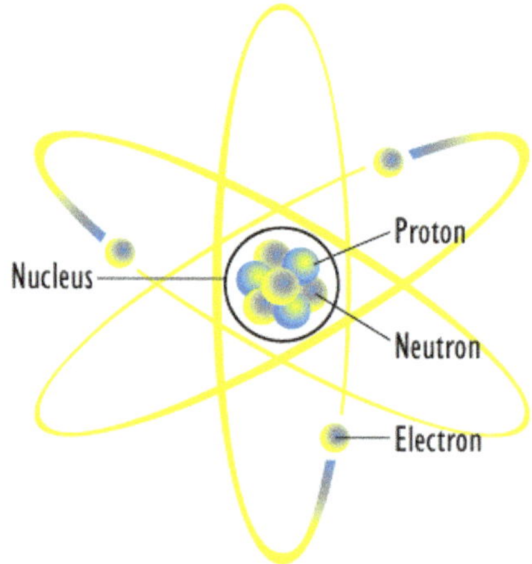

Generic atomic planetary model.

Shield of the U.S. Atomic Energy Commission

Rutherford's model deferred to the idea of many electrons in rings, per Nagaoka. However, once Niels Bohr modified this view into a picture of just a few planet-like electrons for light atoms, the Rutherford-Bohr model caught the imagination of the public. It has since continually been used as a symbol for atoms and even for "atomic" energy (even though this is more properly considered nuclear energy). Examples of its use over the past century include:

- The logo of the United States Atomic Energy Commission, which was in part responsible for its later usage in relation to nuclear fission technology in particular.
- The flag of the International Atomic Energy Agency is a Rutherford atom, enclosed in olive branches.
- The US minor league baseball Albuquerque Isotopes' logo is a Rutherford atom, with the electron orbits forming an A.
- A similar symbol, the Atomic whirl, was chosen as the symbol for the American Atheists, and has come to be used as a symbol of atheism in general.
- The Unicode Miscellaneous Symbols codepoint U+269B (□) uses a Rutherford atom.
- On maps, it is generally used to indicate a nuclear power installation.

Bohr Model

The Rutherford–Bohr model of the hydrogen atom ($Z = 1$) or a hydrogen-like ion ($Z > 1$), where the negatively charged electron confined to an atomic shell encircles a small, positively charged atomic nucleus and where an electron jump between orbits is accompanied by an emitted or absorbed amount of electromagnetic energy (hv). The orbits in which the electron may travel are shown as grey circles; their radius increases as n^2, where n is the principal quantum number. The $3 \rightarrow 2$ transition depicted here produces the first line of the Balmer series, and for hydrogen ($Z = 1$) it results in a photon of wavelength 656 nm (red light).

In atomic physics, the Rutherford–Bohr model or Bohr model, introduced by Niels Bohr and Ernest Rutherford in 1913, depicts the atom as a small, positively charged nucleus surrounded by electrons that travel in circular orbits around the nucleus—similar in structure to the Solar System, but with attraction provided by electrostatic forces rather than gravity. After the cubic model (1902), the plum-pudding model (1904), the Saturnian model (1904), and the Rutherford model (1911) came the Rutherford–Bohr model or just *Bohr model* for short (1913). The improvement to the Rutherford model is mostly a quantum physical interpretation of it. The Bohr model has been superseded, but it helped to lead the way to a modern quantum mechanical model of the atom.

The model's key success lay in explaining the Rydberg formula for the spectral emission lines of atomic hydrogen. While the Rydberg formula had been known experimentally, it did not gain a theoretical underpinning until the Bohr model was introduced. Not only did the Bohr model explain the reason for the structure of the Rydberg formula, it also provided a justification for its empirical results in terms of fundamental physical constants.

The Bohr model is a relatively primitive model of the hydrogen atom, compared to the valence shell atom. As a theory, it can be derived as a first-order approximation of the hydrogen atom using the broader and much more accurate quantum mechanics and thus may be considered to be an obsolete scientific theory. However, because of its simplicity, and its correct results for selected systems (see below for application), the Bohr model is still commonly taught to introduce students to quantum mechanics or energy level diagrams before moving on to the more accurate, but more complex, valence shell atom. A related model was originally proposed by Arthur Erich Haas in 1910, but was rejected. The quantum theory of the period between Planck's discovery of the quantum (1900) and the advent of a full-blown quantum mechanics (1925) is often referred to as the old quantum theory.

Origin

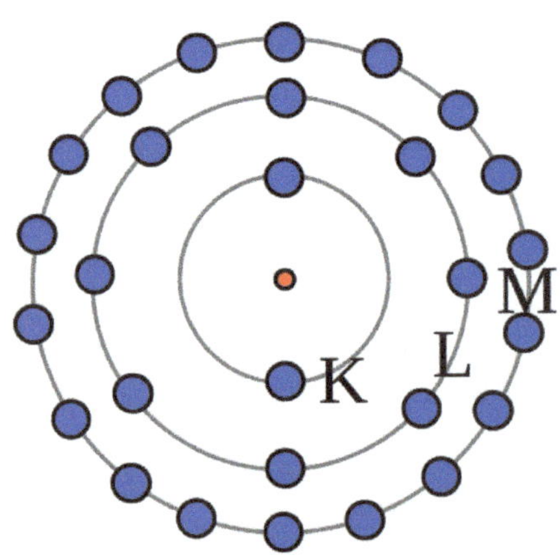

Bohr model showing maximum electrons per shell with shells labeled in X-ray notation

In the early 20th century, experiments by Ernest Rutherford established that atoms consisted of a diffuse cloud of negatively charged electrons surrounding a small, dense, positively charged nucleus. Given this experimental data, Rutherford naturally considered a planetary-model atom, the Rutherford model of 1911 – electrons orbiting a solar nucleus – however, said planetary-model atom has a technical difficulty. The laws of classical mechanics (i.e. the Larmor formula), predict that the electron will release electromagnetic radiation while orbiting a nucleus. Because the electron would lose energy, it would rapidly spiral inwards, collapsing into the nucleus on a timescale of around 16 picoseconds. This atom model is disastrous, because it predicts that all atoms are unstable.

Also, as the electron spirals inward, the emission would rapidly increase in frequency as the orbit got smaller and faster. This would produce a continuous smear, in frequency, of electromagnetic radiation. However, late 19th century experiments with electric discharges have shown that atoms will only emit light (that is, electromagnetic radiation) at certain discrete frequencies.

To overcome this difficulty, Niels Bohr proposed, in 1913, what is now called the *Bohr model of the atom*. He suggested that electrons could only have certain *classical* motions:

1. Electrons in atoms orbit the nucleus.

2. The electrons can only orbit stably, without radiating, in certain orbits (called by Bohr the "stationary orbits") at a certain discrete set of distances from the nucleus. These orbits are associated with definite energies and are also called energy shells or energy levels. In these orbits, the electron's acceleration does not result in radiation and energy loss as required by classical electromagnetics. The Bohr model of an atom was based upon Planck's quantum theory of radiation.

3. Electrons can only gain and lose energy by jumping from one allowed orbit to another, absorbing or emitting electromagnetic radiation with a frequency ν determined by the energy difference of the levels according to the Planck relation:

$$\Delta E = E_2 - E_1 = h\nu \ ,$$

   where h is Planck's constant. The frequency of the radiation emitted at an orbit of period T is as it would be in classical mechanics; it is the reciprocal of the classical orbit period:

$$\nu = \frac{1}{T} \ .$$

The significance of the Bohr model is that the laws of classical mechanics apply to the motion of the electron about the nucleus *only when restricted by a quantum rule*. Although Rule 3 is not completely well defined for small orbits, because the emission process involves two orbits with two different periods, Bohr could determine the energy spacing between levels using Rule 3 and come to an exactly correct quantum rule: the angular momentum L is restricted to be an integer multiple of a fixed unit:

$$L = n\frac{h}{2\pi} = n\hbar$$

where $n = 1, 2, 3, \ldots$ is called the principal quantum number, and $\hbar = h/2\pi$. The lowest value of n is 1; this gives a smallest possible orbital radius of 0.0529 nm known as the Bohr radius. Once an electron is in this lowest orbit, it can get no closer to the proton. Starting from the angular momentum quantum rule, Bohr was able to calculate the energies of the allowed orbits of the hydrogen atom and other hydrogen-like atoms and ions.

Other points are:

1. Like Einstein's theory of the Photoelectric effect, Bohr's formula assumes that during a quantum jump a *discrete* amount of energy is radiated. However, unlike Einstein, Bohr stuck to the *classical* Maxwell theory of the electromagnetic field. Quantization of the electromagnetic field was explained by the discreteness of the atomic energy levels; Bohr did not believe in the existence of photons.

2. According to the Maxwell theory the frequency ν of classical radiation is equal to the rotation frequency ν_{rot} of the electron in its orbit, with harmonics at integer multiples of this frequency. This result is obtained from the Bohr model for jumps between energy levels E_n and E_{n-k} when k is much smaller than n. These jumps reproduce the frequency of the k-th harmonic of orbit n. For sufficiently large values of n (so-called Rydberg states), the two orbits involved in the emission process have nearly the same rotation frequency, so that the classical orbital frequency is not ambiguous. But for small n (or large k), the radiation frequency has no unambiguous classical interpretation. This marks the birth of the correspondence principle, requiring quantum theory to agree with the classical theory only in the limit of large quantum numbers.

3. The Bohr-Kramers-Slater theory (BKS theory) is a failed attempt to extend the Bohr model, which violates the conservation of energy and momentum in quantum jumps, with the conservation laws only holding on average.

Bohr's condition, that the angular momentum is an integer multiple of $\hbar$ was later reinterpreted in 1924 by de Broglie as a standing wave condition: the electron is described by a wave and a whole number of wavelengths must fit along the circumference of the electron's orbit:

$$n\lambda = 2\pi r\ .$$

Substituting de Broglie's wavelength of $\lambda = h/p$ reproduces Bohr's rule. In 1913, however, Bohr justified his rule by appealing to the correspondence principle, without providing any sort of wave interpretation. In 1913, the wave behavior of matter particles such as the electron (i.e., matter waves) was not suspected.

In 1925 a new kind of mechanics was proposed, quantum mechanics, in which Bohr's model of electrons traveling in quantized orbits was extended into a more accurate model of electron motion. The new theory was proposed by Werner Heisenberg. Another form of the same theory, wave mechanics, was discovered by the Austrian physicist Erwin Schrödinger independently, and by different reasoning. Schrödinger employed de Broglie's matter waves, but sought wave solutions of a three-dimensional wave equation describing electrons that were constrained to move about the nucleus of a hydrogen-like atom, by being trapped by the potential of the positive nuclear charge.

Electron Energy Levels

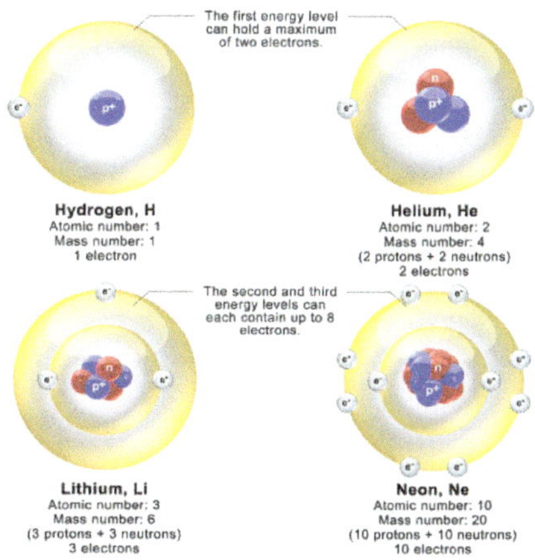

Models depicting electron energy levels in hydrogen, helium, lithium, and neon

The Bohr model gives almost exact results only for a system where two charged points orbit each other at speeds much less than that of light. This not only includes one-electron systems such as the hydrogen atom, singly ionized helium, doubly ionized lithium, but it includes positronium and Rydberg states of any atom where one electron is far away from everything else. It can be used for K-line X-ray transition calculations if other assumptions are added (see Moseley's law below). In high energy physics, it can be used to calculate the masses of heavy quark mesons.

Calculation of the orbits requires two assumptions.

- Classical mechanics

  The electron is held in a circular orbit by electrostatic attraction. The centripetal force is equal to the Coulomb force.

$$\frac{m_e v^2}{r} = \frac{Z k_e e^2}{r^2}$$

  where m_e is the electron's mass, e is the charge of the electron, k_e is Coulomb's constant and Z is the atom's atomic number. It is assumed here that the mass of the nucleus is much larger than the electron mass (which is a good assumption). This equation determines the electron's speed at any radius:

$$v = \sqrt{\frac{Z k_e e^2}{m_e r}}.$$

  It also determines the electron's total energy at any radius:

$$E = \frac{1}{2} m_e v^2 - \frac{Z k_e e^2}{r} = -\frac{Z k_e e^2}{2r}.$$

The total energy is negative and inversely proportional to r. This means that it takes energy to pull the orbiting electron away from the proton. For infinite values of r, the energy is zero, corresponding to a motionless electron infinitely far from the proton. The total energy is half the potential energy, which is also true for noncircular orbits by the virial theorem.

- A quantum rule

The angular momentum $L = m_e vr$ is an integer multiple of $\hbar$:

$$m_e vr = n\hbar$$

How The Equation Was Derived:

If an electron in an atom is moving on an orbit with period T, classically the electromagnetic radiation will repeat itself every orbital period. If the coupling to the electromagnetic field is weak, so that the orbit doesn't decay very much in one cycle, the radiation will be emitted in a pattern which repeats every period, so that the Fourier transform will have frequencies which are only multiples of $1/T$. This is the classical radiation law: the frequencies emitted are integer multiples of $1/T$.

In quantum mechanics, this emission must be in quanta of light, of frequencies consisting of integer multiples of $1/T$, so that classical mechanics is an approximate description at large quantum numbers. This means that the energy level corresponding to a classical orbit of period $1/T$ must have nearby energy levels which differ in energy by h/T, and they should be equally spaced near that level,

$$\Delta E_n = \frac{h}{T(E_n)}.$$

Bohr worried whether the energy spacing $1/T$ should be best calculated with the period of the energy state E_n, or E_{n+1}, or some average—in hindsight, this model is only the leading semiclassical approximation.

Bohr considered circular orbits. Classically, these orbits must decay to smaller circles when photons are emitted. The level spacing between circular orbits can be calculated with the correspondence formula. For a Hydrogen atom, the classical orbits have a period T determined by Kepler's third law to scale as $r^{3/2}$. The energy scales as $1/r$, so the level spacing formula amounts to

$$\Delta E \propto \frac{1}{r^{\frac{3}{2}}} \propto E^{\frac{3}{2}}.$$

It is possible to determine the energy levels by recursively stepping down orbit by orbit, but there is a shortcut.

The angular momentum L of the circular orbit scales as $\sqrt{r}$. The energy in terms of the angular momentum is then

$$E \propto \frac{1}{r} \propto \frac{1}{L^2}.$$

Assuming, with Bohr, that quantized values of L are equally spaced, the spacing between neighboring energies is

$$\Delta E \propto \frac{1}{(L+\hbar)^2} - \frac{1}{L^2} \approx -\frac{2\hbar}{L^3} \propto -E^{\frac{3}{2}}.$$

This is as desired for equally spaced angular momenta. If one kept track of the constants, the spacing would be $\hbar$, so the angular momentum should be an integer multiple of $\hbar$,

$$L = \frac{nh}{2\pi} = n\hbar .$$

This is how Bohr arrived at his model.

Substituting the expression for the velocity gives an equation for r in terms of n:

$$m_e\sqrt{\frac{k_e Z e^2}{m_e r}}\,r = n\hbar$$

so that the allowed orbit radius at any n is:

$$r_n = \frac{n^2\hbar^2}{Zk_e e^2 m_e}$$

The smallest possible value of r in the hydrogen atom (Z=1) is called the Bohr radius and is equal to:

$$= \frac{\quad}{k_e e \ m_e} \approx 5.29\times 10^{\ 11}\mathrm{m}$$

The energy of the n-th level for any atom is determined by the radius and quantum number:

$$E = -\frac{Zk_e e^2}{2r_n} = -\frac{Z^2(k_e e^2)^2 m_e}{2\hbar^2 n^2} \approx \frac{-13.6Z^2}{n^2}\mathrm{eV}$$

An electron in the lowest energy level of hydrogen (n = 1) therefore has about 13.6 eV less energy than a motionless electron infinitely far from the nucleus. The next energy level (n = 2) is −3.4 eV. The third (n = 3) is −1.51 eV, and so on. For larger values of n, these are also the binding energies of a highly excited atom with one electron in a large circular orbit around the rest of the atom. The hydrogen formula also coincides with the Wallis product.

The combination of natural constants in the energy formula is called the Rydberg energy (R_E):

$$R_E = \frac{(k_e e^2)^2 m_e}{2\hbar^2}$$

This expression is clarified by interpreting it in combinations that form more natural units:

$m_e c^2$ is the rest mass energy of the electron (511 keV)

$\frac{k_e e^2}{\hbar c} = \alpha \approx \frac{1}{137}$ is the fine structure constant

$$R_E = \frac{1}{2}(m_e c^2)\alpha^2$$

Since this derivation is with the assumption that the nucleus is orbited by one electron, we can generalize this result by letting the nucleus have a charge $q = Z\,e$ where Z is the atomic number. This will now give us energy levels for hydrogenic atoms, which can serve as a rough order-of-magnitude approximation of the actual energy levels. So for nuclei with Z protons, the energy levels are (to a rough approximation):

$$E_n = -\frac{Z^2 R_{\mathrm{E}}}{n^2}$$

The actual energy levels cannot be solved analytically for more than one electron (see n-body problem) because the electrons are not only affected by the nucleus but also interact with each other via the Coulomb Force.

When $Z = 1/\alpha$ (Z ≈ 137), the motion becomes highly relativistic, and Z^2 cancels the α^2 in R; the orbit energy begins to be comparable to rest energy. Sufficiently large nuclei, if they were stable, would reduce their charge by creating a bound electron from the vacuum, ejecting the positron to infinity. This is the theoretical phenomenon of electromagnetic charge screening which predicts a maximum nuclear charge. Emission of such positrons has been observed in the collisions of heavy ions to create temporary super-heavy nuclei.

The Bohr formula properly uses the reduced mass of electron and proton in all situations, instead of the mass of the electron,

$$m_{\mathrm{red}} = \frac{m_{\mathrm{e}} m_{\mathrm{p}}}{m_{\mathrm{e}} + m_{\mathrm{p}}} = m_{\mathrm{e}} \frac{1}{1 + m_{\mathrm{e}} / m_{\mathrm{p}}}.$$

However, these numbers are very nearly the same, due to the much larger mass of the proton, about 1836.1 times the mass of the electron, so that the reduced mass in the system is the mass of the electron multiplied by the constant 1836.1/(1+1836.1) = 0.99946. This fact was historically important in convincing Rutherford of the importance of Bohr's model, for it explained the fact that the frequencies of lines in the spectra for singly ionized helium do not differ from those of hydrogen by a factor of exactly 4, but rather by 4 times the ratio of the reduced mass for the hydrogen vs. the helium systems, which was much closer to the experimental ratio than exactly 4.

For positronium, the formula uses the reduced mass also, but in this case, it is exactly the electron mass divided by 2. For any value of the radius, the electron and the positron are each moving at half the speed around their common center of mass, and each has only one fourth the kinetic energy. The total kinetic energy is half what it would be for a single electron moving around a heavy nucleus.

$$E_n = \frac{R_{\mathrm{E}}}{2n^2} \quad \text{(positronium)}$$

Rydberg Formula

The Rydberg formula, which was known empirically before Bohr's formula, is seen in Bohr's theory as describing the energies of transitions or quantum jumps between one orbital energy levels. Bohr's formula gives the numerical value of the already-known and measured Rydberg's constant, but in terms of more fundamental constants of nature, including the electron's charge and Planck's constant.

When the electron gets moved from its original energy level to a higher one, it then jumps back each level until it comes to the original position, which results in a photon being emitted. Using the derived formula for the different energy levels of hydrogen one may determine the wavelengths of light that a hydrogen atom can emit.

The energy of a photon emitted by a hydrogen atom is given by the difference of two hydrogen energy levels:

$$E = E_i - E_f = R_{\mathrm{E}}\left(\frac{1}{n_f^2} - \frac{1}{n_i^2}\right)$$

where n_f is the final energy level, and n_i is the initial energy level.

Since the energy of a photon is

$$E = \frac{hc}{\lambda},$$

the wavelength of the photon given off is given by

$$\frac{1}{\lambda} = R\left(\frac{1}{n_f^2} - \frac{1}{n_i^2}\right).$$

This is known as the Rydberg formula, and the Rydberg constant R is R_{E}/hc, or $R_{\mathrm{E}}/2\pi$ in natural units. This formula was known in the nineteenth century to scientists studying spectroscopy, but there was no theoretical explanation for this form or a theoretical prediction for the value of R, until Bohr. In fact, Bohr's derivation of the Rydberg constant, as well as the concomitant agreement of Bohr's formula with experimentally observed spectral lines of the Lyman (n_f=1), Balmer (n_f=2), and Paschen (n_f=3) series, and successful theoretical prediction of other lines not yet observed, was one reason that his model was immediately accepted.

To apply to atoms with more than one electron, the Rydberg formula can be modified by replacing Z with Z–b or n with n–b where b is constant representing a screening effect due to the inner-shell and other electrons (see Electron shell and the later discussion of the "Shell Model of the Atom" below). This was established empirically before Bohr presented his model.

Shell Model of Heavier Atoms

Bohr extended the model of hydrogen to give an approximate model for heavier atoms. This gave a physical picture that reproduced many known atomic properties for the first time.

Heavier atoms have more protons in the nucleus, and more electrons to cancel the charge. Bohr's idea was that each discrete orbit could only hold a certain number of electrons. After that orbit is full, the next level would have to be used. This gives the atom a shell structure, in which each shell corresponds to a Bohr orbit.

This model is even more approximate than the model of hydrogen, because it treats the electrons in each shell as non-interacting. But the repulsions of electrons are taken into account somewhat by the phenomenon of screening. The electrons in outer orbits do not only orbit the nucleus, but they also move around the inner electrons, so the effective charge Z that they feel is reduced by the number of the electrons in the inner orbit.

For example, the lithium atom has two electrons in the lowest 1s orbit, and these orbit at Z=2. Each one sees the nuclear charge of Z=3 minus the screening effect of the other, which crudely reduces the nuclear charge by 1 unit. This means that the innermost electrons orbit at approximately 1/4 the Bohr radius. The outermost electron in lithium orbits at roughly Z=1, since the two inner electrons reduce the nuclear charge by 2. This outer electron should be at nearly one Bohr radius from the nucleus. Because the electrons strongly repel each other, the effective charge description is very approximate; the effective charge Z doesn't usually come out to be an integer. But Moseley's law experimentally probes the innermost pair of electrons, and shows that they do see a nuclear charge of approximately Z−1, while the outermost electron in an atom or ion with only one electron in the outermost shell orbits a core with effective charge Z−k where k is the total number of electrons in the inner shells.

The shell model was able to qualitatively explain many of the mysterious properties of atoms which became codified in the late 19th century in the periodic table of the elements. One property was the size of atoms, which could be determined approximately by measuring the viscosity of gases and density of pure crystalline solids. Atoms tend to get smaller toward the right in the periodic table, and become much larger at the next line of the table. Atoms to the right of the table tend to gain electrons, while atoms to the left tend to lose them. Every element on the last column of the table is chemically inert (noble gas).

In the shell model, this phenomenon is explained by shell-filling. Successive atoms become smaller because they are filling orbits of the same size, until the orbit is full, at which point the next atom in the table has a loosely bound outer electron, causing it to expand. The first Bohr orbit is filled when it has two electrons, which explains why helium is inert. The second orbit allows eight electrons, and when it is full the atom is neon, again inert. The third orbital contains eight again, except that in the more correct Sommerfeld treatment (reproduced in modern quantum mechanics) there are extra "d" electrons. The third orbit may hold an extra 10 d electrons, but these positions are not filled until a few more orbitals from the next level are filled (filling the n=3 d orbitals produces the 10 transition elements). The irregular filling pattern is an effect of interactions between electrons, which are not taken into account in either the Bohr or Sommerfeld models and which are difficult to calculate even in the modern treatment.

Moseley's Law and Calculation of K-alpha X-ray Emission Lines

Niels Bohr said in 1962, "You see actually the Rutherford work [the nuclear atom] was not taken seriously. We cannot understand today, but it was not taken seriously at all. There was no mention of it any place. The great change came from Moseley."

In 1913 Henry Moseley found an empirical relationship between the strongest X-ray line emitted by atoms under electron bombardment (then known as the K-alpha line), and their atomic number Z. Moseley's empiric formula was found to be derivable from Rydberg and Bohr's formula (Moseley actually mentions only Ernest Rutherford and Antonius Van den Broek in terms of models). The two additional assumptions that this X-ray line came from a transition between energy levels with quantum numbers 1 and 2, and , that the atomic number Z when used in the formula for atoms heavier than hydrogen, should be diminished by 1, to $(Z-1)^2$.

Moseley wrote to Bohr, puzzled about his results, but Bohr was not able to help. At that time, he thought that the postulated innermost "K" shell of electrons should have at least four electrons, not the two which would have neatly explained the result. So Moseley published his results without a theoretical explanation.

Later, people realized that the effect was caused by charge screening, with an inner shell containing only 2 electrons. In the experiment, one of the innermost electrons in the atom is knocked out, leaving a vacancy in the lowest Bohr orbit, which contains a single remaining electron. This vacancy is then filled by an electron from the next orbit, which has n=2. But the n=2 electrons see an effective charge of Z−1, which is the value appropriate for the charge of the nucleus, when a single electron remains in the lowest Bohr orbit to screen the nuclear charge +Z, and lower it by −1 (due to the electron's negative charge screening the nuclear positive charge). The energy gained by an electron dropping from the second shell to the first gives Moseley's law for K-alpha lines,

$$E = h\nu = E_i - E_f = R_E (Z-1)^2 \left(\frac{1}{1^2} - \frac{1}{2^2} \right)$$

or

$$f = \nu = R_v \left(\frac{3}{4}\right)(Z-1)^2 = (2.46\times10^{15}\,Hz)(Z-1)^2.$$

Here, $\boldsymbol{R}_v = \boldsymbol{R}_E/\boldsymbol{h}$ is the Rydberg constant, in terms of frequency equal to 3.28 x 10^{15} Hz. For values of Z between 11 and 31 this latter relationship had been empirically derived by Moseley, in a simple (linear) plot of the square root of X-ray frequency against atomic number (however, for silver, Z = 47, the experimentally obtained screening term should be replaced by 0.4). Notwithstanding its restricted validity, Moseley's law not only established the objective meaning of atomic number (see Henry Moseley for detail) but, as Bohr noted, it also did more than the Rydberg derivation to establish the validity of the Rutherford/Van den Broek/Bohr nuclear model of the atom, with atomic number (place on the periodic table) standing for whole units of nuclear charge.

The K-alpha line of Moseley's time is now known to be a pair of close lines, written as ($K\alpha_1$ and $K\alpha_2$) in Siegbahn notation.

Shortcomings

The Bohr model gives an incorrect value $L=\hbar$ for the ground state orbital angular momentum: The angular momentum in the true ground state is known to be zero from experiment. Although mental pictures fail somewhat at these levels of scale, an electron in the lowest modern "orbital" with no orbital momentum, may be thought of as not to rotate "around" the nucleus at all, but merely to go tightly around it in an ellipse with zero area (this may be pictured as "back and forth", without striking or interacting with the nucleus). This is only reproduced in a more sophisticated semiclassical treatment like Sommerfeld's. Still, even the most sophisticated semiclassical model fails to explain the fact that the lowest energy state is spherically symmetric - it doesn't point in any particular direction.

Nevertheless, in the modern *fully quantum treatment in phase space*, the proper deformation (careful full extension) of the semi-classical result adjusts the angular momentum value to the correct effective one. As a consequence, the physical ground state expression is obtained through a shift of the vanishing quantum angular momentum expression, which corresponds to spherical symmetry.

In modern quantum mechanics, the electron in hydrogen is a spherical cloud of probability that grows denser near the nucleus. The rate-constant of probability-decay in hydrogen is equal to the inverse of the Bohr radius, but since Bohr worked with circular orbits, not zero area ellipses, the fact that these two numbers exactly agree is considered a "coincidence". (However, many such coincidental agreements are found between the semiclassical vs. full quantum mechanical treatment of the atom;

these include identical energy levels in the hydrogen atom and the derivation of a fine structure constant, which arises from the relativistic Bohr–Sommerfeld model (see below) and which happens to be equal to an entirely different concept, in full modern quantum mechanics).

The Bohr model also has difficulty with, or else fails to explain:

- Much of the spectra of larger atoms. At best, it can make predictions about the K-alpha and some L-alpha X-ray emission spectra for larger atoms, if *two* additional ad hoc assumptions are made (see Moseley's law above). Emission spectra for atoms with a single outer-shell electron (atoms in the lithium group) can also be approximately predicted. Also, if the empiric electron–nuclear screening factors for many atoms are known, many other spectral lines can be deduced from the information, in similar atoms of differing elements, via the Ritz–Rydberg combination principles (see Rydberg formula). All these techniques essentially make use of Bohr's Newtonian energy-potential picture of the atom.
- the relative intensities of spectral lines; although in some simple cases, Bohr's formula or modifications of it, was able to provide reasonable estimates (for example, calculations by Kramers for the Stark effect).
- The existence of fine structure and hyperfine structure in spectral lines, which are known to be due to a variety of relativistic and subtle effects, as well as complications from electron spin.
- The Zeeman effect – changes in spectral lines due to external magnetic fields; these are also due to more complicated quantum principles interacting with electron spin and orbital magnetic fields.
- The model also violates the uncertainty principle in that it considers electrons to have known orbits and locations, two things which can not be measured simultaneously.
- Doublets and Triplets: Appear in the spectra of some atoms: Very close pairs of lines. Bohr's model cannot say why some energy levels should be very close together.
- Multi-electron Atoms: don't have energy levels predicted by the model. It doesn't work for (neutral) helium.
- A rotating charge, such as the electron classically orbiting around the nucleus, would constantly lose energy in form of electromagnetic radiation (via various mechanisms: dipole radiation, Bremsstrahlung,...). But such radiation is not observed.

Refinements

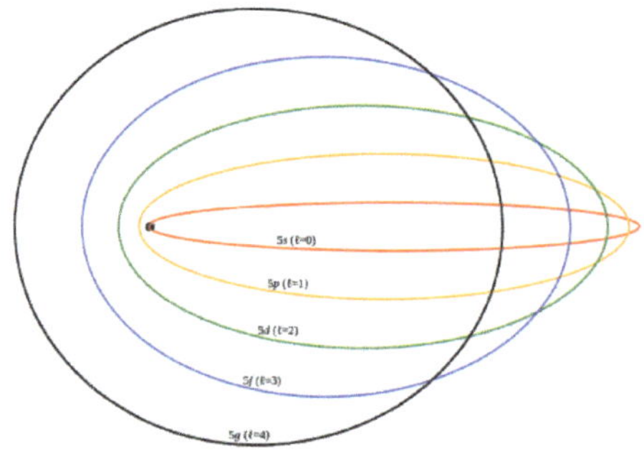

Elliptical orbits with the same energy and quantized angular momentum

Several enhancements to the Bohr model were proposed, most notably the Sommerfeld model or Bohr–Sommerfeld model, which suggested that electrons travel in elliptical orbits around a nucleus instead of the Bohr model's circular orbits. This model supplemented the quantized angular momentum condition of the Bohr model with an additional radial quantization condition, the Sommerfeld–Wilson quantization condition

$$\int_0^T p_r \, dq_r = nh$$

where p_r is the radial momentum canonically conjugate to the coordinate q which is the radial position and T is one full orbital period. The integral is the action of action-angle coordinates. This condition, suggested by the correspondence principle, is the only one possible, since the quantum numbers are adiabatic invariants.

The Bohr–Sommerfeld model was fundamentally inconsistent and led to many paradoxes. The magnetic quantum number measured the tilt of the orbital plane relative to the xy-plane, and it could only take a few discrete values. This contradicted the obvious fact that an atom could be turned this way and that relative to the coordinates without restriction. The Sommerfeld quantization can be performed in different canonical coordinates and sometimes gives different answers. The incorporation of radiation corrections was difficult, because it required finding action-angle coordinates for a combined radiation/atom system, which is difficult when the radiation is allowed to escape. The whole theory did not extend to non-integrable motions, which meant that many systems could not be treated even in principle. In the end, the model was replaced by the modern quantum mechanical treatment of the hydrogen atom, which was first given by Wolfgang Pauli in 1925, using Heisenberg's matrix mechanics. The current picture of the hydrogen atom is based on the atomic orbitals of wave mechanics which Erwin Schrödinger developed in 1926.

However, this is not to say that the Bohr model was without its successes. Calculations based on the Bohr–Sommerfeld model were able to accurately explain a number of more complex atomic spectral effects. For example, up to first-order perturbations, the

Bohr model and quantum mechanics make the same predictions for the spectral line splitting in the Stark effect. At higher-order perturbations, however, the Bohr model and quantum mechanics differ, and measurements of the Stark effect under high field strengths helped confirm the correctness of quantum mechanics over the Bohr model. The prevailing theory behind this difference lies in the shapes of the orbitals of the electrons, which vary according to the energy state of the electron.

The Bohr–Sommerfeld quantization conditions lead to questions in modern mathematics. Consistent semiclassical quantization condition requires a certain type of structure on the phase space, which places topological limitations on the types of symplectic manifolds which can be quantized. In particular, the symplectic form should be the curvature form of a connection of a Hermitian line bundle, which is called a prequantization.

References

- Dijksterhuis, E. The Mechanization of the World Picture. Trans. by C. Dikshoorn. New York: Oxford University Press, 1969. ISBN 0-691-02396-4
- Gangopadhyaya, Mrinalkanti. Indian Atomism: history and sources. Atlantic Highlands, New Jersey: Humanities Press, 1981. ISBN 0-391-02177-X
- Lloyd, G. E. R. Aristotle: The Growth and Structure of his Thought. Cambridge: Cambridge University Press, 1968. ISBN 0-521-09456-9
- McEvilley, Thomas (2002). The Shape of Ancient Thought: Comparative Studies in Greek and Indian Philosophies. New York: Allworth Communications Inc. ISBN 1-58115-203-5.
- Redondi, Pietro. Galileo Heretic. Translated by Raymond Rosenthal. Princeton, NJ: Princeton University Press, 1987. ISBN 0-691-02426-X
- Salpeter, Edwin E. (1996). Lakhtakia, Akhlesh, ed. “Models and Modelers of Hydrogen”. American Journal of Physics. World Scientific. 65 (9): 933. Bibcode:1997AmJPh..65..933L. doi:10.1119/1.18691. ISBN 981-02-2302-1.
- Akhlesh Lakhtakia (Ed.); Salpeter, Edwin E. (1996). “Models and Modelers of Hydrogen”. American Journal of Physics. World Scientific. 65 (9): 933. Bibcode:1997AmJPh..65..933L. doi:10.1119/1.18691. ISBN 981-02-2302-1.
- Nicholas Giordano (1 January 2012). College Physics: Reasoning and Relationships. Cengage Learning. pp. 1051–. ISBN 1-285-22534-1.

4

Atomic Nucleus and its Structure

The atomic nucleus comprises of neutrons and protons, making it positively charged on the whole. The components of the nucleus are held together by a strong nuclear force. This chapter provides comprehensive information on the nuclear makeup of an atom and analyses the nuclear structure and nuclear shell model. This chapter also discusses the semi-empirical mass formula which explains the atomic masses of various elements and is based on the liquid drop model.

Atomic Nucleus

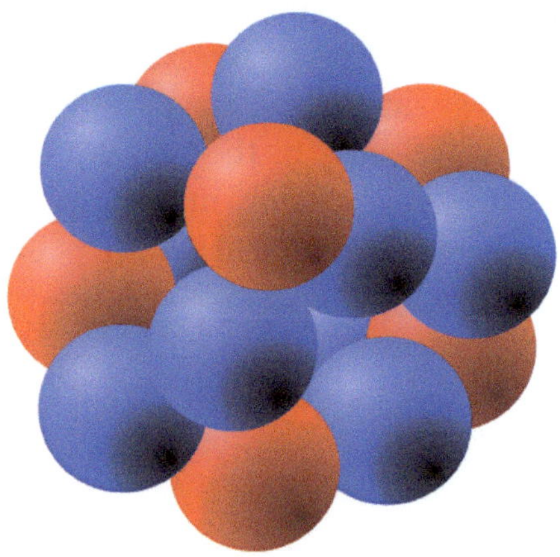

A model of the atomic nucleus showing it as a compact bundle of the two types of nucleons: protons (red) and neutrons (blue). In this diagram, protons and neutrons look like little balls stuck together, but an actual nucleus (as understood by modern nuclear physics) cannot be explained like this, but only by using quantum mechanics. In a nucleus which occupies a certain energy level (for example, the ground state), each nucleon can be said to occupy a range of locations.

The nucleus is the small, dense region consisting of protons and neutrons at the center of an atom, discovered in 1911 by Ernest Rutherford based on the 1909 Geiger–Marsden gold foil experiment. After the discovery of the neutron in 1932, models for a nucleus composed of protons and neutrons were quickly developed by Dmitri Ivanenko and Werner Heisenberg. Almost all of the mass of an atom is located in the nucleus, with a very small contribution from the electron cloud. Protons and neutrons are bound together to form a nucleus by the nuclear force.

The diameter of the nucleus is in the range of 1.75 fm (1.75×10^{-15} m) for hydrogen (the diameter of a single proton) to about 15 fm for the heaviest atoms, such as uranium. These dimensions are much smaller than the diameter of the atom itself (nucleus + electron cloud), by a factor of about 23,000 (uranium) to about 145,000 (hydrogen).

The branch of physics concerned with the study and understanding of the atomic nucleus, including its composition and the forces which bind it together, is called nuclear physics.

Introduction

History

The nucleus was discovered in 1911, as a result of Ernest Rutherford's efforts to test Thomson's "plum pudding model" of the atom. The electron had already been discovered earlier by J.J. Thomson himself. Knowing that atoms are electrically neutral, Thomson postulated that there must be a positive charge as well. In his plum pudding model, Thomson suggested that an atom consisted of negative electrons randomly scattered within a sphere of positive charge. Ernest Rutherford later devised an experiment with his research partner Hans Geiger and with help of Ernest Marsden, that involved the deflection of alpha particles (helium nuclei) directed at a thin sheet of metal foil. He reasoned that if Thomson's model were correct, the positively charged alpha particles would easily pass through the foil with very little deviation in their paths, as the foil should act as electrically neutral if the negative and positive charges are so intimately mixed as to make it appear neutral. To his surprise, many of the particles were deflected at very large angles. Because the mass of an alpha particle is about 8000 times that of an electron, it became apparent that a very strong force must be present if it could deflect the massive and fast moving alpha particles. He realized that the plum pudding model could not be accurate and that the deflections of the alpha particles could only be explained if the positive and negative charges were separated from each other and that the mass of the atom was a concentrated point of positive charge. This justified the idea of a nuclear atom with a dense center of positive charge and mass.

Etymology

The term nucleus is from the Latin word *nucleus*, a diminutive of *nux* ("nut"), meaning the kernel (i.e., the "small nut") inside a watery type of fruit (like a peach). In 1844, Michael Faraday used the term to refer to the "central point of an atom". The modern atomic meaning was proposed by Ernest Rutherford in 1912. The adoption of the term "nucleus" to atomic theory, however, was not immediate. In 1916, for example, Gilbert N. Lewis stated, in his famous article *The Atom and the Molecule*, that "the atom is composed of the *kernel* and an outer atom or *shell*"

Nuclear Makeup

A figurative depiction of the helium-4 atom with the electron cloud in shades of gray. In the nucleus, the two protons and two neutrons are depicted in red and blue. This depiction shows the particles as separate, whereas in an actual helium atom, the protons are superimposed in space and most likely found at the very center of the nucleus, and the same is true of the two neutrons. Thus, all four particles are most likely found in exactly the same space, at the central point. Classical images of separate particles fail to model known charge distributions in very small nuclei. A more accurate image is that the spatial distribution of nucleons in a helium nucleus is much closer to the helium electron cloud shown here, although on a far smaller scale, than to the fanciful nucleus image.

The nucleus of an atom consists of neutrons and protons, which in turn are the manifestation of more elementary particles, called quarks, that are held in association by the nuclear strong force in certain stable combinations of hadrons, called baryons. The nuclear strong force extends far enough from each baryon so as to bind the neutrons and protons together against the repulsive electrical force between the positively charged protons. The nuclear strong force has a very short range, and essentially drops to zero just beyond the edge of the nucleus. The collective action of the positively charged nucleus is to hold the electrically negative charged electrons in their orbits about the nucleus. The collection of negatively charged electrons orbiting the nucleus display an affinity for certain configurations and numbers of electrons that make their orbits stable. Which chemical element an atom represents is determined by the number of protons in the nucleus; the neutral atom will have an equal number of electrons orbiting that nucleus. Individual chemical elements can create more stable electron configurations by combining to share their electrons. It is that sharing of electrons to create stable electronic orbits about the nucleus that appears to us as the chemistry of our macro world.

Protons define the entire charge of a nucleus, and hence its chemical identity. Neutrons are electrically neutral, but contribute to the mass of a nucleus to nearly the same extent as the protons. Neutrons explain the phenomenon of isotopes – varieties of the same chemical element which differ only in their atomic mass, not their chemical action.

Composition and Shape

Protons and neutrons are fermions, with different values of the strong isospin quantum number, so two protons and two neutrons can share the same space wave function since they are not identical quantum entities. They are sometimes viewed as two different quantum states of the same particle, the *nucleon*. Two fermions, such as two protons, or two neutrons, or a proton + neutron (the deuteron) can exhibit bosonic behavior when they become loosely bound in pairs, which have integral spin.

In the rare case of a hypernucleus, a third baryon called a hyperon, containing one or more strange quarks and/or other unusual quark(s), can also share the wave function. However, this type of nucleus is extremely unstable and not found on Earth except in high energy physics experiments.

The neutron has a positively charged core of radius ≈ 0.3 fm surrounded by a compensating negative charge of radius between 0.3 fm and 2 fm. The proton has an approximately exponentially decaying positive charge distribution with a mean square radius of about 0.8 fm.

Nuclei can be spherical, rugby ball-shaped (prolate deformation), discus-shaped (oblate deformation), triaxial (a combination of oblate and prolate deformation) or pear-shaped.

Forces

Nuclei are bound together by the residual strong force (nuclear force). The residual strong force is a minor residuum of the strong interaction which binds quarks together to form protons and neutrons. This force is much weaker *between* neutrons and protons because it is mostly neutralized within them, in the same way that electromagnetic forces *between* neutral atoms (such as van der Waals forces that act between two inert gas atoms) are much weaker than the electromagnetic forces that hold the parts of the atoms together internally (for example, the forces that hold the electrons in an inert gas atom bound to its nucleus).

The nuclear force is highly attractive at the distance of typical nucleon separation, and this overwhelms the repulsion between protons due to the electromagnetic force, thus allowing nuclei to exist. However, the residual strong force has a limited range because it decays quickly with distance (see Yukawa potential); thus only nuclei smaller than a certain size can be completely stable. The largest known completely stable nucleus (i.e. stable to alpha, beta, and gamma decay) is lead-208 which contains a total of 208 nucleons (126 neutrons and 82 protons). Nuclei larger than this maximum are unstable and tend to be increasingly short-lived with larger numbers of nucleons. However, bismuth-209 is also stable to beta decay and has the longest half-life to alpha decay of any known isotope, estimated at a billion times longer than the age of the universe.

The residual strong force is effective over a very short range (usually only a few femtometres (fm); roughly one or two nucleon diameters) and causes an attraction between any pair of nucleons. For example, between protons and neutrons to form [NP] deuteron, and also between protons and protons, and neutrons and neutrons.

Halo Nuclei and Strong Force Range Limits

The effective absolute limit of the range of the strong force is represented by halo nuclei such as lithium-11 or boron-14, in which dineutrons, or other collections of neutrons, orbit at distances of about ten fermis (roughly similar to the 8 fermi radius of the nucleus of uranium-238). These nuclei are not maximally dense. Halo nuclei form at the extreme edges of the chart of the nuclides—the neutron drip line and proton drip line—and are all unstable with short half-lives, measured in milliseconds; for example, lithium-11 has a half-life of 8.8 milliseconds.

Halos in effect represent an excited state with nucleons in an outer quantum shell which has unfilled energy levels "below" it (both in terms of radius and energy). The halo may be made of either neutrons [NN, NNN] or protons [PP, PPP]. Nuclei which have a single neutron halo include ^{11}Be and ^{19}C. A two-neutron halo is exhibited by ^{6}He, ^{11}Li, ^{17}B, ^{19}B and ^{22}C. Two-neutron halo nuclei break into three fragments, never two, and are called *Borromean nuclei* because of this behavior (referring to a system of three interlocked rings in which breaking any ring frees both of the others). ^{8}He and ^{14}Be both exhibit a four-neutron halo. Nuclei which have a proton halo include ^{8}B and ^{26}P. A two-proton halo is exhibited by ^{17}Ne and ^{27}S. Proton halos are expected to be more rare and unstable than the neutron examples, because of the repulsive electromagnetic forces of the excess proton(s).

Nuclear Models

Although the standard model of physics is widely believed to completely describe the composition and behavior of the nucleus, generating predictions from theory is much more difficult than for most other areas of particle physics. This is due to two reasons:

- In principle, the physics within a nucleus can be derived entirely from quantum chromodynamics (QCD). In practice however, current computational and mathematical approaches for solving QCD in low-energy systems such as the nuclei are extremely limited. This is due to the phase transition that occurs between high-energy quark matter and low-energy hadronic matter, which renders perturbative techniques unusable, making it difficult to construct an accurate QCD-derived model of the forces between nucleons. Current approaches are limited to either phenomenological models such as the Argonne v18 potential or chiral effective field theory.

- Even if the nuclear force is well constrained, a significant amount of computational power is required to accurately compute the properties of nuclei *ab initio*. Developments in many-body theory have made this possible for many low mass and relatively stable nuclei, but further improvements in both computational power and mathematical approaches are required before heavy nuclei or highly unstable nuclei can be tackled.

Historically, experiments have been compared to relatively crude models that are necessarily imperfect. None of these models can completely explain experimental data on nuclear structure.

The nuclear radius (R) is considered to be one of the basic quantities that any model must predict. For stable nuclei (not halo nuclei or other unstable distorted nuclei) the nuclear radius is roughly proportional to the cube root of the mass number (A) of the nucleus, and particularly in nuclei containing many nucleons, as they arrange in more spherical configurations:

The stable nucleus has approximately a constant density and therefore the nuclear radius R can be approximated by the following formula,

where A = Atomic mass number (the number of protons Z, plus the number of neutrons N) and $r_0 = 1.25$ fm $= 1.25 \times 10^{-15}$ m. In this equation, the constant r_0 varies by 0.2 fm, depending on the nucleus in question, but this is less than 20% change from a constant.

In other words, packing protons and neutrons in the nucleus gives *approximately* the same total size result as packing hard spheres of a constant size (like marbles) into a tight spherical or almost spherical bag (some stable nuclei are not quite spherical, but are known to be prolate).

Models of nuclear structure include :

Liquid Drop Model

Early models of the nucleus viewed the nucleus as a rotating liquid drop. In this model, the trade-off of long-range electromagnetic forces and relatively short-range nuclear forces, together cause behavior which resembled surface tension forces in liquid drops of different sizes. This formula is successful at explaining many important phenomena of nuclei, such as their changing amounts of binding energy as their size and composition changes (see semi-empirical mass formula), but it does not explain the special stability which occurs when nuclei have special "magic numbers" of protons or neutrons.

The terms in the semi-empirical mass formula, which can be used to approximate the binding energy of many nuclei, are considered as the sum of five types of energies (see below). Then the picture of a nucleus as a drop of incompressible liquid roughly accounts for the observed variation of binding energy of the nucleus:

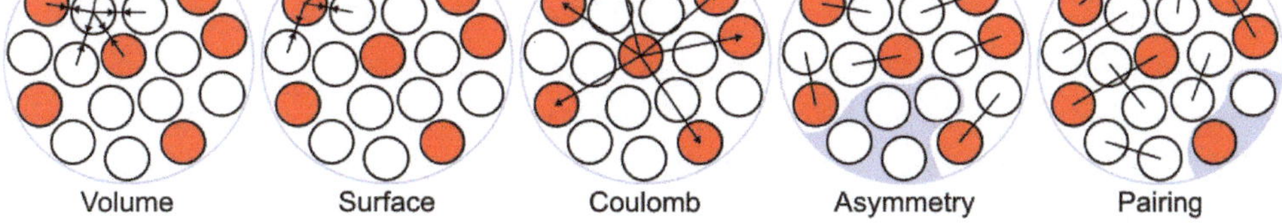

Volume energy. When an assembly of nucleons of the same size is packed together into the smallest volume, each interior nucleon has a certain number of other nucleons in contact with it. So, this nuclear energy is proportional to the volume.

Surface energy. A nucleon at the surface of a nucleus interacts with fewer other nucleons than one in the interior of the nucleus and hence its binding energy is less. This surface energy term takes that into account and is therefore negative and is proportional to the surface area.

Coulomb Energy. The electric repulsion between each pair of protons in a nucleus contributes toward decreasing its binding energy.

Asymmetry energy (also called Pauli Energy). An energy associated with the Pauli exclusion principle. Were it not for the Coulomb energy, the most stable form of nuclear matter would have the same number of neutrons as protons, since unequal numbers of neutrons and protons imply filling higher energy levels for one type of particle, while leaving lower energy levels vacant for the other type.

Pairing energy. An energy which is a correction term that arises from the tendency of proton pairs and neutron pairs to occur. An even number of particles is more stable than an odd number.

Shell Models and Other Quantum Models

A number of models for the nucleus have also been proposed in which nucleons occupy orbitals, much like the atomic orbitals in atomic physics theory. These wave models imagine nucleons to be either sizeless point particles in potential wells, or else probability waves as in the "optical model", frictionlessly orbiting at high speed in potential wells.

In the above models, the nucleons may occupy orbitals in pairs, due to being fermions, which allows to explain even/odd Z and N effects well-known from experiments. The exact nature and capacity of nuclear shells differs from those of electrons in atomic orbitals, primarily because the potential well in which the nucleons move (especially in larger nuclei) is quite different from the central electromagnetic potential well which binds electrons in atoms. Some resemblance to atomic orbital models may be seen in a small atomic nucleus like that of helium-4, in which the two protons and two neutrons separately occupy 1s orbitals analogous to the 1s orbital for the two electrons in the helium atom, and achieve unusual stability for the same reason. Nuclei with 5 nucleons are all extremely unstable and short-lived, yet, helium-3, with 3 nucleons, is very stable even with lack of a closed 1s orbital shell. Another nucleus with 3 nucleons, the triton hydrogen-3 is unstable and will decay into helium-3 when isolated. Weak nuclear stability with 2 nucleons {NP} in the 1s orbital is found in the deuteron hydrogen-2, with only one nucleon in each of the proton and neutron potential wells. While each nucleon is a fermion, the {NP} deuteron is a boson and thus does not follow Pauli Exclusion for close packing within shells. Lithium-6 with 6 nucleons is highly stable without a closed second 1p shell orbital. For light nuclei with total nucleon numbers 1 to 6 only those with 5 do not show some evidence of stability. Observations of beta-stability of light nuclei outside closed shells indicate that nuclear stability is much more complex than simple closure of shell orbitals with magic numbers of protons and neutrons.

For larger nuclei, the shells occupied by nucleons begin to differ significantly from electron shells, but nevertheless, present nuclear theory does predict the magic numbers of filled nuclear shells for both protons and neutrons. The closure of the stable shells predicts unusually stable configurations, analogous to the noble group of nearly-inert gases in chemistry. An example is the stability of the closed shell of 50 protons, which allows tin to have 10 stable isotopes, more than any other element. Similarly, the distance from shell-closure explains the unusual instability of isotopes which have far from stable numbers of these particles, such as the radioactive elements 43 (technetium) and 61 (promethium), each of which is preceded and followed by 17 or more stable elements.

There are however problems with the shell model when an attempt is made to account for nuclear properties well away from closed shells. This has led to complex *post hoc* distortions of the shape of the potential well to fit experimental data, but the question remains whether these mathematical manipulations actually correspond to the spatial deformations in real nuclei. Problems with the shell model have led some to propose realistic two-body and three-body nuclear force effects involving nucleon clusters and then build the nucleus on this basis. Two such cluster models are the Close-Packed Spheron Model of Linus Pauling and the 2D Ising Model of MacGregor.

Consistency Between Models

As with the case of superfluid liquid helium, atomic nuclei are an example of a state in which both (1) "ordinary" particle physical rules for volume and (2) non-intuitive quantum mechanical rules for a wave-like nature apply. In superfluid helium, the helium atoms have volume, and essentially "touch" each other, yet at the same time exhibit strange bulk properties, consistent with a Bose–Einstein condensation. The latter reveals that they also have a wave-like nature and do not exhibit standard fluid properties, such as friction. For nuclei made of hadrons which are fermions, the same type of condensation does not occur, yet nevertheless, many nuclear properties can only be explained similarly by a combination of properties of particles with volume, in addition to the frictionless motion characteristic of the wave-like behavior of objects trapped in Erwin Schrödinger's quantum orbitals.

Nuclear Structure

Understanding the structure of the atomic nucleus is one of the central challenges in nuclear physics.

Models

The Liquid Drop Model

The liquid drop model is one of the first models of nuclear structure, proposed by Carl Friedrich von Weizsäcker in 1935. It describes the nucleus as a semiclassical fluid made up of neutrons and protons, with an internal repulsive electrostatic force proportional to the number of protons. The quantum mechanical nature of these particles appears via the Pauli exclusion principle, which states that no two nucleons of the same kind can be at the same state. Thus the fluid is actually what is known as a Fermi liquid. In this model, the binding energy of a nucleus with Z protons and N neutrons is given by

$$E_B = a_V A - a_S A^{2/3} - a_C \frac{Z^2}{A^{1/3}} - a_A \frac{(N-Z)^2}{A} - \delta(A,Z)$$

where $A = Z + N$ is the total number of nucleons. The terms proportional to A and $A^{2/3}$ represent the volume and surface energy of the liquid drop, the term proportional to Z^2 represents the electrostatic energy, the term proportional to $(N-Z)^2$ represents the Pauli exclusion principle and the last term $\delta(A,Z)$ is the pairing term, which lowers the energy for even numbers of protons or neutrons. The coefficients a_V, a_S, a_C, a_A and the strength of the pairing term may be estimated theoretically, or fit to data. This simple model reproduces the main features of the binding energy of nuclei.

The assumption of nucleus as a drop of Fermi liquid is still widely used in the form of Finite Range Droplet Model (FRDM), due to the possible good reproduction of nuclear binding energy on the whole chart, with the necessary accuracy for predictions of unknown nuclei.

The Shell Model

The expression "shell model" is ambiguous in that it refers to two different eras in the state of the art. It was previously used to describe the existence of nucleon shells in the nucleus according to an approach closer to what is now called mean field theory. Nowadays, it refers to a formalism analogous to the configuration interaction formalism used in quantum chemistry. We shall introduce the latter here.

Introduction to the shell concept

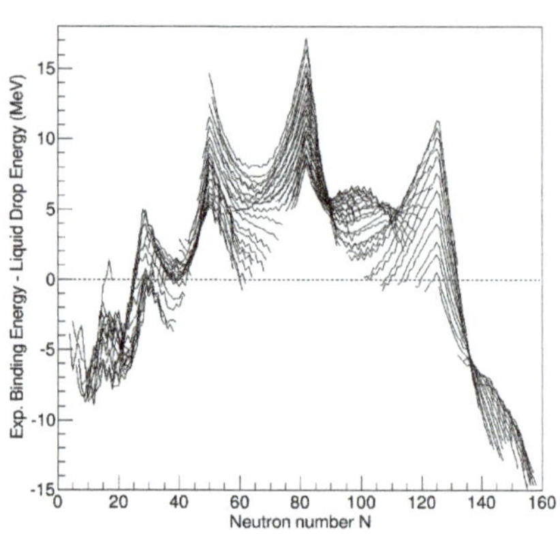

Difference between experimental binding energies and the liquid drop model prediction as a function of neutron number for Z>7

Systematic measurements of the binding energy of atomic nuclei show systematic deviations with respect to those estimated from the liquid drop model. In particular, some nuclei having certain values for the number of protons and/or neutrons are bound more tightly together than predicted by the liquid drop model. These nuclei are called singly/doubly magic. This observation led scientists to assume the existence of a shell structure of nucleons (protons and neutrons) within the nucleus, like that of electrons within atoms.

Indeed, nucleons are quantum objects. Strictly speaking, one should not speak of energies of individual nucleons, because they are all correlated with each other. However, as an approximation one may envision an average nucleus, within which nucleons propagate individually. Owing to their quantum character, they may only occupy *discrete* energy levels. These levels are by no means uniformly distributed; some intervals of energy are crowded, and some are empty, generating a gap in possible energies. A shell is such a set of levels separated from the other ones by a wide empty gap.

The energy levels are found by solving the Schrödinger equation for a single nucleon moving in the average potential generated by all other nucleons. Each level may be occupied by a nucleon, or empty. Some levels accommodate several different quantum states with the same energy; they are said to be *degenerate*. This occurs in particular if the average nucleus has some symmetry.

The concept of shells allows one to understand why some nuclei are bound more tightly than others. This is because two nucleons of the same kind cannot be in the same state (Pauli exclusion principle). So the lowest-energy state of the nucleus is one where nucleons fill all energy levels from the bottom up to some level. A nucleus with full shells is exceptionally stable, as will be explained.

As with electrons in the electron shell model, protons in the outermost shell are relatively loosely bound to the nucleus if there are only few protons in that shell, because they are farthest from the center of the nucleus. Therefore, nuclei which have a full outer proton shell will be more tightly bound and have a higher binding energy than other nuclei with a similar total number of protons. All this is also true for neutrons.

Furthermore, the energy needed to excite the nucleus (i.e. moving a nucleon to a higher, previously unoccupied level) is exceptionally high in such nuclei. Whenever this unoccupied level is the next after a full shell, the only way to excite the nucleus is to raise one nucleon *across the gap*, thus spending a large amount of energy. Otherwise, if the highest occupied energy level lies in a partly filled shell, much less energy is required to raise a nucleon to a higher state in the same shell.

Some evolution of the shell structure observed in stable nuclei is expected away from the valley of stability. For example, observations of unstable isotopes have shown shifting and even a reordering of the single particle levels of which the shell structure is composed. This is sometimes observed as the creation of an island of inversion or in the reduction of excitation energy gaps above the traditional magic numbers.

Basic Hypotheses

Some basic hypotheses are made in order to give a precise conceptual framework to the shell model:

- The atomic nucleus is a quantum n-body system.
- The internal motion of nucleons within the nucleus is non-relativistic, and their behavior is governed by the Schrödinger equation.
- Nucleons are considered to be pointlike, without any internal structure.

Brief Description of the Formalism

The general process used in the shell model calculations is the following. First a Hamiltonian for the nucleus is defined. Usually, for computational practicality, only one- and two-body terms are taken into account in this definition. The interaction is an effective theory : it contains free parameters which have to be fitted with experimental data.

The next step consists in defining a basis of single-particle states, i.e. a set of wavefunctions describing all possible nucleon states. Most of the time, this basis is obtained via a Hartree–Fock computation. With this set of one-particle states, Slater determinants are built, that is, wavefunctions for Z proton variables or N neutron variables, which are antisymmetrized products of single-particle wavefunctions (antisymmetrized meaning that under exchange of variables for any pair of nucleons, the wavefunction only changes sign).

In principle, the number of quantum states available for a single nucleon at a finite energy is finite, say n. The number of nucleons in the nucleus must be smaller than the number of available states, otherwise the nucleus cannot hold all of its nucleons. There are thus several ways to choose Z (or N) states among the n possible. In combinatorial mathematics, the number of choices of Z objects among n is the binomial coefficient CZ n. If n is much larger than Z (or N), this increases roughly like n^Z. Practically, this number becomes so large that every computation is impossible for $A=N+Z$ larger than 8.

To obviate this difficulty, the space of possible single-particle states is divided into a core and a valence shell, by analogy with chemistry. The core is a set of single-particles which are assumed to be inactive, in the sense that they are the well bound lowest-energy states, and that there is no need to reexamine their situation. They do not appear in the Slater determinants, contrary to the states in the valence space, which is the space of all single-particle states *not in the core*, but possibly to be considered in the choice of the build of the (Z-) N-body wavefunction. The set of all possible Slater determinants in the valence space defines a basis for (Z-) N-body states.

The last step consists in computing the matrix of the Hamiltonian within this basis, and to diagonalize it. In spite of the reduction of the dimension of the basis owing to the fixation of the core, the matrices to be diagonalized reach easily dimensions of the order of 10^9, and demand specific diagonalization techniques.

The shell model calculations give in general an excellent fit with experimental data. They depend however strongly on two main factors:

- The way to divide the single-particle space into core and valence.
- The effective nucleon–nucleon interaction.

Mean Field Theories

The Independent-particle Model

The interaction between nucleons, which is a consequence of strong interactions and binds the nucleons within the nucleus, exhibits the peculiar behaviour of having a finite range: it vanishes when the distance between two nucleons becomes too large; it is attractive at medium range, and repulsive at very small range. This last property correlates with the Pauli exclusion principle according to which two fermions (nucleons are fermions) cannot be in the same quantum state. This results in a very large mean free path predicted for a nucleon within the nucleus.

The main idea of the Independent Particle approach is that a nucleon moves inside a certain potential well (which keeps it bound to the nucleus) independently from the other nucleons. This amounts to replacing a *N*-body problem (*N* particles interacting) by *N* single-body problems. This essential simplification of the problem is the cornerstone of mean field theories. These are also widely used in atomic physics, where electrons move in a mean field due to the central nucleus and the electron cloud itself.

The independent particle model and mean field theories (we shall see that there exist several variants) have a great success in describing the properties of the nucleus starting from an effective interaction or an effective potential, thus are a basic part of atomic nucleus theory. One should also notice that they are modular enough, in that it is quite easy to extend the model to introduce effects such as nuclear pairing, or collective motions of the nucleon like rotation, or vibration, adding the corresponding energy terms in the formalism. This implies that in many representation the mean field is only a starting point for a more complete description which introduces correlations reproducing properties like collective excitations and nucleon transfer.

Nuclear Potential and Effective Interaction

A large part of the practical difficulties met in mean field theories is the definition (or calculation) of the potential of the mean field itself. One can very roughly distinguish between two approaches:

- The phenomenological approach is a parameterization of the nuclear potential by an appropriate mathematical function. Historically, this procedure was applied with the greatest success by Sven Gösta Nilsson, who used as a potential a (deformed) harmonic oscillator potential. The most recent parameterizations are based on more realistic functions, which account more accurately for scattering experiments, for example. In particular the form known as the Woods Saxon potential can be mentioned.

- The self-consistent or Hartree–Fock approach aims to deduce mathematically the nuclear potential from an effective nucleon–nucleon interaction. This technique implies a resolution of the Schrödinger equation in an iterative fashion, starting from an ansatz wavefunction and improving it variationally, since the potential depends there upon the wavefunctions to be determined. The latter are written as Slater determinants.

In the case of the Hartree–Fock approaches, the trouble is not to find the mathematical function which describes best the nuclear potential, but that which describes best the nucleon–nucleon interaction. Indeed, in contrast with atomic physics where the interaction is known (it is the Coulomb interaction), the nucleon–nucleon interaction within the nucleus is not known analytically.

There are two main reasons for this fact. First, the strong interaction acts essentially among the quarks forming the nucleons. The nucleon–nucleon interaction in vacuum is a mere *consequence* of the quark–quark interaction. While the latter is well understood in the framework of the Standard Model at high energies, it is much more complicated in low energies due to color confinement and asymptotic freedom. Thus there is yet no fundamental theory allowing one to deduce the nucleon–nucleon interaction from the quark–quark interaction. Furthermore, even if this problem were solved, there would remain a large difference between the ideal (and conceptually simpler) case of two nucleons interacting in vacuum, and that of these nucleons interacting in the nuclear matter. To go further, it was necessary to invent the concept of effective interaction. The latter is basically a mathematical function with several arbitrary parameters, which are adjusted to agree with experimental data.

Most modern interaction are zero-range so they act only when the two nucleons are in contact, as introduced by Tony Skyrme.

The Self-consistent Approaches of the Hartree–Fock Type

In the Hartree–Fock approach of the n-body problem, the starting point is a Hamiltonian containing n kinetic energy terms, and potential terms. As mentioned before, one of the mean field theory hypotheses is that only the two-body interaction is to be taken into account. The potential term of the Hamiltonian represents all possible two-body interactions in the set of n fermions. It is the first hypothesis.

The second step consists in assuming that the wavefunction of the system can be written as a Slater determinant of one-particle spin-orbitals. This statement is the mathematical translation of the independent-particle model. This is the second hypothesis.

There remains now to determine the components of this Slater determinant, that is, the individual wavefunctions of the nucleons. To this end, it is assumed that the total wavefunction (the Slater determinant) is such that the energy is minimum. This is the third hypothesis.

Technically, it means that one must compute the mean value of the (known) two-body Hamiltonian on the (unknown) Slater determinant, and impose that its mathematical variation vanishes. This leads to a set of equations where the unknowns are the individual wavefunctions: the Hartree–Fock equations. Solving these equations gives the wavefunctions and individual energy levels of nucleons, and so the total energy of the nucleus and its wavefunction.

This short account of the Hartree–Fock method explains why it is called also the variational approach. At the beginning of the calculation, the total energy is a "function of the individual wavefunctions" (a so-called functional), and everything is then made in order to optimize the choice of these wavefunctions so that the functional has a minimum – hopefully absolute, and not only local. To be more precise, there should be mentioned that the energy is a functional of the density, defined as the sum of the individual squared wavefunctions. Let us note also that the Hartree–Fock method is also used in atomic physics and condensed matter physics as Density Functional Theory, DFT.

The process of solving the Hartree–Fock equations can only be iterative, since these are in fact a Schrödinger equation in which the potential depends on the density, that is, precisely on the wavefunctions to be determined. Practically, the algorithm is started with a set of individual grossly reasonable wavefunctions (in general the eigenfunctions of a harmonic oscillator). These allow to compute the density, and therefrom the Hartree–Fock potential. Once this done, the Schrödinger equation is solved anew, and so on. The calculation stops – convergence is reached – when the difference among wavefunctions, or energy levels, for two successive iterations is less than a fixed value. Then the mean field potential is completely determined, and the Hartree–Fock equations become standard Schrödinger equations. The corresponding Hamiltonian is then called the Hartree–Fock Hamiltonian.

The Relativistic Mean Field Approaches

Born first in the 1970s with the works of John Dirk Walecka on quantum hadrodynamics, the relativistic models of the nucleus were sharpened up towards the end of the 1980s by P. Ring and coworkers. The starting point of these approaches is the relativistic quantum field theory. In this context, the nucleon interactions occur via the exchange of virtual particles called mesons. The idea is, in a first step, to build a Lagrangian containing these interaction terms. Second, by an application of the least action principle, one gets a set of equations of motion. The real particles (here the nucleons) obey the Dirac equation, whilst the virtual ones (here the mesons) obey the Klein–Gordon equations.

In view of the non-perturbative nature of strong interaction, and also in view of the fact that the exact potential form of this interaction between groups of nucleons is relatively badly known, the use of such an approach in the case of atomic nuclei requires drastic approximations. The main simplification consists in replacing in the equations all field terms (which are operators in the mathematical sense) by their mean value (which are functions). In this way, one gets a system of coupled integro-differential equations, which can be solved numerically, if not analytically.

The Interacting Boson Model

The interacting boson model (IBM) is a model in nuclear physics in which nucleons are represented as pairs, each of them acting as a boson particle, with integral spin of 0, 2 or 4. This makes calculations feasible for larger nuclei. There are several branches of this model - in one of them (IBM-1) one can group all types of nucleons in pairs, in others (for instance - IBM-2) one considers protons and neutrons in pairs separately.

Spontaneous Breaking of Symmetry in Nuclear Physics

One of the focal points of all physics is symmetry. The nucleon–nucleon interaction and all effective interactions used in practice have certain symmetries. They are invariant by translation (changing the frame of reference so that directions are not altered), by rotation (turning the frame of reference around some axis), or parity (changing the sense of axes) in the sense that the interaction does not change under any of these operations. Nevertheless, in the Hartree–Fock approach, solutions which are not invariant under such a symmetry can appear. One speaks then of spontaneous symmetry breaking.

Qualitatively, these spontaneous symmetry breakings can be explained in the following way : in the mean field theory, the nucleus is described as a set of independent particles. Most additional correlations among nucleons which do not enter the mean field are neglected. They can appear however by a breaking of the symmetry of the mean field Hamiltonian, which is only approximate. If the density used to start the iterations of the Hartree–Fock process breaks certain symmetries, the final Hartree–Fock Hamiltonian may break these symmetries, if it is advantageous to keep these broken from the point of view of the total energy.

It may also converge towards a symmetric solution. In any case, if the final solution breaks the symmetry, for example, the rotational symmetry, so that the nucleus appears not to be spherical, but elliptic, all configurations deduced from this deformed nucleus by a rotation are just as good solutions for the Hartree–Fock problem. The ground state of the nucleus is then *degenerate.*

A similar phenomenon happens with the nuclear pairing, which violates the conservation of the number of baryons (see below).

Extensions of the Mean Field Theories

Nuclear Pairing Phenomenon

The most common extension to mean field theory is the nuclear pairing. Nuclei with an even number of nucleons are systematically more bound than those with an odd one. This implies that each nucleon binds with another one to form a pair, consequently the system cannot be described as independent particles subjected to a common a mean field. When the nucleus has an even number of protons and neutrons, each one of them finds a partner. To excite such a system, one must at least use such an energy as to break a pair. Conversely, in the case of odd number of protons or neutrons, there exists an unpaired nucleon, which needs less energy to be excited.

This phenomenon is closely analogous to that of Type 1 superconductivity in solid state physics. The first theoretical description of nuclear pairing was proposed at the end of the 1950s by Aage Bohr, Ben Mottelson, and David Pines (which contributed to the reception of the Nobel Prize in Physics in 1975 by Bohr and Mottelson). It was close to the BCS theory of Bardeen, Cooper and Schrieffer, which accounts for metal superconductivity. Theoretically, the pairing phenomenon as described by the BCS theory combines with the mean field theory: nucleons are both subject to the mean field potential and to the pairing interaction.

The Hartree–Fock–Bogolyubov (HFB) method is a more sophisticated approach, enabling one to consider the pairing and mean field interactions consistently on equal footing. HFB is now the de facto standard in the mean field treatment of nuclear systems.

Symmetry Restoration

Peculiarity of mean field methods is the calculation of nuclear property by explicit symmetry breaking. The calculation of the mean field with self-consistent methods (e.g. Hartree-Fock), breaks rotational symmetry, and the calculation of pairing property breaks particle-number.

Several techniques for symmetry restoration by projecting on good quantum numbers have been developed.

Particle Vibrations Coupling

Mean field methods (eventually considering symmetry restoration) are a good approximation for the ground state of the system, even postulating a system of independent particles. Higher-order corrections consider the fact that the particles interact together by the means of correlation. These correlations can be introduced taking into account the coupling of independent particle degrees of freedom, low-energy collective excitation of systems with even number of protons and neutrons.

In these way excited states can be reproduced by the means of Random Phase Approximation (RPA), and eventually consistently calculating also corrections to the ground state (e.g. by the means of Nuclear Field Theory).

Semi-empirical Mass Formula

In nuclear physics, the semi-empirical mass formula (SEMF) (sometimes also called Weizsäcker's formula, or the Bethe–Weizsäcker formula, or the Bethe–Weizsäcker mass formula to distinguish it from the Bethe–Weizsäcker process) is used to approximate the mass and various other properties of an atomic nucleus from its number of protons and neutrons. As the name suggests, it is based partly on theory and partly on empirical measurements. The theory is based on the liquid drop model proposed by George Gamow, which can account for most of the terms in the formula and gives rough estimates for the values of the coefficients. It was first formulated in 1935 by German physicist Carl Friedrich von Weizsäcker, and although refinements have been made to the coefficients over the years, the structure of the formula remains the same today.

The SEMF gives a good approximation for atomic masses and several other effects, but does not explain the appearance of magic numbers of protons and neutrons, and the extra binding-energy and measure of stability that are associated with these numbers of nucleons.

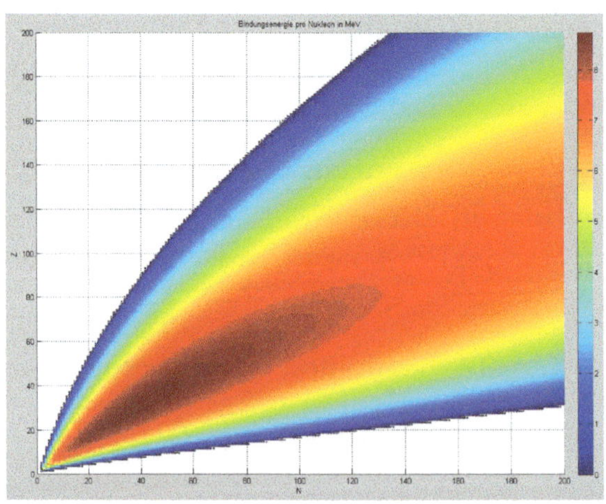

A graphical representation of the semi-empirical binding energy formula. The binding energy per nucleon in MeV (highest numbers in dark red, in excess of 8.5 MeV per nucleon) is plotted for various nuclides as a function of Z, the atomic number (on the y-axis), vs. N, the neutron number (on the x-axis). The highest binding energies are seen for $Z = 26$ (iron).

The Liquid Drop Model and its Analysis

The liquid drop model in nuclear physics treats the nucleus as drop of incompressible nuclear fluid of very high density. It was first proposed by George Gamow and then developed by Niels Bohr and John Archibald Wheeler. The nucleus is made of nucleons (protons and neutrons), which are held together by the nuclear force (a residual effect of the strong force). This is very similar to the structure of spherical liquid drop made of microscopic molecules. This is a crude model that does not explain all the properties of the nucleus, but does explain the spherical shape of most nuclei. It also helps to predict the nuclear binding energy and to assess how much is available for consumption.

Mathematical analysis of the theory delivers an equation which attempts to predict the binding energy of a nucleus in terms of the numbers of protons and neutrons it contains. This equation has five terms on its right hand side. These correspond to the cohesive binding of all the nucleons by the nuclear force, a surface energy term, the electrostatic mutual repulsion of the protons, an asymmetry term (derivable from the protons and neutrons occupying independent quantum momentum states) and a pairing term (partly derivable from the protons and neutrons occupying independent quantum spin states).

If we consider the sum of the following five types of energies, then the picture of a nucleus as a drop of incompressible liquid roughly accounts for the observed variation of binding energy of the nucleus:

Volume energy. When an assembly of nucleons of the same size is packed together into the smallest volume, each interior nucleon has a certain number of other nucleons in contact with it. So, this nuclear energy is proportional to the volume.

Surface energy. A nucleon at the surface of a nucleus interacts with fewer other nucleons than one in the interior of the nucleus and hence its binding energy is less. This surface energy term takes that into account and is therefore negative and is proportional to the surface area.

Coulomb Energy. The electric repulsion between each pair of protons in a nucleus contributes toward decreasing its binding energy.

Asymmetry energy (also called Pauli Energy). An energy associated with the Pauli exclusion principle. Were it not for the Coulomb energy, the most stable form of nuclear matter would have the same number of neutrons as protons, since unequal numbers of neutrons and protons imply filling higher energy levels for one type of particle, while leaving lower energy levels vacant for the other type.

Pairing energy. An energy which is a correction term that arises from the tendency of proton pairs and neutron pairs to occur. An even number of particles is more stable than an odd number.

The Formula

In the following formulae, let A be the total number of nucleons, Z the number of protons, and N the number of neutrons, so that $A=Z+N$.

The mass of an atomic nucleus is given by

$$m = Zm_p + Nm_n - \frac{E_B}{c^2}$$

where m_p and m_n are the rest mass of a proton and a neutron, respectively, and E_B is the binding energy of the nucleus. The semi-empirical mass formula states that the binding energy will take the following form:

$$E_B = a_V A - a_S A^{2/3} - a_C \frac{Z^2}{A^{1/3}} - a_A \frac{(A-2Z)^2}{A} \pm \delta(A,Z)$$

Each of the terms in this formula has a theoretical basis, as will be explained below. The coefficients a_V, a_S, a_C, a_A and a coefficient that appears in the formula for $\delta(A,Z)$ are determined empirically.

Terms

Volume Term

The term $a_V A$ is known as the *volume term*. The volume of the nucleus is proportional to A, so this term is proportional to the volume, hence the name.

The basis for this term is the strong nuclear force. The strong force affects both protons and neutrons, and as expected, this term is independent of Z. Because the number of pairs that can be taken from A particles is $\frac{A(A-1)}{2}$, , one might expect a term proportional to A^2. However, the strong force has a very limited range, and a given nucleon may only interact strongly with its nearest neighbors and next nearest neighbors. Therefore, the number of pairs of particles that actually interact is roughly proportional to A, giving the volume term its form.

The coefficient a_V is smaller than the binding energy of the nucleons to their neighbours E_b, which is of order of 40 MeV. This is because the larger the number of nucleons in the nucleus, the larger their kinetic energy is, due to the Pauli exclusion principle. If one treats the nucleus as a Fermi ball of A nucleons, with equal numbers of protons and neutrons, then the total kinetic energy is $\frac{3}{5} A\grave{o}_F$, with $\grave{o}_F$ the Fermi energy which is estimated as 28 MeV. Thus the expected value of a_V in this model is $E_b - \frac{3}{5}\epsilon_F \sim 17$ MeV, , not far from the measured value.

Surface Term

The term $a_S A^{2/3}$ is known as the *surface term*. This term, also based on the strong force, is a correction to the volume term.

The volume term suggests that each nucleon interacts with a constant number of nucleons, independent of A. While this is very nearly true for nucleons deep within the nucleus, those nucleons on the surface of the nucleus have fewer nearest neighbors, justifying this correction. This can also be thought of as a surface tension term, and indeed a similar mechanism creates surface tension in liquids.

If the volume of the nucleus is proportional to A, then the radius should be proportional to $A^{1/3}$ and the surface area to $A^{2/3}$. This explains why the surface term is proportional to $A^{2/3}$. It can also be deduced that a_S should have a similar order of magnitude as a_V.

Coulomb Term

The term $a_C \frac{Z(Z-1)}{A^{1/3}}$ or $a_C Z^2 A1>$ is known as the *Coulomb* or *electrostatic term*.

The basis for this term is the electrostatic repulsion between protons. To a very rough approximation, the nucleus can be considered a sphere of uniform charge density. The potential energy of such a charge distribution can be shown to be

$$E = \frac{3}{5}\left(\frac{1}{4\pi\grave{o}_0}\right)\frac{Q^2}{R}$$

where Q is the total charge and R is the radius of the sphere. Identifying Q with Ze, and noting as above that the radius is proportional to $A^{1/3}$, we get close to the form of the Coulomb term. However, because electrostatic repulsion will only exist for more than one proton, becomes $Z(Z-1)$. The value of a_C can be approximately calculated using the equation above:

Empirical nuclear radius:

$$R \approx r_0 A^{\frac{1}{3}}.$$

Quantum charge integers:

$$Q = Ze$$

$$Z^2 \approx Z(Z-1).$$

Integration by substitution:

$$E = \frac{3}{5}\left(\frac{1}{4\pi\epsilon_0}\right)\frac{Q^2}{R} = \frac{3}{5}\left(\frac{1}{4\pi\epsilon_0}\right)\frac{(Ze)^2}{(r_0 A^{\frac{1}{3}})} = \frac{3e^2Z^2}{20\pi\epsilon_0 r_0 A^{\frac{1}{3}}} \approx \frac{3e^2Z(Z-1)}{20\pi\epsilon_0 r_0 A^{\frac{1}{3}}} = a_C\frac{Z(Z-1)}{A^{1/3}}$$

Potential energy of charge distribution:

$$E = \frac{3e^2Z(Z-1)}{20\pi\epsilon_0 r_0 A^{\frac{1}{3}}}$$

Electrostatic Coulomb constant:

$$a_C = \frac{3e^2}{20\pi\epsilon_0 r_0}$$

The value of a_C using the fine structure constant:

$$a_C = \frac{3}{5}\left(\frac{\hbar c\alpha}{r_0}\right) = \frac{3}{5}\left(\frac{R_P}{r_0}\right)\alpha m_p c^2$$

where α is the fine structure constant and $r_0 A^{1/3}$ is the radius of a nucleus, giving r_0 to be approximately 1.25 femtometers. R_P is the proton Compton radius and m_p the proton mass. This gives a_C an approximate theoretical value of 0.691 MeV, not far from the measured value.

$$a_C = 0.691\,\text{MeV}$$

Asymmetry Term

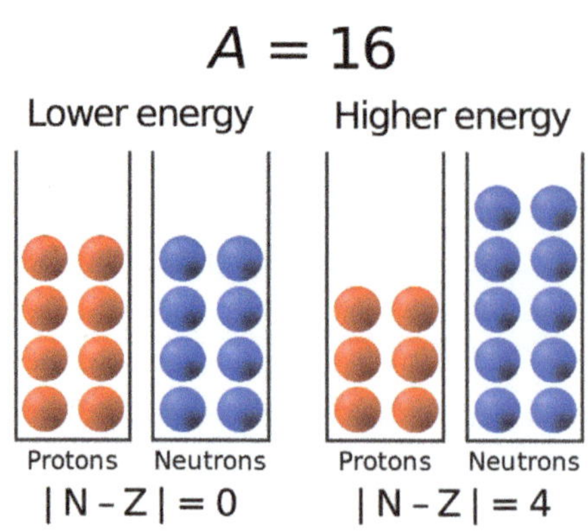

The term $a_A\frac{(A-2Z)^2}{A}$ or $4a_A((A >$ is known as the *asymmetry term*. Note that as $A = N + Z$, the parenthesized expression can be rewritten as $(N - Z)$. The form $(A - 2Z)$ is used to keep the dependence on *A* explicit, as will be important for a number of uses of the formula.

The theoretical justification for this term is more complex. The Pauli exclusion principle states that no two identical fermions can occupy exactly the same quantum state in an atom. At a given energy level, there are only finitely many quantum states available for particles. What this means in the nucleus is that as more particles are "added", these particles must occupy higher energy levels, increasing the total energy of the nucleus (and decreasing the binding energy). Note that this effect is not based on any of the fundamental forces (gravitational, electromagnetic, etc.), only the Pauli exclusion principle.

Protons and neutrons, being distinct types of particles, occupy different quantum states. One can think of two different "pools" of states, one for protons and one for neutrons. Now, for example, if there are significantly more neutrons than protons in a nucleus, some of the neutrons will be higher in energy than the available states in the proton pool. If we could move some particles from the neutron pool to the proton pool, in other words change some neutrons into protons, we would significantly decrease the energy. The imbalance between the number of protons and neutrons causes the energy to be higher than it needs to be, *for a given number of nucleons*. This is the basis for the asymmetry term.

The actual form of the asymmetry term can again be derived by modelling the nucleus as a Fermi ball of protons and neutrons. Its total kinetic energy is

$$E_k = \frac{3}{5}(N_p \varepsilon_{Fp} + N_n \varepsilon_{Fn})$$

where N_p, N_n are the numbers of protons and neutrons and $\grave{o}_{Fp}$, ϵ_{Fn} are their Fermi energies. Since the latter are proportional to $N_p{}^{2/3}$ and $N_n 2 >,$, respectively, one gets

$E_k = C(N_p^{5/3} + N_n^{5/3})$ for some constant *C*.

The leading expansion in the difference $N_n - N_p$ is then

$$E_k = \frac{C}{2^{2/3}}\left((N_p + N_n)^{5/3} + \frac{5}{9}\frac{(N_n - N_p)^2}{(N_p + N_n)^{1/3}}\right) + O((N_n - N_p)^2).$$

At the zeroth order expansion the kinetic energy is just the Fermi energy $\epsilon_F \equiv \epsilon_{Fp} = \epsilon_{Fn}$ multiplied by $\frac{3}{5}(N_p + N_n)^{2/3}$. Thus we get

$$E_k = \frac{3}{5}\varepsilon_F(N_p + N_n) + \frac{1}{3}\varepsilon_F\frac{(N_n - N_p)^2}{(N_p + N_n)} + O((N_n - N_p)^4) = \frac{3}{5}\varepsilon_F A + \frac{1}{3}\varepsilon_F\frac{(A-2Z)^2}{A} + O((A-2Z)^4).$$

The first term contributes to the volume term in the semi-empirical mass formula, and the second term is minus the asymmetry term (remember the kinetic energy contributes to the total binding energy with a *negative* sign).

ϵ_F is 38 MeV, so calculating a_A from the equation above, we get only half the measured value. The discrepancy is explained by our model not being accurate: nucleons in fact interact with each other, and are not spread evenly across the nucleus. For example, in the shell model, a proton and a neutron with overlapping wavefunctions will have a greater strong interaction between them and stronger binding energy. This makes it energetically favourable (i.e. having lower energy) for protons and neutrons to have the same quantum numbers (other than isospin), and thus increase the energy cost of asymmetry between them.

One can also understand the asymmetry term intuitively, as follows. It should be dependent on the absolute difference $|N-Z|$, , and the form $(A-2Z)^2$ is simple and differentiable, which is important for certain applications of the formula. In addition, small differences between Z and N do not have a high energy cost. The A in the denominator reflects the fact that a given difference $|N-Z|$ is less significant for larger values of A.

Pairing Term

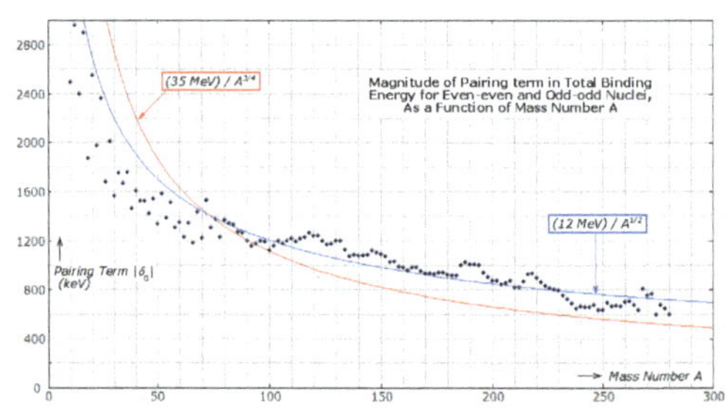

Magnitude of the pairing term in the total binding energy for even-even and odd-odd nuclei, as a function of mass number. Two fits are shown (blue and red line). The pairing term (positive for even-even and negative for odd-odd nuclei) was derived from the binding energy data in: G. Audi et al., 'The AME2012 atomic mass evaluation', in Chinese Physics C 36 (2012/12) pp. 1287-1602.

The term $\delta(A,Z)$ is known as the *pairing term* (possibly also known as the pairwise interaction). This term captures the effect of spin-coupling. It is given by:

$$\delta(A,Z)=\begin{cases}+\delta_0 & Z,N \text{ even } (A \text{ even})\\ 0 & A \text{ odd}\\ -\delta_0 & Z,N \text{ odd } (A \text{ even})\end{cases}$$

where δ_0 is found empirically to have a value of about 1000 keV, slowly decreasing with mass number A. The dependence on mass number is commonly parametrized as

$$\delta_0=\frac{a_P}{A^{k_P}}.$$

The value of the exponent k_p is determined from experimental binding energy data. In the past its value was often assumed to be -3/4, but modern experimental data indicate that a value of -1/2 is nearer the mark:

$$\delta_0 = \frac{a_P}{A^{1/2}} \text{ or } \delta_0 = \frac{a_P'}{A^{3/4}}..$$

Due to the Pauli exclusion principle the nucleus would have a lower energy if the number of protons with spin up were equal to the number of protons with spin down. This is also true for neutrons. Only if both Z and N are even can both protons and neutrons have equal numbers of spin up and spin down particles. This is a similar effect to the asymmetry term.

The factor A^{-k_P} is not easily explained theoretically. The Fermi ball calculation we have used above, based on the liquid drop model but neglecting interactions, will give an A^{-1} dependence, as in the asymmetry term. This means that the actual effect for large nuclei will be larger than expected by that model. This should be explained by the interactions between nucleons; For example, in the shell model, two protons with the same quantum numbers (other than spin) will have completely overlapping wavefunctions and will thus have greater strong interaction between them and stronger binding energy. This makes it energetically favourable (i.e. having lower energy) for protons to pair in pairs of opposite spin. The same is true for neutrons.

Calculating the Coefficients

The semi-empirical mass formula provides a good fit to heavier nuclei, and a poor fit to very light nuclei, especially ^{4}He. This is because the formula does not consider the internal shell structure of the nucleus. For light nuclei, it is usually better to use a model that takes this structure into account.

Examples for Consequences of the Formula

By maximizing $E_b(A,Z)$ with respect to Z, we find the best neutrons to protons ratio N/Z for a given atomic weight A. We get

$$N/Z \approx 1 + \frac{a_C}{2a_A} A^{2/3}.$$

This is roughly 1 for light nuclei, but for heavy nuclei the ratio grows in good agreement with nature.

By substituting the above value of Z back into E_b one obtains the binding energy as a function of the atomic weight, $E_b(A)$. Maximizing $E_b(A)/A$ with respect to A gives the nucleus which is most strongly bound, i.e. most stable. The value we get is $A = 63$ (copper), close to the measured values of $A = 62$ (nickel) and $A = 58$ (iron).

Nuclear Shell Model

In nuclear physics and nuclear chemistry, the nuclear shell model is a model of the atomic nucleus which uses the Pauli exclusion principle to describe the structure of the nucleus in terms of energy levels. The first shell model was proposed by Dmitry Ivanenko (together with E. Gapon) in 1932. The model was developed in 1949 following independent work by several physicists, most notably Eugene Paul Wigner, Maria Goeppert Mayer and J. Hans D. Jensen, who shared the 1963 Nobel Prize in Physics for their contributions.

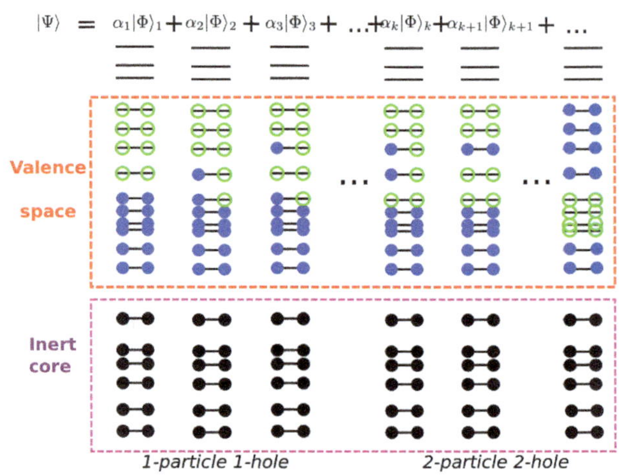

Partly filled valence orbitals for both neutrons and protons appear at energies over the filled inert core orbitals, in the shell model of the atomic nucleus

The shell model is partly analogous to the atomic shell model which describes the arrangement of electrons in an atom, in that a filled shell results in greater stability. When adding nucleons (protons or neutrons) to a nucleus, there are certain points where the binding energy of the next nucleon is significantly less than the last one. This observation, that there are certain magic numbers of nucleons: 2, 8, 20, 28, 50, 82, 126 which are more tightly bound than the next higher number, is the origin of the shell model.

The shells for protons and for neutrons are independent of each other. Therefore, one can have "magic nuclei" where one nucleon type or the other is at a magic number, and "doubly magic nuclei", where both are. Due to some variations in orbital filling, the upper magic numbers are 126 and, speculatively, 184 for neutrons but only 114 for protons, playing a role in the search for the so-called island of stability. Some semimagic numbers have been found, notably Z=40 giving nuclear shell filling for the various elements; 16 may also be a magic number.

In order to get these numbers, the nuclear shell model starts from an average potential with a shape something between the square well and the harmonic oscillator. To this potential a spin orbit term is added. Even so, the total perturbation does not coincide with experiment, and an empirical spin orbit coupling must be added with at least two or three different values of its coupling constant, depending on the nuclei being studied.

Nevertheless, the magic numbers of nucleons, as well as other properties, can be arrived at by approximating the model with a three-dimensional harmonic oscillator plus a spin-orbit interaction. A more realistic but also complicated potential is known as Woods Saxon potential.

Igal Talmi developed a method to obtain the information from experimental data and use it to calculate and predict energies which have not been measured. This method has been successfully used by many nuclear physicists and has led to deeper understanding of nuclear structure. The theory which gives a good description of these properties was developed. This description turned out to furnish the shell model basis of the elegant and successful interacting boson model.

Deformed Harmonic Oscillator Approximated Model

Consider a three-dimensional harmonic oscillator. This would give, for example, in the first two levels (*"l"* is angular momentum)

level n	l	m_l	m_s
0	0	0	$1/2$
			$-1/2$
1	1	1	$1/2$
			$-1/2$
		0	$1/2$
			$-1/2$
		−1	$1/2$
			$-1/2$

We can imagine ourselves building a nucleus by adding protons and neutrons. These will always fill the lowest available level. Thus the first two protons fill level zero, the next six protons fill level one, and so on. As with electrons in the periodic table, protons in the outermost shell will be relatively loosely bound to the nucleus if there are only few protons in that shell, because they are farthest from the center of the nucleus. Therefore nuclei which have a full outer proton shell will have a higher binding energy than other nuclei with a similar total number of protons. All this is true for neutrons as well.

This means that the magic numbers are expected to be those in which all occupied shells are full. We see that for the first two numbers we get 2 (level 0 full) and 8 (levels 0 and 1 full), in accord with experiment. However the full set of magic numbers does not turn out correctly. These can be computed as follows:

In a three-dimensional harmonic oscillator the total degeneracy at level n is $\frac{(n+1)(n+2)}{2}$. Due to the spin, the degeneracy is doubled and is $(n+1)(n+2)$..

Thus the magic numbers would be

$$\sum_{n=0}^{k}(n+1)(n+2) = \frac{(k+1)(k+2)(k+3)}{3}$$

for all integer k. This gives the following magic numbers: 2,8,20,40,70,112..., which agree with experiment only in the first three entries. These numbers are twice the tetrahedral numbers (1,4,10,20,35,56...) from the Pascal Triangle.

In particular, the first six shells are:

- level 0: 2 states ($l = 0$) = 2.
- level 1: 6 states ($l = 1$) = 6.
- level 2: 2 states ($l = 0$) + 10 states ($l = 2$) = 12.
- level 3: 6 states ($l = 1$) + 14 states ($l = 3$) = 20.
- level 4: 2 states ($l = 0$) + 10 states ($l = 2$) + 18 states ($l = 4$) = 30.
- level 5: 6 states ($l = 1$) + 14 states ($l = 3$) + 22 states ($l = 5$) = 42.

where for every l there are $2l+1$ different values of m_l and 2 values of m_s, giving a total of $4l+2$ states for every specific level.

These numbers are twice the values of triangular numbers from the Pascal Triangle: 1,3,6,10,15,21....

Including a Spin-orbit Interaction

We next include a spin-orbit interaction. First we have to describe the system by the quantum numbers j, m_j and parity instead of l, m_l and m_s, as in the hydrogen-like atom. Since every even level includes only even values of l, it includes only states of even (positive) parity; Similarly every odd level includes only states of odd (negative) parity. Thus we can ignore parity in counting states. The first six shells, described by the new quantum numbers, are

- level 0 (n=0): 2 states ($j = 1/2$). Even parity.
- level 1 (n=1): 2 states ($j = 1/2$) + 4 states ($j = 3/2$) = 6. Odd parity.
- level 2 (n=2): 2 states ($j = 1/2$) + 4 states ($j = 3/2$) + 6 states ($j = 5/2$) = 12. Even parity.

- level 3 (n=3): 2 states (j = ½) + 4 states (j = 3/2) + 6 states (j = 5/2) + 8 states (j = 7/2) = 20. Odd parity.
- level 4 (n=4): 2 states (j = ½) + 4 states (j = 3/2) + 6 states (j = 5/2) + 8 states (j = 7/2) + 10 states (j = 9/2) = 30. Even parity.
- level 5 (n=5): 2 states (j = ½) + 4 states (j = 3/2) + 6 states (j = 5/2) + 8 states (j = 7/2) + 10 states (j = 9/2) + 12 states (j = 11/2) = 42. Odd parity.

where for every j there are $2j$+1 different states from different values of m_j.

Due to the spin-orbit interaction the energies of states of the same level but with different j will no longer be identical. This is because in the original quantum numbers, when is parallel to , the interaction energy is positive; and in this case $j = l + s = l + ½$. When is anti-parallel to (i.e. aligned oppositely), the interaction energy is negative, and in this case j=l–s=l–½. Furthermore, the strength of the interaction is roughly proportional to l.

For example, consider the states at level 4:

- The 10 states with j = 9/2 come from l = 4 and s parallel to l. Thus they have a positive spin-orbit interaction energy.
- The 8 states with j = 7/2 came from l = 4 and s anti-parallel to l. Thus they have a negative spin-orbit interaction energy.
- The 6 states with j = 5/2 came from l = 2 and s parallel to l. Thus they have a positive spin-orbit interaction energy. However its magnitude is half compared to the states with j = 9/2.
- The 4 states with j = 3/2 came from l = 2 and s anti-parallel to l. Thus they have a negative spin-orbit interaction energy. However its magnitude is half compared to the states with j = 7/2.
- The 2 states with j = ½ came from l = 0 and thus have zero spin-orbit interaction energy.

Deforming the Potential

The harmonic oscillator potential $V(r) = \mu\omega^2 r^2 / 2$ grows infinitely as the distance from the center r goes to infinity. A more realistic potential, such as Woods Saxon potential, would approach a constant at this limit. One main consequence is that the average radius of nucleons' orbits would be larger in a realistic potential; This leads to a reduced term $\hbar^2 l(l+1)/2mr^2$ in the Laplace operator of the Hamiltonian. Another main difference is that orbits with high average radii, such as those with high n or high l, will have a lower energy than in a harmonic oscillator potential. Both effects lead to a reduction in the energy levels of high l orbits.

Predicted Magic Numbers

Together with the spin-orbit interaction, and for appropriate magnitudes of both effects, one is led to the following qualitative picture: At all levels, the highest j states have their energies shifted downwards, especially for high n (where the highest j is high). This is both due to the negative spin-orbit interaction energy and to the reduction in energy resulting from deforming the potential to a more realistic one. The second-to-highest j states, on the contrary, have their energy shifted up by the first effect and down by the second effect, leading to a small overall shift. The shifts in the energy of the highest j states can thus bring the energy of states of one level to be closer to the energy of states of a lower level. The "shells" of the shell model are then no longer identical to the levels denoted by n, and the magic numbers are changed.

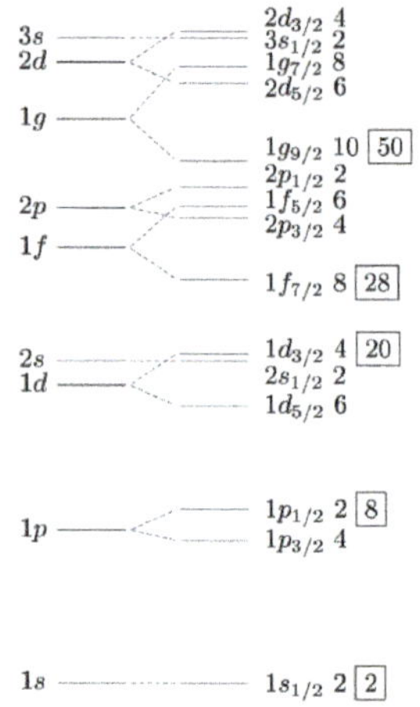

Low-lying energy levels in a single-particle shell model with an oscillator potential (with a small negative l2 term) without spin-orbit (left) and with spin-orbit (right) interaction. The number to the right of a level indicates its degeneracy, (2j+1). The boxed integers indicate the magic numbers.

We may then suppose that the highest j states for $n = 3$ have an intermediate energy between the average energies of $n = 2$ and $n = 3$, and suppose that the highest j states for larger n (at least up to $n = 7$) have an energy closer to the average energy of $n-1$. Then we get the following shells (see the figure)

- 1st shell: 2 states ($n = 0, j = 1/2$).
- 2nd shell: 6 states ($n = 1, j = 1/2$ or $3/2$).
- 3rd shell: 12 states ($n = 2, j = 1/2, 3/2$ or $5/2$).
- 4th shell: 8 states ($n = 3, j = 7/2$).
- 5th shell: 22 states ($n = 3, j = 1/2, 3/2$ or $5/2$; $n = 4, j = 9/2$).
- 6th shell: 32 states ($n = 4, j = 1/2, 3/2, 5/2$ or $7/2$; $n = 5, j = 11/2$).

- 7th shell: 44 states ($n = 5, j = {}^1\!/_2, {}^3\!/_2, {}^5\!/_2, {}^7\!/_2$ or ${}^9\!/_2$; $n = 6, j = {}^{13}\!/_2$).
- 8th shell: 58 states ($n = 6, j = {}^1\!/_2, {}^3\!/_2, {}^5\!/_2, {}^7\!/_2, {}^9\!/_2$ or ${}^{11}\!/_2$; $n = 7, j = {}^{15}\!/_2$).

and so on.

The magic numbers are then

- 2
- 8=2+6
- 20=2+6+12
- 28=2+6+12+8
- 50=2+6+12+8+22
- 82=2+6+12+8+22+32
- 126=2+6+12+8+22+32+44
- 184=2+6+12+8+22+32+44+58

and so on. This gives all the observed magic numbers, and also predicts a new one (the so-called *island of stability*) at the value of 184 (for protons, the magic number 126 has not been observed yet, and more complicated theoretical considerations predict the magic number to be 114 instead).

Another way to predict magic (and semi-magic) numbers is by laying out the idealized filling order (with spin-orbit splitting but energy levels not overlapping). For consistency s is split into j = ½ and j = -½ components with 2 and 0 members respectively. Taking leftmost and rightmost total counts within sequences marked bounded by / here gives the magic and semi-magic numbers.

- s(2,0)/p(4,2)> 2,2/6,8, so (semi)magic numbers 2,2/6,8
- d(6,4):s(2,0)/f(8,6):p(4,2)> 14,18:20,20/28,34:38,40, so 14,20/28,40
- g(10,8):d(6,4):s(2,0)/h(12,10):f(8,6):p(4,2)> 50,58,64,68,70,70/82,92,100,106,110,112, so 50,70/82,112
- i(14,12):g(10,8):d(6,4):s(2,0)/j(16,14):h(12,10):f(8,6):p(4,2)> 126,138,148,156,162,166,168,168/184,198,210,220,228,234,238,240, so 126,168/184,240

The rightmost predicted magic numbers of each pair within the quartets bisected by / are double tetrahedral numbers from the Pascal Triangle: 2,8,20,40,70,112,168,240 are 2x 1,4,10,20,35,56,84,120..., and the leftmost members of the pairs differ from the rightmost by double triangular numbers: 2-2=0, 8-6=2, 20-14=6, 40-28=12, 70-

50=20, 112-82=30, 168-126=42, 240-184=56, where 0,2,6,12,20,30,42,56... are 2x 0,1,3,6,10,15,21,28....

Other Properties of Nuclei

This model also predicts or explains with some success other properties of nuclei, in particular spin and parity of nuclei ground states, and to some extent their excited states as well. Take 17O as an example — its nucleus has eight protons filling the three first proton 'shells', eight neutrons filling the three first neutron 'shells', and one extra neutron. All protons in a complete proton shell have total angular momentum zero, since their angular momenta cancel each other; The same is true for neutrons. All protons in the same level (n) have the same parity (either +1 or −1), and since the parity of a pair of particles is the product of their parities, an even number of protons from the same level (n) will have +1 parity. Thus the total angular momentum of the eight protons and the first eight neutrons is zero, and their total parity is +1. This means that the spin (i.e. angular momentum) of the nucleus, as well as its parity, are fully determined by that of the ninth neutron. This one is in the first (i.e. lowest energy) state of the 4th shell, which is a d-shell (l = 2), and since $p = (-1)^l$, this gives the nucleus an overall parity of +1. This 4th d-shell has a $j = 5/2$, thus the nucleus of 17O is expected to have positive parity and total angular momentum $5/2$, which indeed it has.

The rules for the ordering of the nucleus shells are similar to Hund's Rules of the atomic shells, however, unlike its use in atomic physics the completion of a shell is not signified by reaching the next n, as such the shell model cannot accurately predict the order of excited nuclei states, though it is very successful in predicting the ground states. The order of the first few terms are listed as follows: 1s, 1p$3/2$, 1p$1/2$, 1d$5/2$, 2s, 1d$3/2$... For further clarification on the notation refer to the article on the Russell-Saunders term symbol.

For nuclei farther from the magic numbers one must add the assumption that due to the relation between the strong nuclear force and angular momentum, protons or neutrons with the same n tend to form pairs of opposite angular momenta. Therefore a nucleus with an even number of protons and an even number of neutrons has 0 spin and positive parity. A nucleus with an even number of protons and an odd number of neutrons (or vice versa) has the parity of the last neutron (or proton), and the spin equal to the total angular momentum of this neutron (or proton). By "last" we mean the properties coming from the highest energy level.

In the case of a nucleus with an odd number of protons and an odd number of neutrons, one must consider the total angular momentum and parity of both the last neutron and the last proton. The nucleus parity will be a product of theirs, while the nucleus spin will be one of the possible results of the sum of their angular momenta (with other possible results being excited states of the nucleus).

The ordering of angular momentum levels within each shell is according to the principles described above - due to spin-orbit interaction, with high angular momentum states having their energies shifted downwards due to the deformation of the potential (i.e. moving from a harmonic oscillator potential to a more realistic one). For nucleon pairs, however, it is often energetically favorable to be at high angular momentum, even if its energy level for a single nucleon would be higher. This is due to the relation between angular momentum and the strong nuclear force.

Nuclear magnetic moment is partly predicted by this simple version of the shell model. The magnetic moment is calculated through j, l and s of the "last" nucleon, but nuclei are not in states of well defined l and s. Furthermore, for odd-odd nuclei, one has to consider the two "last" nucleons, as in deuterium. Therefore one gets several possible answers for the nuclear magnetic moment, one for each possible combined l and s state, and the real state of the nucleus is a superposition of them. Thus the real (measured) nuclear magnetic moment is somewhere in between the possible answers.

The electric dipole of a nucleus is always zero, because its ground state has a definite parity, so its matter density (ψ^2, where ψ is the wavefunction) is always invariant under parity. This is usually the situations with the atomic electric dipole as well.

Higher electric and magnetic multipole moments cannot be predicted by this simple version of the shell model, for the reasons similar to those in the case of deuterium.

Alpha Particle Model

A model derived from the nuclear shell model is the alpha particle model developed by Henry Margenau, Edward Teller, J. K. Pering, T.H. Skyrme.

Nuclear Force

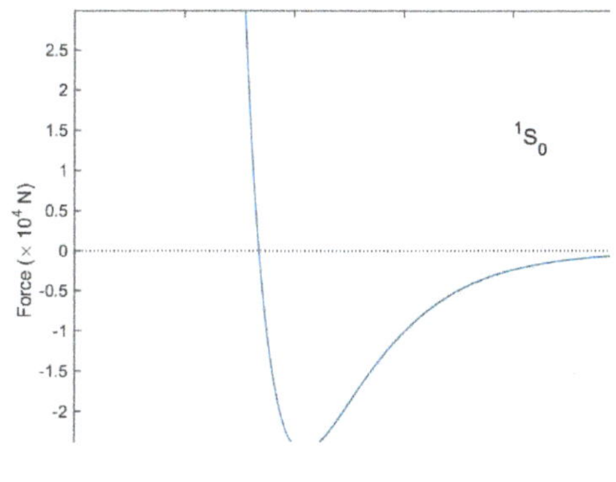

Force (in units of 10,000 N) between two nucleons as a function of distance as computed from the Reid potential (1968). The spins of the neutron and proton are aligned, and they are in the S angular momentum state. The attractive (negative) force has a maximum at a distance of about 1 fm with a force of about 25,000 N. Particles much closer than a distance of 0.8 fm experience a large repulsive (positive) force. Particles separated by a distance greater than 1 fm are still attracted (Yukawa potential), but the force falls as an exponential function of distance.

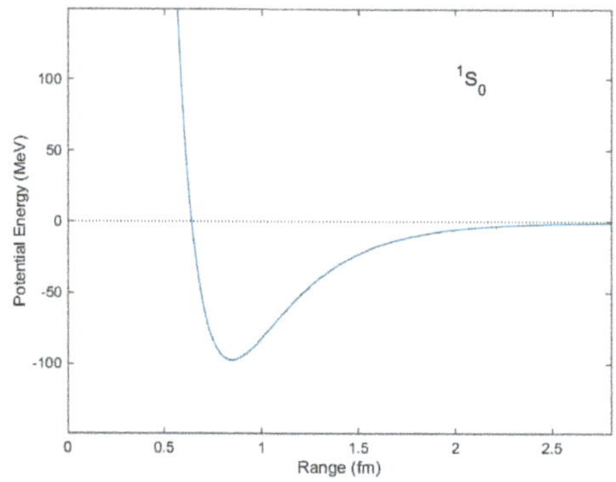

Corresponding potential energy (in units of MeV) of two nucleons as a function of distance as computed from the Reid potential. The potential well is a minimum at a distance of about 0.8 fm. With this potential nucleons can become bound with a negative "binding energy."

The nuclear force (or nucleon–nucleon interaction or residual strong force) is the force between protons and neutrons, subatomic particles that are collectively called nucleons. The nuclear force is responsible for binding protons and neutrons into atomic nuclei. Neutrons and protons are affected by the nuclear force almost identically. Since protons have charge +1 *e*, they experience a strong electric field repulsion (following Coulomb's law) that tends to push them apart, but at short range the attractive nuclear force overcomes the repulsive electromagnetic force. The mass of a nucleus is less than the sum total of the individual masses of the protons and neutrons which form it. The difference in mass between bound and unbound nucleons is known as the mass defect. Energy is released when some large nuclei break apart, and it is this energy that is used in nuclear power and nuclear weapons.

The nuclear force is powerfully attractive between nucleons at distances of about 1 femtometer (fm, or 1.0×10^{-15} metres) between their centers, but rapidly decreases to insignificance at distances beyond about 2.5 fm. At distances less than 0.7 fm, the nuclear force becomes repulsive. This repulsive component is responsible for the physical size of nuclei, since the nucleons can come no closer than the force allows. By comparison, the size of an atom, measured in angstroms (Å, or 1.0×10^{-10} m), is five orders of magnitude larger. The nuclear force is not simple, however, since it depends on the nucleon spins, has a tensor component, and may depend on the relative momentum of the nucleons.

A quantitative description of the nuclear force relies on partially empirical equations that model the internucleon potential energies, or potentials. (Generally, forces within a system of particles can be more simply modeled by describing the system's potential energy; the negative gradient of a potential is equal to the vector force.) The constants for the equations are phenomenological, that is, determined by fitting the equations to experimental data. The internucleon potentials attempt to describe the properties of nucleon–nucleon interaction. Once determined, any given potential can be used in, e.g., the Schrödinger equation to determine the quantum mechanical properties of the nucleon system.

The discovery of the neutron in 1932 revealed that atomic nuclei were made of protons and neutrons, held together by an attractive force. By 1935 the nuclear force was conceived to be transmitted by particles called mesons. This theoretical development included a description of the Yukawa potential, an early example of a nuclear potential. Mesons, predicted by theory, were discovered experimentally in 1947. By the 1970s, the quark model had been developed, by which the mesons and nucleons were viewed as composed of quarks and gluons. By this new model, the nuclear force, resulting from the exchange of mesons between neighboring nucleons, is a residual effect of the strong force.

Description

The nuclear force is only felt between hadrons (particles composed of quarks). At small separations between nucleons (less than ~ 0.7 fm between their centers, depending upon spin alignment) the force becomes repulsive, which keeps the nucleons at a certain average separation, even if they are of different types. This repulsion arises from the Pauli exclusion force for identical nucleons (such as two neutrons or two protons). A Pauli exclusion force also occurs between quarks of the same type within nucleons, when the nucleons are different (a proton and a neutron, for example).

Field Strength

At distances larger than 0.7 fm the force becomes attractive between spin-aligned nucleons, becoming maximal at a center–center distance of about 0.9 fm. Beyond this distance the force drops exponentially, until beyond about 2.0 fm separation, the force is negligible. Nucleons have a radius of about 0.8 fm.

At short distances (less than 1.7 fm or so), the attractive nuclear force is stronger than the repulsive Coulomb force between protons; it thus overcomes the repulsion of protons within the nucleus. However, the Coulomb force between protons has a much greater range as it varies as the inverse square of the charge separation, and Coulomb repulsion thus becomes the only significant force between protons when their separation exceeds about 2 to 2.5 fm.

The nuclear force has a spin-dependent component. The force is stronger for particles with their spins aligned than for those with their spins anti-aligned. If two particles are the same, such as two neutrons or two protons, the force is not enough to bind the particles, since the spin vectors of two particles of the same type must point in opposite directions when the particles are near each other and are (save for spin) in the same quantum state. This requirement for fermions stems from the Pauli exclusion principle. For fermion particles of different types, such as a proton and neutron, particles may be close to each other and have aligned spins without violating the Pauli exclusion principle, and the nuclear force may bind them (in this case, into a deuteron), since the nuclear force is much stronger for spin-aligned particles. But if the particles' spins are anti-aligned the nuclear force is too weak to bind them, even if they are of different types.

The nuclear force also has a tensor component which depends on the interaction between the nucleon spins and the angular momentum of the nucleons, leading to deformation from a simple spherical shape.

Nuclear Binding

To disassemble a nucleus into unbound protons and neutrons requires work against the nuclear force. Conversely, energy is released when a nucleus is created from free nucleons or other nuclei: the nuclear binding energy. Because of mass–energy equivalence (i.e. Einstein's famous formula $E = mc^2$), releasing this energy causes the mass of the nucleus to be lower than the total mass of the individual nucleons, leading to the so-called "mass defect".

The nuclear force is nearly independent of whether the nucleons are neutrons or protons. This property is called *charge independence*. The force depends on whether the spins of the nucleons are parallel or antiparallel, as it has a non-central or *tensor* component. This part of the force does not conserve orbital angular momentum, which under the action of central forces is conserved.

The symmetry resulting in the strong force, proposed by Werner Heisenberg, is that protons and neutrons are identical in every respect, other than their charge. This is not completely true, because neutrons are a tiny bit heavier, but it is an approximate symmetry. Protons and neutrons are therefore viewed as the same particle, but with different isospin quantum number . The strong force is invariant under SU(2) transformations, just as are particles with intrinsic spin. Isospin and intrinsic spin are related under this SU(2) symmetry group. There are only strong attractions when the total isospin is 0, which is confirmed by experiment.

Our understanding of the nuclear force is obtained by scattering experiments and the binding energy of light nuclei.

The nuclear force occurs by the exchange of virtual light mesons, such as the virtual pions, as well as two types of virtual mesons with spin (vector mesons), the rho mesons and the omega mesons. The vector mesons account for the spin-dependence of the nuclear force in this "virtual meson" picture.

The nuclear force is distinct from what historically was known as the weak nuclear force. The weak interaction is one of the four fundamental interactions, and plays a role in such processes as beta decay. The weak force plays no role in the interaction of nucleons, though it is responsible for the decay of neutrons to protons and vice versa.

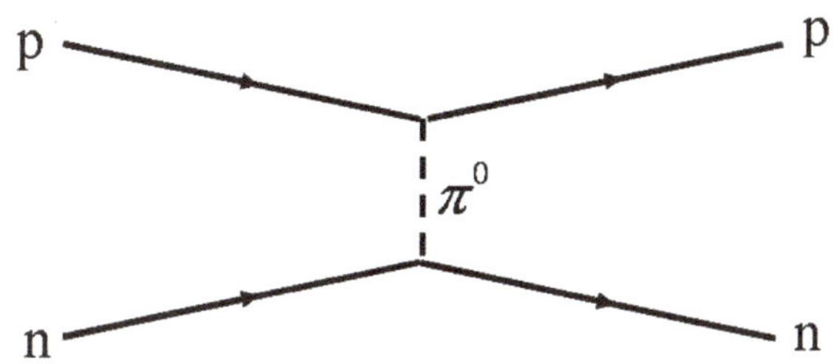

A Feynman diagram of a strong proton–neutron interaction mediated by a neutral pion. Time proceeds from left to right.

History

The nuclear force has been at the heart of nuclear physics ever since the field was born in 1932 with the discovery of the neutron by James Chadwick. The traditional goal of nuclear physics is to understand the properties of atomic nuclei in terms of the 'bare' interaction between pairs of nucleons, or nucleon–nucleon forces (NN forces).

Within months after the discovery of the neutron, Werner Heisenberg and Dmitri Ivanenko had proposed proton–neutron models for the nucleus. Heisenberg approached the description of protons and neutrons in the nucleus through quantum mechanics, an approach that was not at all obvious at the time. Heisenberg's theory for protons and neutrons in the nucleus was a "major step toward understanding the nucleus as a quantum mechanical system." Heisenberg introduced the first theory of nuclear exchange forces that bind the nucleons. He considered protons and neutrons to be different quantum states of the same particle, i.e., nucleons distinguished by the value of their nuclear isospin quantum numbers.

One of the earliest models for the nucleus was the liquid drop model developed in the 1930s. One property of nuclei is that the average binding energy per nucleon is approximately the same for all stable nuclei, which is similar to a liquid drop. The liquid drop model treated the nucleus as a drop of incompressible nuclear fluid, with nucleons behaving like molecules in a liquid. The model was first proposed by George Gamow and then developed by Niels Bohr, Werner Heisenberg and Carl Friedrich von Weizsäcker. This crude model did not explain all the properties of the nucleus, but it did explain the spherical shape of most nuclei. The model also gave good predictions for the nuclear binding energy of nuclei.

In 1934, Hideki Yukawa made the earliest attempt to explain the nature of the nuclear force. According to his theory, massive bosons (mesons) mediate the interaction between two nucleons. Although, in light of quantum chromodynamics (QCD), meson theory is no longer perceived as fundamental, the meson-exchange concept (where hadrons are treated as elementary particles) continues to represent the best working model for a quantitative *NN* potential. The Yukawa potential (also called a screened Coulomb potential) is a potential of the form

where g is a magnitude scaling constant, i.e., the amplitude of potential, is the Yukawa particle mass, r is the radial distance to the particle. The potential is monotone increasing, implying that the force is always attractive. The constants are determined empirically. The Yukawa potential depends only on the distance between particles, r, hence it models a central force.

Throughout the 1930s a group at Columbia University led by I. I. Rabi developed magnetic resonance techniques to determine the magnetic moments of nuclei. These measurements led to the discovery in 1939 that the deuteron also possessed an electric quadrupole moment. This electrical property of the deuteron had been interfering with the measurements by the Rabi group. The deuteron, composed of a proton and a neutron, is one of the simplest nuclear systems. The discovery meant that the physical shape of the deuteron was not symmetric, which provided valuable insight into the nature of the nuclear force binding nucleons. In particular, the result showed that the nuclear force was not a central force, but had a tensor character. Hans Bethe identified the discovery of the deuteron's quadrupole moment as one of the important events during the formative years of nuclear physics.

Historically, the task of describing the nuclear force phenomenologically was formidable. The first semi-empirical quantitative models came in the mid-1950s, such as the Woods–Saxon potential (1954). There was substantial progress in experiment and theory related to the nuclear force in the 1960s and 1970s. One influential model was the Reid potential (1968). In recent years, experimenters have concentrated on the subtleties of the nuclear force, such as its charge dependence, the precise value of the πNN coupling constant, improved phase shift analysis, high-precision *NN* data, high-precision *NN* potentials, *NN* scattering at intermediate and high energies, and attempts to derive the nuclear force from QCD.

The Nuclear Force as a Residual of the Strong Force

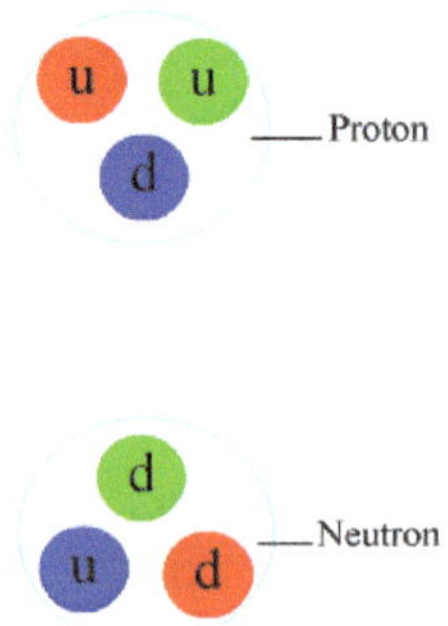

An animation of the interaction. The colored double circles are gluons. Anticolors are shown as per this diagram (larger version).

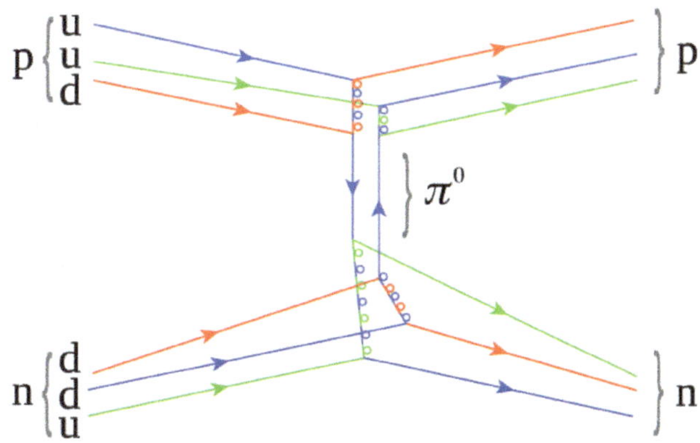

The same diagram as that above with the individual quark constituents shown, to illustrate how the *fundamental* strong interaction gives rise to the **nuclear force**. Straight lines are quarks, while multi-colored loops are gluons (the carriers of the fundamental force). Other gluons, which bind together the proton, neutron, and pion "in-flight," are not shown.

The nuclear force is a residual effect of the more fundamental strong force, or strong interaction. The strong interaction is the attractive force that binds the elementary particles called quarks together to form the nucleons (protons and neutrons) themselves. This more powerful force is mediated by particles called gluons. Gluons hold quarks together with a force like that of electric charge, but of far greater strength. Quarks, gluons and their dynamics are mostly confined within nucleons, but residual influences extend slightly beyond nucleon boundaries to give rise to the nuclear force.

The nuclear forces arising between nucleons are analogous to the forces in chemistry between neutral atoms or molecules called London forces. Such forces between atoms are much weaker than the attractive electrical forces that hold the atoms themselves together (i.e., that bind electrons to the nucleus), and their range between atoms is shorter, because they arise from small separation of charges inside the neutral atom. Similarly, even though nucleons are made of quarks in combinations which cancel most gluon forces (they are "color neutral"), some combinations of quarks and gluons nevertheless leak away from nucleons, in the form of short-range nuclear force fields that extend from one nucleon to another nearby nucleon. These nuclear forces are very weak compared to direct gluon forces ("color forces" or strong forces) inside nucleons, and the nuclear forces extend only over a few nuclear diameters, falling exponentially with distance. Nevertheless, they are strong enough to bind neutrons and protons over short distances, and overcome the electrical repulsion between protons in the nucleus.

Sometimes, the nuclear force is called the residual strong force, in contrast to the strong interactions which arise from QCD. This phrasing arose during the 1970s when QCD was being established. Before that time, the *strong nuclear force* referred to the inter-nucleon potential. After the verification of the quark model, *strong interaction* has come to mean QCD.

Nucleon–nucleon Potentials

Two-nucleon systems such as the deuteron, the nucleus of a deuterium atom, as well as proton–proton or neutron–proton scattering are ideal for studying the *NN* force. Such systems can be described by attributing a *potential* (such as the Yukawa potential) to the nucleons and using the potentials in a Schrödinger equation. The form of the potential is derived phenomenologically (by measurement), although for the long-range interaction, meson-exchange theories help to construct the potential. The parameters of the potential are determined by fitting to experimental data such as the deuteron binding energy or *NN* elastic scattering cross sections (or, equivalently in this context, so-called *NN* phase shifts).

The most widely used *NN* potentials are the Paris potential, the Argonne AV18 potential , the CD-Bonn potential and the Nijmegen potentials.

A more recent approach is to develop effective field theories for a consistent description of nucleon–nucleon and three-nucleon forces. Quantum hadrodynamics is an effective field theory of the nuclear force, comparable to QCD for color interactions and QED for electromagnetic interactions. Additionally, chiral symmetry breaking can be analyzed in terms of an effective field theory (called chiral perturbation theory) which allows perturbative calculations of the interactions between nucleons with pions as exchange particles.

From Nucleons to Nuclei

The ultimate goal of nuclear physics would be to describe all nuclear interactions from the basic interactions between nucleons. This is called the *microscopic* or *ab initio* approach of nuclear physics. There are two major obstacles to overcome before this dream can become reality:

- Calculations in many-body systems are difficult and require advanced computation techniques.
- There is evidence that three-nucleon forces (and possibly higher multi-particle interactions) play a significant role. This means that three-nucleon potentials must be included into the model.

This is an active area of research with ongoing advances in computational techniques leading to better first-principles calculations of the nuclear shell structure. Two- and three-nucleon potentials have been implemented for nuclides up to $A = 12$.

Nuclear Potentials

A successful way of describing nuclear interactions is to construct one potential for the whole nucleus instead of considering all its nucleon components. This is called the *mac-*

roscopic approach. For example, scattering of neutrons from nuclei can be described by considering a plane wave in the potential of the nucleus, which comprises a real part and an imaginary part. This model is often called the optical model since it resembles the case of light scattered by an opaque glass sphere.

Nuclear potentials can be *local* or *global*: local potentials are limited to a narrow energy range and/or a narrow nuclear mass range, while global potentials, which have more parameters and are usually less accurate, are functions of the energy and the nuclear mass and can therefore be used in a wider range of applications.

Nuclear Magnetic Moment

The nuclear magnetic moment is the magnetic moment of an atomic nucleus and arises from the spin of the protons and neutrons. It is mainly a magnetic dipole moment; the quadrupole moment does cause some small shifts in the hyperfine structure as well. All nuclei that have nonzero spin also possess a nonzero magnetic moment and vice versa, although the connection between the two quantities is not straightforward or easy to calculate.

The nuclear magnetic moment varies from isotope to isotope of an element. For a nucleus of which the numbers of protons and of neutrons are *both* even in its ground state (i.e. lowest energy state), the nuclear spin and magnetic moment are both always zero. In cases with odd numbers of either or both protons and neutrons, the nucleus often has nonzero spin and magnetic moment. The nuclear magnetic moment is not sum of nucleon magnetic moments, this property being assigned to the tensorial character of the nuclear force, such as in the case of the most simple nucleus where both proton and neutron appear, namely deuterium nucleus, deuteron.

Measurement Methods

The methods for measuring nuclear magnetic moments can be divided into two broad groups in regard to the interaction with internal or external applied fields. Generally the methods based on external fields are more accurate.

Shell Model

According to the shell model, protons or neutrons tend to form pairs of opposite total angular momentum. Therefore the magnetic moment of a nucleus with even numbers of each protons and neutrons is zero, while that of a nucleus with an odd number of protons and even number of neutrons (or vice versa) will have to be that of the remaining unpaired nucleon. For a nucleus with odd numbers of each protons and neutrons, the total magnetic moment will be some combination of the magnetic moments of both of the "last", unpaired proton and neutron.

The magnetic moment is calculated through j, l and s of the unpaired nucleon, but nuclei are not in states of well defined l and s. Furthermore, for odd–odd nuclei, there are two unpaired nucleons to be considered, as in deuterium. There is consequently a value for the nuclear magnetic moment associated with each possible l and s state combination, and the actual state of the nucleus is a superposition of these. Thus the real (measured) nuclear magnetic moment is between the values associated with the "pure" states, though it may be close to one or the other (as in deuterium).

G-factors

The values of $g^{(l)}$ and $g^{(s)}$ are known as the g-factors of the nucleons.

The measured values of $g^{(l)}$ for the neutron and the proton are according to their electric charge. Thus, in units of nuclear magneton, $g^{(l)} = 0$ for the neutron and $g^{(l)} = 1$ for the proton.

The measured values of $g^{(s)}$ for the neutron and the proton are twice their magnetic moment (either the neutron magnetic moment or the proton magnetic moment). In nuclear magneton units, $g^{(s)} = -3.8263$ for the neutron and $g^{(s)} = 5.5858$ for the proton.

References

- Miller A. I. Early Quantum Electrodynamics: A Sourcebook, Cambridge University Press, Cambridge, 1995, ISBN 0521568919.
- Fernandez, Bernard & Ripka, Georges (2012). "Nuclear Theory After the Discovery of the Neutron". Unravelling the Mystery of the Atomic Nucleus: A Sixty Year Journey 1896 — 1956. Springer. p. 263. ISBN 9781461441809.
- Sitenko, A.G. & Tartakovskiĭ, V.K. (1997). Theory of Nucleus: Nuclear Structure and Nuclear Interaction. Kluwer Academic. p. 3. ISBN 0-7923-4423-5.
- Serway, Raymond; Vuille, Chris; Faughn, Jerry (2009). College Physics (8th ed.). Belmont, CA: Brooks/Cole, Cengage Learning. p. 915. ISBN 9780495386933.
- Broglia, Ricardo A.; Zelevinsky, Vladimir (2013). Fifty Years of Nuclear BCS: Pairing in Finite Systems. World Scientific. ISBN 978-981-4412-48-3.
- Freedman, R.; Young, H. (2004). Sears and Zemanskey's University Physics with Modern Physics (11th ed.). pp. 1633–1634. ISBN 0-8053-8768-4.
- Choppin, G.; Liljenzin, J.-O.; Rydberg, J. (2002). "Nuclear Mass and Stability" (PDF). Radiochemistry and Nuclear Chemistry (3rd ed.). Butterworth-Heinemann. pp. 41–57. ISBN 978-0-7506-7463-8.
- Povh, B.; Rith, K.; Scholz, C.; Zetsche, F. (2002). Particles and Nuclei: An Introduction to the Physical Concepts. Berlin: Springer-Verlag. p. 73. ISBN 978-3-540-43823-6.
- Miller A. I. Early Quantum Electrodynamics: A Sourcebook, Cambridge University Press, Cambridge, 1995, ISBN 0521568919.
- Brown, L.M.; Rechenberg, H. (1996). The Origin of the Concept of Nuclear Forces. Bristol and Philadelphia: Institute of Physics Publishing. ISBN 0750303735.

5

Spectral Line: A Comprehensive Study

Spectral lines help identify molecules and atoms as each atom has its own unique spectral line. This enables one to identify the chemical composition of various materials and is of particular use in identifying the atomic and molecular structure of stars and other planets. This chapter studies spectral lines, emission spectrum, Doppler broadening, Zeeman effect, Stark effect, forbidden mechanism, selection rule among others.

Spectral Line

Continuous spectrum.

Emission lines.

Absorption lines.

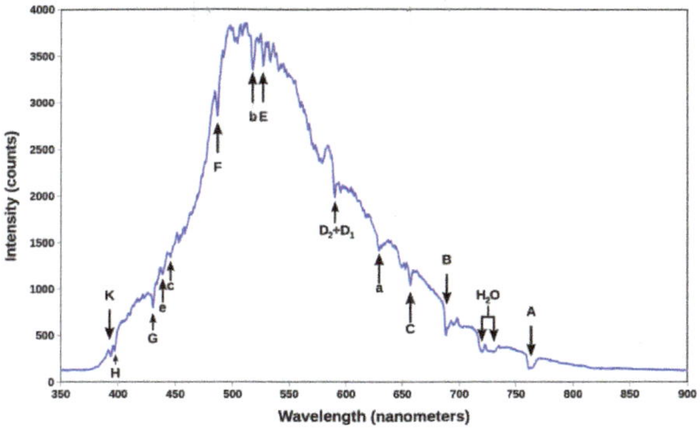

Absorption lines for air, under indirect illumination, with the direct light source not visible, so that the gas is not directly between source and detector. Here, Fraunhofer lines in sunlight and Rayleigh scattering of this sunlight is the "source." This is the spectrum of a blue sky somewhat close to the horizon, pointing east at around 3 or 4 pm (i.e., Sun in the West) on a clear day.

A spectral line is a dark or bright line in an otherwise uniform and continuous spectrum, resulting from emission or absorption of light in a narrow frequency range, compared with the nearby frequencies. Spectral lines are often used to identify atoms and molecules from their characteristic spectral lines. These "fingerprints" can be compared to the previously collected "fingerprints" of atoms and molecules, and are thus used to identify the atomic and molecular components of stars and planets which would otherwise be impossible.

Types of Line Spectra

Continuous spectrum of an incandescent lamp (mid) and discrete spectrum lines of a fluorescent lamp (bottom).

Spectral lines are the result of interaction between a quantum system (usually atoms, but sometimes molecules or atomic nuclei) and a single photon. When a photon has about the right amount of energy to allow a change in the energy state of the system (in the case of an atom this is usually an electron changing orbitals), the photon is absorbed. Then it will be spontaneously re-emitted, either in the same frequency as the original or in a cascade, where the sum of the energies of the photons emitted will be equal to the energy of the one absorbed (assuming the system returns to its original state).

A spectral line may be observed either as an emission line or an absorption line. Which type of line is observed depends on the type of material and its temperature relative to another emission source. An absorption line is produced when photons from a hot, broad spectrum source pass through a cold material. The intensity of light, over a narrow frequency range, is reduced due to absorption by the material and re-emission in random directions. By contrast, a bright, emission line is produced when photons from a hot material are detected in the presence of a broad spectrum from a cold source. The intensity of light, over a narrow frequency range, is increased due to emission by the material.

Spectral lines are highly atom-specific, and can be used to identify the chemical composition of any medium capable of letting light pass through it (typically gas is used). Several elements were discovered by spectroscopic means, such as helium, thallium, and cerium. Spectral lines also depend on the physical conditions of the gas, so they are widely used to determine the chemical composition of stars and other celestial bodies that cannot be analyzed by other means, as well as their physical conditions.

Mechanisms other than atom-photon interaction can produce spectral lines. Depending on the exact physical interaction (with molecules, single particles, etc.), the frequency of the involved photons will vary widely, and lines can be observed across the electromagnetic spectrum, from radio waves to gamma rays.

Nomenclature

Strong spectral lines in the visible part of the spectrum often have a unique Fraunhofer line designation, such as K for a line at 393.366 nm emerging from singly ionized Ca^+, though some of the Fraunhofer "lines" are blends of multiple lines from several different species. In other cases the lines are designated according to the level of ionization by adding a Roman numeral to the designation of the chemical element, so that Ca^+ also has the designation Ca II. Neutral atoms are denoted with the roman number I, singly ionized atoms with II, and so on, so that for example Fe IX (IX, roman 9) represents eight times ionized iron. More detailed designations usually include the line wavelength and may include a multiplet number (for atomic lines) or band designation (for molecular lines). Many spectral lines of atomic hydrogen also have designations within their respective series, such as the Lyman series or Balmer series.

Line Broadening and Shift

A spectral line extends over a range of frequencies, not a single frequency (i.e., it has a nonzero linewidth). In addition, its center may be shifted from its nominal central wavelength. There are several reasons for this broadening and shift. These reasons may be divided into two broad categories – broadening due to local conditions and broadening due to extended conditions. Broadening due to local conditions is due to effects which hold in a small region around the emitting element, usually small enough to assure local thermodynamic equilibrium. Broadening due to extended conditions may result from changes to the spectral distribution of the radiation as it traverses its path to the observer. It also may result from the combining of radiation from a number of regions which are far from each other.

Broadening Due to Local Effects

Natural Broadening

The uncertainty principle relates the lifetime of an excited state (due to spontaneous radiative decay or the Auger process) with the uncertainty of its energy. A short lifetime will have a large energy uncertainty and a broad emission. This broadening effect results in an unshifted Lorentzian profile. The natural broadening can be experimentally altered only to the extent that decay rates can be artificially suppressed or enhanced.

Thermal Doppler Broadening

The atoms in a gas which are emitting radiation will have a distribution of velocities. Each photon emitted will be "red"- or "blue"-shifted by the Doppler effect depending on the velocity of the atom relative to the observer. The higher the temperature of the gas, the wider the distribution of velocities in the gas. Since the spectral line is a combination of all of the emitted radiation, the higher the temperature of the gas, the broader the spectral line emitted from that gas. This broadening effect is described by a Gaussian profile and there is no associated shift.

Pressure Broadening

The presence of nearby particles will affect the radiation emitted by an individual particle. There are two limiting cases by which this occurs:

- *Impact pressure broadening* or *collisional broadening*: The collision of other particles with the emitting particle interrupts the emission process, and by shortening the characteristic time for the process, increases the uncertainty in the energy emitted (as occurs in natural broadening). The duration of the collision is much shorter than the lifetime of the emission process. This effect depends on both the density and the temperature of the gas. The broadening effect is described by a Lorentzian profile and there may be an associated shift.

- *Quasistatic pressure broadening*: The presence of other particles shifts the energy levels in the emitting particle, thereby altering the frequency of the emitted radiation. The duration of the influence is much longer than the lifetime of the emission process. This effect depends on the density of the gas, but is rather insensitive to temperature. The form of the line profile is determined by the functional form of the perturbing force with respect to distance from the perturbing particle. There may also be a shift in the line center. The general expression for the lineshape resulting from quasistatic pressure broadening is a 4-parameter generalization of the Gaussian distribution known as a stable distribution.

Pressure broadening may also be classified by the nature of the perturbing force as follows:

- *Linear Stark broadening* occurs via the linear Stark effect, which results from the interaction of an emitter with an electric field, causing a shift in energy that is linear in the field strength. $(\Delta E \sim 1/r^2)$
- *Resonance broadening* occurs when the perturbing particle is of the same type as the emitting particle, which introduces the possibility of an energy exchange process. $(\Delta E \sim 1/r^3)$
- *Quadratic Stark broadening* occurs via the quadratic Stark effect, which results from the interaction of an emitter with an electric field, causing a shift in energy that is quadratic in the field strength. $(\Delta E \sim 1/r^4)$
- *Van der Waals broadening* occurs when the emitting particle is being perturbed by van der Waals forces. For the quasistatic case, a van der Waals profile is often useful in describing the profile. The energy shift as a function of distance is given in the wings by e.g. the Lennard-Jones potential. $(\Delta E \sim 1/r^6)$

Inhomogeneous Broadening

Inhomogeneous broadening is a general term for broadening because some emitting particles are in a different local environment from others, and therefore emit at a different frequency. This term is used especially for solids, where surfaces, grain boundaries, and stoichiometry variations can create a variety of local environments for a given atom to occupy. In liquids, the effects of inhomogeneous broadening is sometimes reduced by a process called *motional narrowing*.

Broadening Due to Non-local Effects

Certain types of broadening are the result of conditions over a large region of space rather than simply upon conditions that are local to the emitting particle.

Opacity Broadening

Electromagnetic radiation emitted at a particular point in space can be reabsorbed as it travels through space. This absorption depends on wavelength. The line is broadened because the photons at the line wings have a smaller reabsorption probability than the photons at the line center. Indeed, the reabsorption near the line center may be so great as to cause a self reversal in which the intensity at the center of the line is less than in the wings. This process is also sometimes called self-absorption.

Macroscopic Doppler Broadening

Radiation emitted by a moving source is subject to Doppler shift due to a finite line-of-sight velocity projection. If different parts of the emitting body have different velocities (along the line of sight), the resulting line will be broadened, with the line width proportional to the width of the velocity distribution. For example, radiation emitted from a distant rotating body, such as a star, will be broadened due to the line-of-sight variations in velocity on opposite sides of the star. The greater the rate of rotation, the broader the line. Another example is an imploding plasma shell in a Z-pinch.

Combined Effects

Each of these mechanisms can act in isolation or in combination with others. Assuming each effect is independent, the observed line profile is a convolution of the line profiles of each mechanism. For example, a combination of the thermal Doppler broadening and the impact pressure broadening yields a Voigt profile.

However, the different line broadening mechanisms are not always independent. For example, the collisional effects and the motional Doppler shifts can act in a coherent manner, resulting under some conditions even in a collisional *narrowing*, known as the Dicke effect.

Emission Spectrum

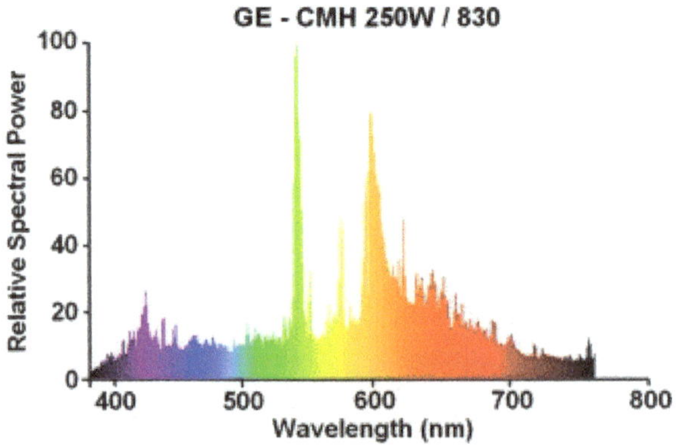

Emission spectrum of a metal halide lamp.

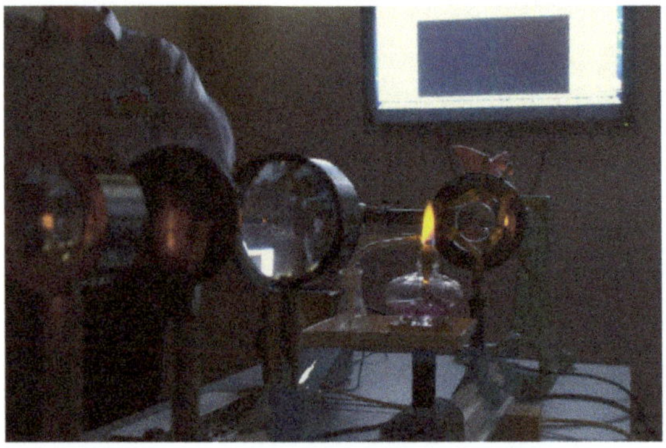

A demonstration of the 589 nm D_2 (left) and 590 nm D_1 (right) emission sodium D lines using a wick with salt water in a flame

The emission spectrum of a chemical element or chemical compound is the spectrum of frequencies of electromagnetic radiation emitted due to an atom or molecule making a transition from a high energy state to a lower energy state. The photon energy of the emitted photon is equal to the energy difference between the two states. There are many possible electron transitions for each atom, and each transition has a specific energy difference. This collection of different transitions, leading to different radiated wavelengths, make up an emission spectrum. Each element's emission spectrum is unique. Therefore, spectroscopy can be used to identify the elements in matter of unknown composition. Similarly, the emission spectra of molecules can be used in chemical analysis of substances.

Emission

In physics, emission is the process by which a higher energy quantum mechanical state of a particle becomes converted to a lower one through the emission of a photon, resulting in the production of light. The frequency of light emitted is a function of the energy of the transition. Since energy must be conserved, the energy difference between the two states equals the energy carried off by the photon. The energy states of the transitions can lead to emissions over a very large range of frequencies. For example, visible light is emitted by the coupling of electronic states in atoms and molecules (then the phenomenon is called fluorescence or phosphorescence). On the other hand, nuclear shell transitions can emit high energy gamma rays, while nuclear spin transitions emit low energy radio waves.

The emittance of an object quantifies how much light is emitted by it. This may be related to other properties of the object through the Stefan–Boltzmann law. For most substances, the amount of emission varies with the temperature and the spectroscopic composition of the object, leading to the appearance of color temperature and emission lines. Precise measurements at many wavelengths allow the identification of a substance via emission spectroscopy.

Emission of radiation is typically described using semi-classical quantum mechanics: the particle's energy levels and spacings are determined from quantum mechanics, and light is treated as an oscillating electric field that can drive a transition if it is in resonance with the system's natural frequency. The quantum mechanics problem is treated using time-dependent perturbation theory and leads to the general result known as Fermi's golden rule. The description has been superseded by quantum electrodynamics, although the semi-classical version continues to be more useful in most practical computations.

Origins

When the electrons in the atom are excited, for example by being heated, the additional energy pushes the electrons to higher energy orbitals. When the electrons fall back down and leave the excited state, energy is re-emitted in the form of a photon. The wavelength (or equivalently, frequency) of the photon is determined by the difference in energy between the two states. These emitted photons form the element's spectrum.

The fact that only certain colors appear in an element's atomic emission spectrum means that only certain frequencies of light are emitted. Each of these frequencies are related to energy by the formula:

$$E_{\text{photon}} = h\nu, ,$$

where E_{photon} is the energy of the photon, ν is its frequency, and h is Planck's constant. This concludes that only photons with specific energies are emitted by the atom. The principle of the atomic emission spectrum explains the varied colors in neon signs, as well as chemical flame test results (described below).

The frequencies of light that an atom can emit are dependent on states the electrons can be in. When excited, an electron moves to a higher energy level or orbital. When the electron falls back to its ground level the light is emitted.

Emission spectrum of hydrogen

The above picture shows the visible light emission spectrum for hydrogen. If only a single atom of hydrogen were present, then only a single wavelength would be observed at a given instant. Several of the possible emissions are observed because the sample contains many hydrogen atoms that are in different initial energy states and reach different final energy states. These different combinations lead to simultaneous emissions at different wavelengths.

Emission spectrum of iron

Radiation from Molecules

As well as the electronic transitions discussed above, the energy of a molecule can also change via rotational, vibrational, and vibronic (combined vibrational and electronic) transitions. These energy transitions often lead to closely spaced groups of many different spectral lines, known as spectral bands. Unresolved band spectra may appear as a spectral continuum.

Emission Spectroscopy

Light consists of electromagnetic radiation of different wavelengths. Therefore, when the elements or their compounds are heated either on a flame or by an electric arc they emit energy in the form of light. Analysis of this light, with the help of a spectroscope gives us a discontinuous spectrum. A spectroscope or a spectrometer is an instrument which is used for separating the components of light, which have different wavelengths. The spectrum appears in a series of lines called the line spectrum. This line spectrum is called an atomic spectrum when it originates from an atom in elemental form. Each element has a different atomic spectrum. The production of line spectra by the atoms of an element indicate that an atom can radiate only a certain amount of energy. This leads to the conclusion that bound electrons cannot have just any amount of energy but only a certain amount of energy.

The emission spectrum can be used to determine the composition of a material, since it is different for each element of the periodic table. One example is astronomical spectroscopy: identifying the composition of stars by analysing the received light. The emission spectrum characteristics of some elements are plainly visible to the naked eye when these elements are heated. For example, when platinum wire is dipped into a strontium nitrate solution and then inserted into a flame, the strontium atoms emit a red color. Similarly, when copper is inserted into a flame, the flame becomes green. These definite characteristics allow elements to be identified by their atomic emission spectrum. Not all emitted lights are perceptible to the naked eye, as the spectrum also includes ultraviolet rays and infrared lighting. An emission is formed when an excited gas is viewed directly through a spectroscope.

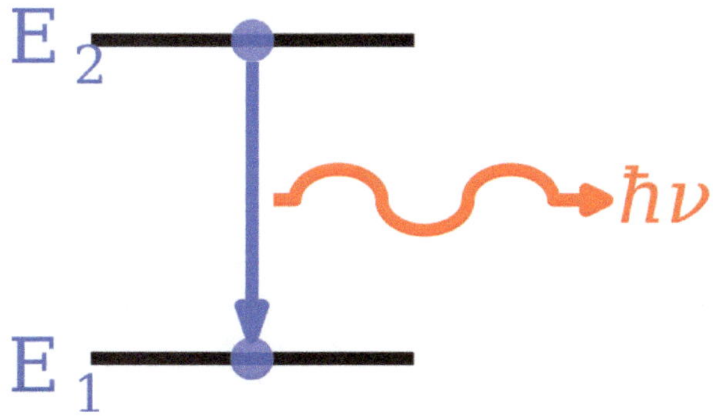

Schematic diagram of spontaneous emission

Emission spectroscopy is a spectroscopic technique which examines the wavelengths of photons emitted by atoms or molecules during their transition from an excited state to a lower energy state. Each element emits a characteristic set of discrete wavelengths according to its electronic structure, and by observing these wavelengths the elemental composition of the sample can be determined. Emission spectroscopy developed in the late 19th century and efforts in theoretical explanation of atomic emission spectra eventually led to quantum mechanics.

There are many ways in which atoms can be brought to an excited state. Interaction with electromagnetic radiation is used in fluorescence spectroscopy, protons or other heavier particles in Particle-Induced X-ray Emission and electrons or X-ray photons in Energy-dispersive X-ray spectroscopy or X-ray fluorescence. The simplest method is to heat the sample to a high temperature, after which the excitations are produced by collisions between the sample atoms. This method is used in flame emission spectroscopy, and it was also the method used by Anders Jonas Ångström when he discovered the phenomenon of discrete emission lines in the 1850s.

Although the emission lines are caused by a transition between quantized energy states and may at first look very sharp, they do have a finite width, i.e. they are composed of more than one wavelength of light. This spectral line broadening has many different causes.

Emission spectroscopy is often referred to as optical emission spectroscopy because of the light nature of what is being emitted.

History

Emission lines from hot gases were first discovered by Ångström, and the technique was further developed by David Alter, Gustav Kirchhoff and Robert Bunsen.

See the history of spectroscopy for details.

Experimental Technique in Flame Emission Spectroscopy

The solution containing the relevant substance to be analysed is drawn into the burner and dispersed into the flame as a fine spray. The solvent evaporates first, leaving finely divided solid particles which move to the hottest region of the flame where gaseous atoms and ions are produced. Here electrons are excited as described above. It is common for a monochromator to be used to allow for easy detection.

On a simple level, flame emission spectroscopy can be observed using just a flame and samples of metal salts. This method of qualitative analysis is called a flame test. For example, sodium salts placed in the flame will glow yellow from sodium ions, while strontium (used in road flares) ions color it red. Copper wire will create a blue colored flame, however in the presence of chloride gives green (molecular contribution by CuCl).

Emission Coefficient

Emission coefficient is a coefficient in the power output per unit time of an electromagnetic source, a calculated value in physics. The emission coefficient of a gas varies with the wavelength of the light. It has units of $ms^{-3}sr^{-1}$. It is also used as a measure of environmental emissions (by mass) per MWh of electricity generated, see: Emission factor.

Scattering of Light

In Thomson scattering a charged particle emits radiation under incident light. The particle may be an ordinary atomic electron, so emission coefficients have practical applications.

If $X\, dV\, d\Omega\, d\lambda$ is the energy scattered by a volume element dV into solid angle $d\Omega$ between wavelengths λ and $\lambda+d\lambda$ per unit time then the Emission coefficient is X.

The values of X in Thomson scattering can be predicted from incident flux, the density of the charged particles and their Thomson differential cross section (area/solid angle).

Spontaneous Emission

A warm body emitting photons has a monochromatic emission coefficient relating to its temperature and total power radiation. This is sometimes called the second Einstein coefficient, and can be deduced from quantum mechanical theory.

Selection Rule

In physics and chemistry, a selection rule, or transition rule, formally constrains the possible transitions of a system from one quantum state to another. Selection rules have been derived for electromagnetic transitions in molecules, in atoms, in atomic nuclei, and so on. The selection rules may differ according to the technique used to observe the transition.

In the following, mainly atomic and molecular transitions are considered.

Overview

In quantum mechanics the basis for a spectroscopic selection rule is the value of the transition moment integral

$$\int \psi_1^* \mu \psi_2 d\tau,,$$

where ψ_1 and ψ_2 are the wave functions of the two states involved in the transition and μ is the transition moment operator. If the value of this integral is zero the transition is forbidden. In practice, the integral itself does not need to be calculated to determine a selection rule. It is sufficient to determine the symmetry of transition moment function, $\psi_1^* \mu \psi_2$. If the symmetry of this function spans the totally symmetric representation of the point group to which the atom or molecule belongs then its value is (in general) not zero and the transition is allowed. Otherwise, the transition is forbidden.

The transition moment integral is zero if the transition moment function, $\psi_1^* \mu \psi_2,$, is anti-symmetric or odd, i.e. $y(x) = -y(-x)$ holds. The symmetry of the transition moment function is the direct product of the parities of its three components. The symmetry characteristics of each component can be obtained from standard character tables. Rules for obtaining the symmetries of a direct product can be found in texts on character tables.

Symmetry characteristics of transition moment operator		
Transition type	μ transforms as	**note**
Electric dipole	x, y, z	Optical spectra
Electric quadrupole	x^2, y^2, z^2, xy, xz, yz	Constraint $x^2 + y^2 + z^2 = 0$
Electric polarizability	x^2, y^2, z^2, xy, xz, yz	Raman spectra
Magnetic dipole	R_x, R_y, R_z	Optical spectra (weak)

Examples

Electronic Spectra

The Laporte rule is a selection rule formally stated as follows: In a centrosymmetric environment, transitions between like atomic orbitals such as *s-s*,*p-p*, *d-d*, or *f-f*, transitions are forbidden. The Laporte rule applies to electric dipole transitions, so the operator has *u* symmetry (meaning *ungerade*, odd). *p* orbitals also have *u* symmetry, so the symmetry of the transition moment function is given by the triple product $u \times u \times u$, which has *u* symmetry. The transitions are therefore forbidden. Likewise, *d* orbitals have *g* symmetry, so the triple product $g \times u \times g$ also has *u* symmetry and the transition is forbidden.

The wave function of a single electron is the product of a space-dependent wave function and a spin wave function. Spin is directional and can be said to have odd parity. It follows that transitions in which the spin "direction" changes are forbidden. In formal terms, only states with the same total spin quantum number are "spin-allowed". In crystal field theory, *d-d* transitions that are spin-forbidden are much weaker than spin-allowed transitions. Both can be observed, in spite of the Laporte rule, because the actual transitions are coupled to vibrations that are anti-symmetric and have the same symmetry as the dipole moment operator.

Vibrational Spectra

In vibrational spectroscopy, transitions are observed between different vibrational states. In a fundamental vibration, the molecule is excited from its ground state ($v = 0$) to the first excited state ($v = 1$). The symmetry of the ground-state wave function is the same as that of the molecule. It is, therefore, a basis for the totally symmetric representation in the point group of the molecule. It follows that, for a vibrational transition to be allowed, the symmetry of the excited state wave function must be the same as the symmetry of the transition moment operator.

In infrared spectroscopy, the transition moment operator transforms as either x and/or y and/or z. The excited state wave function must also transform as at least one of these vectors. In Raman spectroscopy, the operator transforms as one of the second-order terms in the right-most column of the character table, below.

Character table for the T_d point group							
	E	$8C_3$	$3C_2$	$6S_4$	$6\sigma_d$		
A_1	1	1	1	1	1		$x^2 + y^2 + z^2$
A_2	1	1	1	-1	-1		
E	2	-1	2	0	0		$(2z^2 - x^2 - y^2, x^2 - y^2)$
T_1	3	0	-1	1	-1	(R_x, R_y, R_z)	
T_2	3	0	-1	-1	1	(x, y, z)	(xy, xz, yz)

The molecule methane, CH_4, may be used as an example to illustrate the application of these principles. The molecule is tetrahedral and has T_d symmetry. The vibrations of methane span the representations $A_1 + E + 2T_2$. Examination of the character table shows that all four vibrations are Raman-active, but only the T_2 vibrations can be seen in the infrared spectrum.

In the harmonic approximation, it can be shown that overtones are forbidden in both infrared and Raman spectra. However, when anharmonicity is taken into account, the transitions are weakly allowed.

Rotational Spectra

The selection rule for rotational transitions, derived from the symmetries of the rotational wave functions in a rigid rotor, is $\Delta J = \pm 1$, where J is a rotational quantum number.

Coupled Transitions

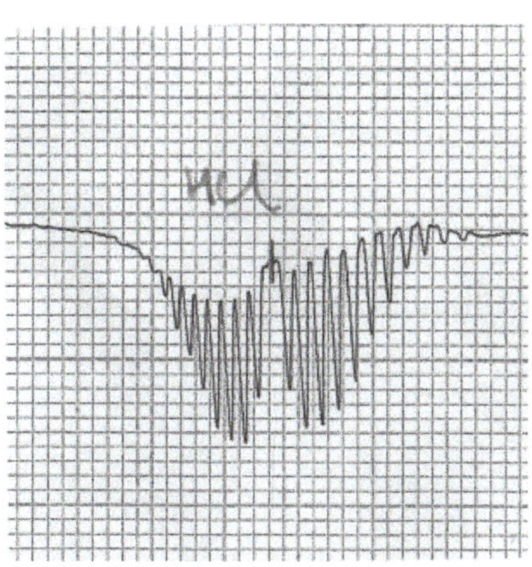

The infrared spectrum of HCl gas

There are many types of coupled transition such as are observed in vibration-rotation spectra. The excited-state wave function is the product of two wave functions such as vibrational and rotational. The general principle is that the symmetry of the excited state is obtained as the direct product of the symmetries of the component wave functions. In rovibronic transitions, the excited states involve three wave functions.

The infrared spectrum of hydrogen chloride gas shows rotational fine structure superimposed on the vibrational spectrum. This is typical of the infrared spectra of heteronuclear diatomic molecules. It shows the so-called *P* and *R* branches. The *Q* branch, located at the vibration frequency, is absent. Symmetric top molecules display the *Q* branch. This follows from the application of selection rules.

Resonance Raman spectroscopy involves a kind of vibronic coupling. It results in much-increased intensity of fundamental and overtone transitions as the vibrations "steal" intensity from an allowed electronic transition. In spite of appearances, the selection rules are the same as in Raman spectroscopy.

Angular Momentum

In general, electric (charge) radiation or magnetic (current, magnetic moment) radiation can be classified into multipoles Eλ (electric) or Mλ (magnetic) of order 2^λ, e.g., E1 for electric dipole, E2 for quadrupole, or E3 for octupole. In transitions where the change in angular momentum between the initial and final states makes several multipole radiations possible, usually the lowest-order multipoles are overwhelmingly more likely, and dominate the transition.

The emitted particle carries away an angular momentum λ, which for the photon must be at least 1, since it is a vector particle (i.e., it has $J^P = 1^-$). Thus, there is no E0 (electric monopoles) or M0 (magnetic monopoles, which do not seem to exist) radiation.

Since the total angular momentum has to be conserved during the transition, we have that

$$\mathbf{J}_\mathrm{i} = \mathbf{J}_\mathrm{f} + \lambda$$

where $\| \ddot{\mathbf{e}} \| = \sqrt{\lambda(\lambda+1)}\hbar$, and its z-projection is given by $\lambda_z = \mu\hbar$; $\mathbf{J}_\mathrm{i}$ and $\mathbf{J}_\mathrm{f}$ are, respectively, the initial and final angular momenta of the atom. The corresponding quantum numbers λ and μ (z-axis angular momentum) must satisfy

$$|J_\mathrm{i} - J_\mathrm{f}| \le \lambda \le J_\mathrm{i} + J_\mathrm{f}$$

and

$$\mu = M_\mathrm{i} - M_\mathrm{f}.$$

Parity is also preserved. For electric multipole transitions

$$\pi(\mathrm{E}\lambda) = \pi_{\mathrm{i}}\pi_{\mathrm{f}} = (-1)^{\lambda}$$

while for magnetic multipoles

$$\pi(\mathrm{M}\lambda) = \pi_{\mathrm{i}}\pi_{\mathrm{f}} = (-1)^{\lambda+1}.$$

Thus, parity does not change for E-even or M-odd multipoles, while it changes for E-odd or M-even multipoles.

These considerations generate different sets of transitions rules depending on the multipole order and type. The expression *forbidden transitions* is often used; this does not mean that these transitions cannot occur, only that they are *electric-dipole-forbidden*. These transitions are perfectly possible; they merely occur at a lower rate. If the rate for an E1 transition is non-zero, the transition is said to be permitted; if it is zero, then M1, E2, etc. transitions can still produce radiation, albeit with much lower transitions rates. These are the so-called forbidden transitions. The transition rate decreases by a factor of about 1000 from one multipole to the next one, so the lowest multipole transitions are most likely to occur.

Semi-forbidden transitions (resulting in so-called intercombination lines) are electric dipole (E1) transitions for which the selection rule that the spin does not change is violated. This is a result of the failure of LS coupling.

Summary Table

$J = L + S$ is the total angular momentum, L is the Azimuthal quantum number, S is the Spin quantum number, and M_J is the secondary total angular momentum quantum number. Which transitions are allowed is based on the Hydrogen-like atom. The symbol $\not\leftrightarrow$ is used to indicate a forbidden transition.

Surface

In surface vibrational spectroscopy, the *surface selection rule* is applied to identify the peaks observed in vibrational spectra. When a molecule is adsorbed on a substrate, the molecule induces opposite image charges in the substrate. The dipole moment of the molecule and the image charges perpendicular to the surface reinforce each other. In contrast, the dipole moments of the molecule and the image charges parallel to the surface cancel out. Therefore, only molecular vibrational peaks giving rise to a dynamic dipole moment perpendicular to the surface will be observed in the vibrational spectrum.

Forbidden Mechanism

In spectroscopy, a forbidden mechanism (forbidden transition or forbidden line) is a spectral line associated with absorption or emission of light by atomic nuclei, atoms, or molecules undergoes a transition that is not allowed by a particular selection rule but is allowed if the approximation associated with that rule is not made. For example, in a situation where, according to usual approximations (such as the electric-dipole approximation for the interaction with light), the process cannot happen, but at a higher level of approximation (*e.g.* magnetic dipole, or, electric quadrupole) the process is allowed but at a much lower rate.

An example is phosphorescent glow in the dark materials, which absorb light and form an excited state whose decay involves a spin flip, and is therefore forbidden by electric dipole transitions. The result is emission of light slowly over minutes or hours.

Although the transitions are nominally forbidden, there is a small probability of their spontaneous occurrence, should an atomic nucleus, atom or molecule be raised to an excited state. More precisely, there is a certain probability that such an excited entity will make a forbidden transition to a lower energy state per unit time; by definition, this probability is much lower than that for any transition permitted or allowed by the selection rules. Therefore, if a state can de-excite via a permitted transition (or otherwise, e.g. via collisions) it will almost certainly do so before any transition occurs via a forbidden route. Nevertheless, most forbidden transitions are only relatively unlikely: states that can only decay in this way (so-called meta-stable states) usually have lifetimes on the order milliseconds to seconds, compared to less than a microsecond for decay via permitted transitions. In some radioactive decay systems, multiple levels of forbiddenness can stretch life times by many orders of magnitude for each additional unit by which the system changes beyond what is most allowed under the selection rules. Such excited states can last years, or even for many billions of years (too long to have been measured).

In Radioactive Decay

Gamma Decay

The most common mechanism for suppression of the rate of gamma decay of excited atomic nuclei, and thus make possible the existence of a metastable isomer for the nucleus, is lack of a decay route for the excited state that will change nuclear angular momentum (along any given direction) by the most common (allowed) amount of 1 quantum unit $\hbar$ of spin angular momentum. Such a change is necessary to emit a gamma photon, which has a spin of 1 unit in this system. Integral changes of 2, 3, 4, and more units in angular momentum are possible (the emitted photons carry off the additional angular momentum), but changes of more than 1 unit are known as forbidden transitions. Each degree of forbiddenness (additional unit of spin change larger than 1, that the emitted gamma ray must carry) inhibits decay rate by about 5 orders of magnitude. The highest known spin change of 8 units occurs in the decay of Ta-180m, which suppresses its decay by a factor of 10^{35} from that associated with 1 unit, so that instead of a natural gamma decay half life of 10^{-12} seconds, it has a half life of more than 10^{23} seconds, or at least 3×10^{15} years, and thus has yet to be observed to decay.

Although gamma decays with nuclear angular momentum changes of 2, 3, 4, etc., are forbidden, they are only relatively forbidden, and do proceed, but with a slower rate than the normal allowed change of 1 unit. However, gamma emission is absolutely forbidden when the nucleus begins in a zero-spin state, as such an emission would not conserve angular momentum. These transitions cannot occur by gamma decay, but must proceed by another route, such as beta decay in some cases, or internal conversion where beta decay is not favored.

Beta Decay

Beta decays here are classified according to the *L*-value of the emitted radiation. Unlike gamma decays, beta decays may proceed from a nucleus with a spin of zero and even parity, to a nucleus also with a spin of zero and even parity (Fermi transition). This is possible because the electron and neutrino emitted may be of opposing spin (giving a radiation total angular momentum of zero), thus preserving angular momentum of the initial state even if the nucleus remains at spin-zero before and after emission. This type of emission is super-allowed meaning that it is the most rapid type of beta decay in nuclei that are susceptible to a change in proton/neutron ratios that accompanies a beta decay process.

The next possible total angular momentum of the electron and neutrino emitted in beta decay is a combined spin of 1 (electron and neutrino spinning in the same direction), and is allowed. This type of emission (Gamow-Teller transition) changes nuclear spin by a unit of 1 to compensate. States involving higher angular momenta of the emitted radiation (2, 3, 4, etc.) are forbidden, and are ranked in degree of forbiddenness by their increasing angular momentum.

Specifically, when $L > 0$, the decay is referred to as forbidden. Nuclear selection rules require L-values greater than two to be accompanied by changes in both nuclear spin (J) and parity (π). The selection rules for the *L*th forbidden transitions are:

$$\text{Ä}J = L-1, L, L+1; \text{Ä}\pi = (-1)^L,$$

where $\Delta\pi = 1$ or -1 corresponds to no parity change or parity change, respectively. As noted, the special case of a Fermi $0^+ \rightarrow 0^+$ transition (which in gamma decay is absolutely forbidden) is referred to as super-allowed for beta decay, and proceeds very quickly if beta decay is possible. The following table lists the ΔJ and $\Delta\pi$ values for the first few values of L:

Forbiddenness	ΔJ	$\Delta\pi$
Superallowed	$0^+ \rightarrow 0^+$	no
Allowed	0, 1	no
First forbidden	0, 1, 2	yes
Second forbidden	1, 2, 3	no
Third forbidden	2, 3, 4	yes

As with gamma decay, each degree of increasing forbiddenness increases the half life of the beta decay process involved by a factor of about 4 to 5 orders of magnitude.

Double beta decay has been observed in the laboratory, e. g. in 82Se. Geochemical experiments have also found this rare type of forbidden decay in several isotopes. with mean half lives over 10^{18} yr .

In Solid-state Physics

Forbidden transitions in rare earth atoms such as erbium and neodymium make them useful as dopants for solid-state lasing media. In such media, the atoms are held in a matrix which keeps them from de-exciting by collision, and the long half life of their excited states makes them easy to optically pump to create a large population of excited atoms. Neodymium doped glass derives its unusual coloration from forbidden *f-f* transitions within the neodymium atom, and is used in extremely high power solid state lasers. Bulk semiconductor transitions can also be forbidden by symmetry, which change the functional form of the absorption spectrum, as can be shown in a Tauc plot.

In Astrophysics and Atomic Physics

Forbidden emission lines have been observed in extremely low-density gases and plasmas, either in outer space or in the extreme upper atmosphere of the Earth. In space environments, densities may be only a few atoms per cubic centimetre, making atomic collisions unlikely. Under such conditions, once an atom or molecule has been excited for any reason into a meta-stable state, then it is almost certain to decay by emitting a forbidden-line photon. Since meta-stable states are rather common, forbidden transitions account for a significant percentage of the photons emitted by the ultra-low density gas in space. Forbidden transitions in highly charged ions resulting in the emission of visible, vacuum-ultraviolet, soft x-ray and x-ray photons are routinely observed in certain laboratory devices such as electron beam ion traps and ion storage rings, where in both cases residual gas densities are sufficiently low for forbidden line emission to occur before atoms are collisionally de-excited. Using laser spectroscopy techniques, forbidden transitions are used to stabilize atomic clocks and quantum clocks that have the highest accuracies currently available.

Forbidden lines of nitrogen ([N II] at 654.8 and 658.4 nm), sulfur ([S II] at 671.6 and 673.1 nm), and oxygen ([O II] at 372.7 nm, and [O III] at 495.9 and 500.7 nm) are commonly observed in astrophysical plasmas. These lines are important to the energy balance of such things as planetary nebulae and H II regions. The forbidden 21-cm hydrogen line is particularly important for radio astronomy as it allows very cold neutral hydrogen gas to be seen.

Notation

Forbidden line transitions are noted by placing square brackets around the atomic or molecular species in question, e.g. [O III] or [S II].

Doppler Broadening

In atomic physics, Doppler broadening is the broadening of spectral lines due to the Doppler effect caused by a distribution of velocities of atoms or molecules. Different velocities of the emitting particles result in different Doppler shifts, the cumulative effect of which is the line broadening. This resulting line profile is known as a Doppler profile. A particular case is the thermal Doppler broadening due to the thermal motion of the particles. Then, the broadening depends only on the frequency of the spectral line, the mass of the emitting particles, and their temperature, and therefore can be used for inferring the temperature of an emitting body.

Saturated absorption spectroscopy, also known as Doppler-free spectroscopy, can be used to find the true frequency of an atomic transition without cooling a sample down to temperatures at which the Doppler broadening is minimal.

Derivation

When thermal motion causes a particle to move towards the observer, the emitted radiation will be shifted to a higher frequency. Likewise, when the emitter moves away, the frequency will be lowered. For non-relativistic thermal velocities, the Doppler shift in frequency will be:

$$f = f_0\left(1+\frac{v}{c}\right)$$

where f is the observed frequency, f_0 is the rest frequency, v is the velocity of the emitter towards the observer, and c is the speed of light.

Since there is a distribution of speeds both toward and away from the observer in any volume element of the radiating body, the net effect will be to broaden the observed line. If $P_v(v)dv$ is the fraction of particles with velocity component v to $v+dv$ along a line of sight, then the corresponding distribution of the frequencies is

$$P_f(f)df = P_v(v_f)\frac{dv}{df}df,$$

where $v_f = c\left(\frac{f}{f_0}-1\right)$ is the velocity towards the observer corresponding to the shift of the rest frequency f_0 to f. Therefore,

$$P_f(f)df = \frac{c}{f_0}P_v\left(c\left(\frac{f}{f_0}-1\right)\right)df.$$

We can also express the broadening in terms of the wavelength λ. Recalling that in the non-relativistic limit $\frac{\lambda-\lambda_0}{\lambda_0} \approx -\frac{f-f_0}{f_0}$, we obtain

$$P_\lambda(\lambda)d\lambda = \frac{c}{\lambda_0} P_v\left(c\left(1-\frac{\lambda}{\lambda_0}\right)\right)d\lambda.$$

In the case of the thermal Doppler broadening, the velocity distribution is given by the Maxwell distribution

$$P_v(v)dv = \sqrt{\frac{m}{2\pi kT}}\exp\left(-\frac{mv^2}{2kT}\right)dv,$$

where m is the mass of the emitting particle, T is the temperature and k is the Boltzmann constant.

Then,

$$P_f(f)df = \left(\frac{c}{f_0}\right)\sqrt{\frac{m}{2\pi kT}}\exp\left(-\frac{m\left[c\left(\frac{f}{f_0}-1\right)\right]^2}{2kT}\right)df.$$

We can simplify this expression as

$$P_f(f)df = \sqrt{\frac{mc^2}{2\pi kTf_0^{\,2}}}\exp\left(-\frac{mc^2(f-f_0)^2}{2kTf_0^{\,2}}\right)df,$$

which we immediately recognize as a Gaussian profile with the standard deviation

$$\sigma_f = \sqrt{\frac{kT}{mc^2}}f_0$$

and full width at half maximum (FWHM)

$$\Delta f_{\text{FWHM}} = \sqrt{\frac{8kT\ln 2}{mc^2}}f_0.$$

Applications and Caveats

In astronomy and plasma physics, the thermal Doppler broadening is one of the explanations for the broadening of spectral lines, and as such gives an indication for the temperature of observed material. Other causes of velocity distributions may exist, though, for example due to turbulent motion. For a fully developed turbulence, the resulting line profile is generally very difficult to distinguish from the thermal one. Another cause could be a large range of *macroscopic* velocities resulting, e.g., from the receding and approaching portions of a rapidly spinning accretion disk. Finally, there are many other factors which can also broaden the lines. For example, a sufficiently high particle number density may lead to significant Stark broadening.

Doppler broadening can also be used to determine the velocity distribution of a gas given its absorption spectrum. In particular, this has been used to determine the velocity distribution of interstellar gas clouds.

Doppler broadening has also been used as a design consideration in high temperature nuclear reactors. In principle, as the reactor fuel heats up, the neutron absorption spectrum will broaden due to the relative thermal motion of the fuel atoms with respect to the neutrons. Given the shape of the neutron absorption spectrum, this has the result of reducing neutron absorption cross section, reducing the likelihood of absorption and fission. The end result is that reactors designed to take advantage of doppler broadening will decrease their reactivity as temperature increases, creating a passive safety measure. This tends to be more relevant to gas cooled reactors as other mechanisms are dominant in water cooled reactors.

Zeeman Effect

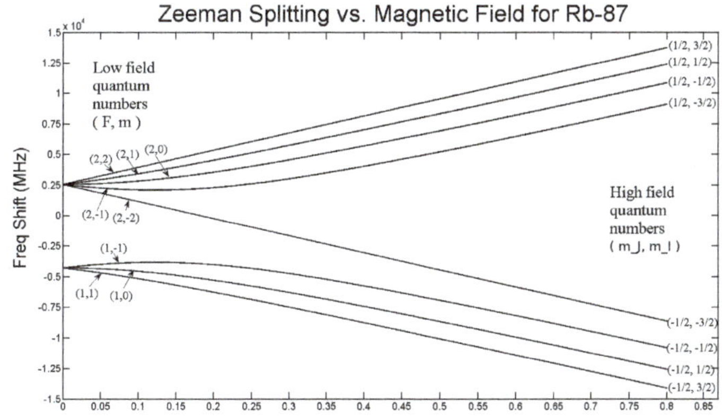

Zeeman splitting of the 5s level of Rb-87, including fine structure and hyperfine structure splitting. Here F = J + I, where I is the nuclear spin. (for Rb-87, I = 3/2)

The Zeeman effect (/ˈzeɪmən/; Dutch pronunciation: [ˈzeːmɑn]), named after the Dutch physicist Pieter Zeeman, is the effect of splitting a spectral line into several components in the presence of a static magnetic field. It is analogous to the Stark effect, the splitting of a spectral line into several components in the presence of an electric field. Also similar to the Stark effect, transitions between different components have, in general, different intensities, with some being entirely forbidden (in the dipole approximation), as governed by the selection rules.

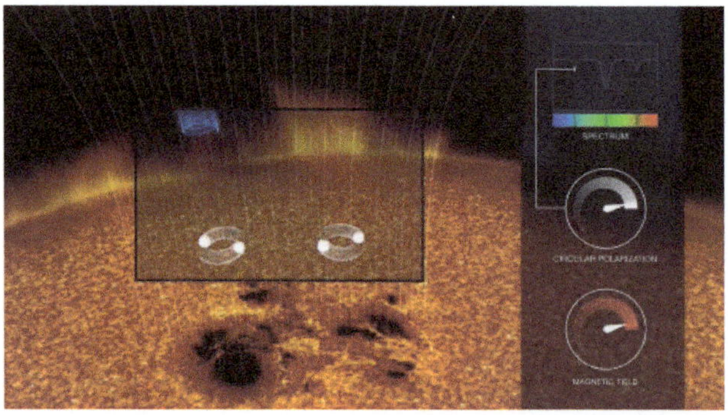

This animation shows what happens as a sunspot (or starspot) forms and the magnetic field increases in strength. The light emerging from the spot starts to demonstrate the Zeeman effect.The dark spectra lines in the spectrum of the emitted light split into three components and the strength of the circular polarisation in parts of the spectrum increases significantly. This polarisation effect is a powerful tool for astronomers to detect and measure stellar magnetic fields.

Since the distance between the Zeeman sub-levels is a function of the magnetic field, this effect can be used to measure the magnetic field, e.g. that of the Sun and other stars or in laboratory plasmas. The Zeeman effect is very important in applications such as nuclear magnetic resonance spectroscopy, electron spin resonance spectroscopy, magnetic resonance imaging (MRI) and Mössbauer spectroscopy. It may also be utilized to improve accuracy in atomic absorption spectroscopy. A theory about the magnetic sense of birds assumes that a protein in the retina is changed due to the Zeeman effect.

When the spectral lines are absorption lines, the effect is called inverse Zeeman effect.

Nomenclature

Historically, one distinguishes between the normal and an anomalous Zeeman effect (discovered by Thomas Preston in Dublin, Ireland). The anomalous effect appears on transitions where the net spin of the electrons is an odd half-integer, so that the number of Zeeman sub-levels is even. It was called "anomalous" because the electron spin had not yet been discovered, and so there was no good explanation for it at the time that Zeeman observed the effect.

At higher magnetic fields the effect ceases to be linear. At even higher field strength, when the strength of the external field is comparable to the strength of the atom's internal field, electron coupling is disturbed and the spectral lines rearrange. This is called the Paschen-Back effect.

In the modern scientific literature, these terms are rarely used, with a tendency to use just the "Zeeman effect".

Theoretical Presentation

The total Hamiltonian of an atom in a magnetic field is

$$H = H_0 + V_M,$$

where H_0 is the unperturbed Hamiltonian of the atom, and V_M is the perturbation due to the magnetic field:

$$V_M = -\vec{\mu} \cdot \vec{B},$$

where $\vec{\mu}$ is the magnetic moment of the atom. The magnetic moment consists of the electronic and nuclear parts; however, the latter is many orders of magnitude smaller and will be neglected here. Therefore,

$$\vec{\mu} \approx -\frac{\mu_B g \vec{J}}{\hbar},$$

where μ_B is the Bohr magneton, $\vec{J}$ is the total electronic angular momentum, and g is the Landé g-factor. A more accurate approach is to take into account that the operator of the magnetic moment of an electron is a sum of the contributions of the orbital angular momentum $\vec{L}$ and the spin angular momentum $\vec{L}$, with each multiplied by the appropriate gyromagnetic ratio:

$$\vec{\mu} = -\frac{\mu_B (g_l \vec{L} + g_s \vec{S})}{\hbar},$$

where $g_l = 1$ and $g_s \approx 2.0023192$ (the latter is called the anomalous gyromagnetic ratio; the deviation of the value from 2 is due to the effects of quantum electrodynamics). In the case of the LS coupling, one can sum over all electrons in the atom:

$$g\vec{J} = \sum_i (g_l \vec{l_i} + g_s \vec{s_i}) = (g_l \vec{L} + g_s \vec{S}),$$

where $\vec{L}$ and $\vec{S}$ are the total orbital momentum and spin of the atom, and averaging is done over a state with a given value of the total angular momentum.

If the interaction term V_M is small (less than the fine structure), it can be treated as a perturbation; this is the Zeeman effect proper. In the Paschen-Back effect, described below, V_M exceeds the LS coupling significantly (but is still small compared to H_0). In ultrastrong magnetic fields, the magnetic-field interaction may exceed H_0, in which case the atom can no longer exist in its normal meaning, and one talks about Landau levels instead. There are, of course, intermediate cases which are more complex than these limit cases.

Weak Field (Zeeman Effect)

If the spin-orbit interaction dominates over the effect of the external magnetic field, H_0 and $\vec{S}$ are not separately conserved, only the total angular momentum $\vec{J} = \vec{L} + \vec{S}$ is.

The spin and orbital angular momentum vectors can be thought of as precessing about the (fixed) total angular momentum vector $\vec{J}$. The (time-)"averaged" spin vector is then the projection of the spin onto the direction of $\vec{J}$:

$$\vec{S}_{avg} = \frac{(\vec{S}\cdot\vec{J})}{J^2}\vec{J}$$

and for the (time-)"averaged" orbital vector:

$$\vec{L}_{avg} = \frac{(\vec{L}\cdot\vec{J})}{J^2}\vec{J}.$$

Thus,

$$\langle\ V_M\ \rangle = \frac{\mu_B}{\hbar}\vec{J}\left(g_L\frac{\vec{L}\cdot\vec{J}}{J^2} + g_S\frac{\vec{S}\cdot\vec{J}}{J^2}\right)\cdot\vec{B}.$$

Using $\vec{L} = \vec{J} - \vec{S}$ and squaring both sides, we get

$$\vec{S}\cdot\vec{J} = \frac{1}{2}(J^2 + S^2 - L^2) = \frac{\hbar^2}{2}[j(j+1) - l(l+1) + s(s+1)],$$

and: using $\vec{S} = \vec{J} - \vec{L}$ and squaring both sides, we get

$$\vec{L}\cdot\vec{J} = \frac{1}{2}(J^2 - S^2 + L^2) = \frac{\hbar^2}{2}[j(j+1) + l(l+1) - s(s+1)].$$

Combining everything and taking $J_z = \hbar m_j$, we obtain the magnetic potential energy of the atom in the applied external magnetic field,

$$\begin{aligned} V_M &= \mu_B B m_j\left[g_L\frac{j(j+1)+l(l+1)-s(s+1)}{2j(j+1)} + g_S\frac{j(j+1)-l(l+1)+s(s+1)}{2j(j+1)}\right] \\ &= \mu_B B m_j\left[1 + (g_S - 1)\frac{j(j+1)-l(l+1)+s(s+1)}{2j(j+1)}\right], \\ &= \mu_B B m_j g_j \end{aligned}$$

where the quantity in square brackets is the Landé g-factor g_J of the atom ($g_L = 1$ and $g_S \approx 2$) and m_j is the z-component of the total angular momentum. For a single electron above filled shells $s = 1/2$ and $j = l \pm s$, the Landé g-factor can be simplified into:

$$g_j = 1 \pm \frac{g_S - 1}{2l+1}$$

Taking V_m to be the perturbation, the Zeeman correction to the energy is

$$E_Z^{(1)} = \left\langle nljm_j \mid H_Z^{'} \mid nljm_j \right\rangle = \langle V_M \rangle_\Psi = \mu_B g_J B_{ext} m_j$$

Example: Lyman Alpha Transition in Hydrogen

The Lyman alpha transition in hydrogen in the presence of the spin-orbit interaction involves the transitions

$$2P_{1/2} \to 1S_{1/2} \text{ and } 2P_{3/2} \to 1S_{1/2}.$$

In the presence of an external magnetic field, the weak-field Zeeman effect splits the $1S_{1/2}$ and $2P_{1/2}$ levels into 2 states each ($m_j = 1/2, -1/2$) and the $2P_{3/2}$ level into 4 states ($m_j = 3/2, 1/2, -1/2, -3/2$). The Landé g-factors for the three levels are:

$g_J = 2$ for $1S1>$ (j=1/2, l=0)

$g_J = 2/3$ for $2P_{1/2}$ (j=1/2, l=1)

$g_J = 4/3$ for $2P_{3/2}$ (j=3/2, l=1).

Note in particular that the size of the energy splitting is different for the different orbitals, because the g_J values are different. On the left, fine structure splitting is depicted. This splitting occurs even in the absence of a magnetic field, as it is due to spin-orbit coupling. Depicted on the right is the additional Zeeman splitting, which occurs in the presence of magnetic fields.

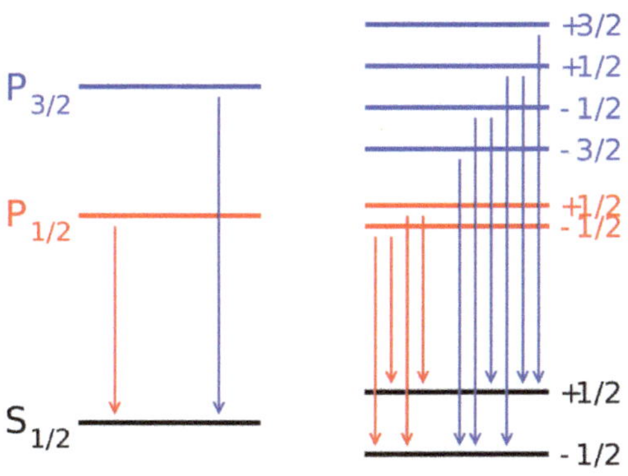

Strong Field (Paschen-Back Effect)

The Paschen-Back effect is the splitting of atomic energy levels in the presence of a strong magnetic field. This occurs when an external magnetic field is sufficiently large to disrupt the coupling between orbital ($\vec{L}$) and spin ($\vec{S}$) angular momenta. This effect is the strong-field limit of the Zeeman effect. When $s = 0$, the two effects are equivalent. The effect was named after the German physicists Friedrich Paschen and Ernst E. A. Back.

When the magnetic-field perturbation significantly exceeds the spin-orbit interaction, one can safely assume $[H_0, S] = 0.$. This allows the expectation values of L_z and S_z to be easily evaluated for a state $|\psi$. The energies are simply

$$E_z = \psi\left|H_0 + \frac{B_z \mu_B}{\hbar}(L_z + g_s S_z)\right|\psi = E_0 + B_z \mu_B (m_l + g_s m_s).$$

The above may be read as implying that the LS-coupling is completely broken by the external field. However m_l and m_s are still "good" quantum numbers. Together with the selection rules for an electric dipole transition, i.e., $\Delta s = 0, \Delta m_s = 0, \Delta l = \pm 1, \Delta m_l = 0, \pm 1$ this allows to ignore the spin degree of freedom altogether. As a result, only three spectral lines will be visible, corresponding to the $\text{Ä}m_l = 0, \pm 1$ selection rule. The splitting $\text{Ä}E = B\mu_B \text{Ä}m_l$ is *independent* of the unperturbed energies and electronic configurations of the levels being considered. It should be noted that in general (if $s \neq 0$), these three components are actually groups of several transitions each, due to the residual spin-orbit coupling.

In general, one must now add spin-orbit coupling and relativistic corrections (which are of the same order, known as 'fine structure') as a perturbation to these 'unperturbed' levels. First order perturbation theory with these fine-structure corrections yields the following formula for the Hydrogen atom in the Paschen–Back limit:

$$E_{z+fs} = E_z + \frac{\alpha^2}{2n^3}\left\{\frac{3}{4n} - \left[\frac{l(l+1) - m_l m_s}{l(l+1/2)(l+1)}\right]\right\}.$$

Intermediate Field For j = 1/2

In the magnetic dipole approximation, the Hamiltonian which includes both the hyperfine and Zeeman interactions is

$$H = hA\vec{I} \cdot \vec{J} - \vec{\mu} \cdot \vec{B}$$

$$H = hA\vec{I} \cdot \vec{J} + \mu_B (g_J \vec{J} + g_I \vec{I}) \cdot \vec{B}$$

To arrive at the Breit-Rabi formula we will include the hyperfine structure (interaction between the electron's spin and the magnetic moment of the nucleus), which is governed by the quantum number $F \equiv |\vec{F}| = |\vec{J} + \vec{I}|$, where $\vec{I}$ is the spin angular momentum operator of the nucleus. Alternatively, the derivation could be done with J only. The constant A is known as the zero field hyperfine constant and is given in units of Hertz. μ_B is the Bohr magneton. $\hbar\vec{J}$ and $\hbar\vec{I}$ are the electron and nuclear angular momentum operators. g_J and g_F can be found via a classical vector coupling model or a more detailed quantum mechanical calculation to be:

$$g_J = g_L \frac{J(J+1)+L(L+1)-S(S+1)}{2J(J+1)} + g_S \frac{J(J+1)-L(L+1)+S(S+1)}{2J(J+1)}$$

$$g_F = g_J \frac{F(F+1)+J(J+1)-I(I+1)}{2F(F+1)} + g_I \frac{F(F+1)-J(J+1)+I(I+1)}{2F(F+1)}$$

As discussed, in the case of weak magnetic fields, the Zeeman interaction can be treated as a perturbation to the $|F, m_f$ basis. In the high field regime, the magnetic field becomes so large that the Zeeman effect will dominate, and we must use a more complete basis of $|I, J, m_I, m_J$ or just $|m_I, m_J\rangle$ since I and J will be constant within a given level.

To get the complete picture, including intermediate field strengths, we must consider eigenstates which are superpositions of the $|F, m_F$ and $|m_I, m_J$ basis states. For $J = 1/2$, the Hamiltonian can be solved analytically, resulting in the Breit-Rabi formula. Notably, the electric quadrupole interaction is zero for $L = 0$ ($J = 1/2$), so this formula is fairly accurate.

To solve this system, we note that at all times, the total angular momentum projection $m_F = m_J + m_I$ will be conserved. Furthermore, since $J = 1/2$ between states m_J will change between only $\pm 1/2$. Therefore, we can define a good basis as:

$$|\pm \equiv | m_J = \pm 1/2, m_I = m_F \mp 1/2$$

We now utilize quantum mechanical ladder operators, which are defined for a general angular momentum operator L as

$$|\pm \equiv | m_J = \pm 1/2, m_I = m_F \mp 1/2$$

These ladder operators have the property

$$L_\pm | L, m_L\rangle = \sqrt{(L \mp m_L)(L \pm m_L + 1)}\, | L, m_L \pm 1\rangle$$

as long as m_L lies in the range $-L, \ldots\ldots, L$ (otherwise, they return zero). Using ladder operators $J_\pm$ and $I_\pm$ We can rewrite the Hamiltonian as

$$H = hAI_z J_z + \frac{hA}{2}(J_+ I_- + J_- I_+) + \mu_B B(g_J J_z + g_I I_Z)$$

Now we can determine the matrix elements of the Hamiltonian:

$$\langle \pm | H | \pm \rangle = -\frac{1}{4} hA + \mu_B B g_I m_F \pm \frac{1}{2}(hAm_F + \mu_B B(g_J - g_I))$$

$$\langle \pm | H | \mp \rangle = \frac{1}{2} hA \sqrt{(I + 1/2)^2 - m_F^2}$$

Solving for the eigenvalues of this matrix, (as can be done by hand, or more easily, with a computer algebra system) we arrive at the energy shifts:

$$\Delta E_{F=I\pm 1/2} = -\frac{h\Delta W}{2(2I+1)} + {}_{B}g_I m_F B \pm \frac{h\Delta W}{2}\sqrt{1+\frac{m\ x}{I+1/2}+x}$$

$$x \equiv \frac{\mu_B B(g_J - g_I)}{h\Delta W} \qquad \Delta W = A\left(I+\frac{1}{2}\right)$$

where ÄW is the splitting (in units of Hz) between two hyperfine sublevels in the absence of magnetic field B,

x is referred to as the 'field strength parameter' (Note: for $m=-(I+1/2)$ the square root is an exact square, and should be interpreted as $+(1-x)$). This equation is known as the Breit-Rabi formula and is useful for systems with one valence electron in an s ($J=1/2$) level.

Note that index F in Ä$E_{F=I\pm 1/2}$ should be considered not as total angular momentum of the atom but as *asymptotic total angular momentum*. It is equal to total angular momentum only if $B=0$ otherwise eigenvectors corresponding different eigenvalues of the Hamiltonian are the superpositions of states with different F but equal m_F (the only exceptions are $|F=I+1/2, m_F=\pm F\rangle$).

Applications

Astrophysics

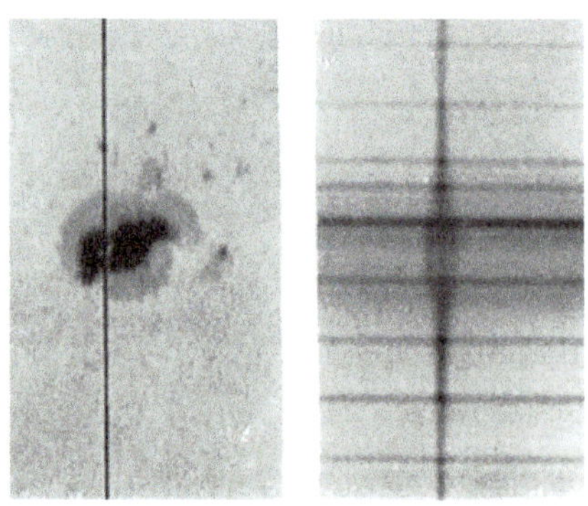

Zeeman effect on a sunspot spectral line

George Ellery Hale was the first to notice the Zeeman effect in the solar spectra, indicating the existence of strong magnetic fields in sunspots. Such fields can be quite high, on the order of 0.1 tesla or higher. Today, the Zeeman effect is used to produce magnetograms showing the variation of magnetic field on the sun.

Laser Cooling

The Zeeman effect is utilized in many laser cooling applications such as a magneto-optical trap and the Zeeman slower.

Stark Effect

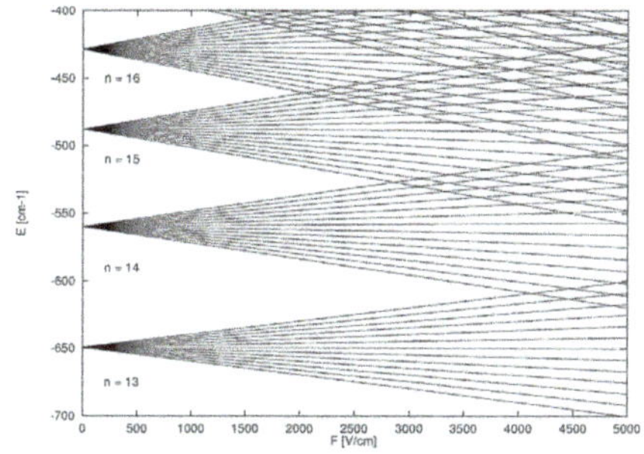

Computed regular (non-chaotic) Rydberg atom energy level spectra of hydrogen in an electric field near $n = 15$ for magnetic quantum number $m = 0$. Each n level consists of $n - 1$ degenerate sublevels; application of an electric field breaks the degeneracy. Note that energy levels can cross due to underlying symmetries of dynamical motion.

The Stark effect is the shifting and splitting of spectral lines of atoms and molecules due to presence of an external electric field. The amount of splitting or shifting is called the Stark splitting or Stark shift. In general, one distinguishes first- and second-order Stark effects. The first-order effect is linear in the applied electric field, while the second-order effect is quadratic in the field.

The Stark effect is responsible for the pressure broadening (Stark broadening) of spectral lines by charged particles. When the split/shifted lines appear in absorption, the effect is called the inverse Stark effect.

The Stark effect is the electric analogue of the Zeeman effect where a spectral line is split into several components due to the presence of a magnetic field.

The Stark effect can be explained with fully quantum-mechanical approaches, but it has also been a fertile testing ground for semiclassical methods.

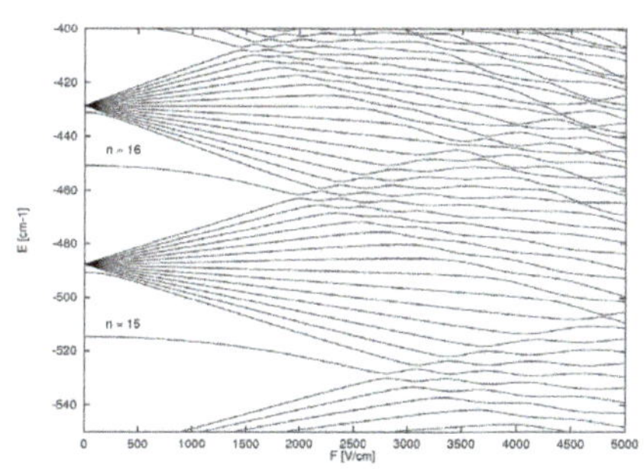

Computed chaotic Rydberg atom energy level spectra of lithium in an electric field near $n = 15$ for $m = 0$. Note that energy levels cannot cross due to the ionic core (and resulting quantum defect) breaking symmetries of dynamical motion.

History

The effect is named after Johannes Stark, who discovered it in 1913. It was independently discovered in the same year by the Italian physicist Antonino Lo Surdo, and in Italy it is thus sometimes called the Stark–Lo Surdo effect. The discovery of this effect contributed importantly to the development of quantum theory.

Inspired by the magnetic Zeeman effect, and especially by Lorentz's explanation of it, Woldemar Voigt performed classical mechanical calculations of quasi-elastically bound electrons in an electric field. By using experimental indices of refraction he gave an estimate of the Stark splittings. This estimate was a few orders of magnitude too low. Not deterred by this prediction, Stark undertook measurements on excited states of the hydrogen atom and succeeded in observing splittings.

By the use of the Bohr–Sommerfeld ("old") quantum theory, Paul Epstein and Karl Schwarzschild were independently able to derive equations for the linear and quadratic Stark effect in hydrogen. Four years later, Hendrik Kramers derived formulas for intensities of spectral transitions. Kramers also included the effect of fine structure, which includes corrections for relativistic kinetic energy and coupling between electron spin and orbit. The first quantum mechanical treatment (in the framework of Heisenberg's matrix mechanics) was by Wolfgang Pauli. Erwin Schrödinger discussed at length the Stark effect in his third paper on quantum theory (in which he introduced his perturbation theory), once in the manner of the 1916 work of Epstein (but generalized from the old to the new quantum theory) and once by his (first-order) perturbation approach. Finally, Epstein reconsidered the linear and quadratic Stark effect from the point of view of the new quantum theory. He derived equations for the line intensities which were a decided improvement over Kramers' results obtained by the old quantum theory.

While first-order perturbation effects for the Stark effect in hydrogen are in agreement for the Bohr–Sommerfeld model and the quantum-mechanical theory of the atom, higher-order effects are not. Measurements of the Stark effect under high field strengths confirmed the correctness of the quantum theory over the Bohr model.

Mechanism

Overview

An electric field pointing from left to right, for example, tends to pull nuclei to the right and electrons to the left. In another way of viewing it, if an electronic state has its electron disproportionately to the left, its energy is lowered, while if it has the electron disproportionately to the right, its energy is raised.

Other things equal, the effect of the electric field is greater for outer electron shells, because the electron is more distant from the nucleus, so it travels farther left and farther right.

The Stark effect can lead to splitting of degenerate energy levels. For example, in the Bohr model, an electron has the same energy whether it is in the 2s state or any of the 2p states. However, in an electric field, there will be hybrid orbitals (also called quantum superpositions) of the 2s and 2p states where the electron tends to be to the left, which will acquire a lower energy, and other hybrid orbitals where the electron tends to be to the right, which will acquire a higher energy. Therefore, the formerly degenerate energy levels will split into slightly lower and slightly higher energy levels.

Classical Electrostatics

The Stark effect originates from the interaction between a charge distribution (atom or molecule) and an external electric field. Before turning to quantum mechanics we describe the interaction classically and consider a continuous charge distribution ρ(r). If this charge distribution is non-polarizable its interaction energy with an external electrostatic potential V(r) is

$$E_{\text{int}} = \int \rho(\mathbf{r}) V(\mathbf{r}) d\mathbf{r}^3 .$$

If the electric field is of macroscopic origin and the charge distribution is microscopic, it is reasonable to assume that the electric field is uniform over the charge distribution. That is, V is given by a two-term Taylor expansion,

$$V(\mathbf{r}) = V(\mathbf{0}) - \sum_{i=1}^{3} r_i F_i \text{, with the electric field: } Fi \equiv -\left(\frac{\partial V}{\partial r_i}\right)\bigg|_{\mathbf{0}},$$

where we took the origin o somewhere within ρ. Setting V as the zero energy, the interaction becomes

$$E_{\text{int}} = -\sum_{i=1}^{3} F_i \int \rho(\mathbf{r}) r_i d\mathbf{r} \equiv -\sum_{i=1}^{3} F_i \mu_i = -\mathbf{F} \cdot \mu$$

Here we have introduced the dipole moment μ of ρ as an integral over the charge distribution. In case ρ consists of N point charges q_j this definition becomes a sum

$$\mu \equiv \sum_{j=1}^{N} q_j \mathbf{r}_j ..$$

Perturbation Theory

Astronomical perturbation applied to a classical hydrogen atom produces a distortion of the electron orbit in a direction perpendicular to the applied electric field. This effect can be shown without perturbation theory using the relation between the angular momentum and the Laplace–Runge–Lenz vector. Using the Laplace-Runge-Lenz approach, one can see both the transverse distortion and the usual Stark effect. The transverse distortion is not mentioned in most textbooks. This approach can also lead to an *exactly solvable* approximate model Hamiltonian for an atom in a strong oscillatory field. "There are few exactly-solvable problems in quantum mechanics, and even fewer with a time-dependent Hamiltonian."

Turning now to quantum mechanics an atom or a molecule can be thought of as a collection of point charges (electrons and nuclei), so that the second definition of the dipole applies. The interaction of atom or molecule with a uniform external field is described by the operator

$$V_{\text{int}} = -\mathbf{F} \cdot \boldsymbol{\mu}$$

This operator is used as a perturbation in first- and second-order perturbation theory to account for the first- and second-order Stark effect.

First Order

Let the unperturbed atom or molecule be in a g-fold degenerate state with orthonormal zeroth-order state functions $\psi_1^0, \ldots, \psi_g^0$. (Non-degeneracy is the special case $g = 1$). According to perturbation theory the first-order energies are the eigenvalues of the g x g matrix with general element

$$(\mathbf{V}_{\text{int}})_{kl} = \langle \psi_k^0 \| V_{\text{int}} \| \psi_l^0 \rangle = -\mathbf{F} \cdot \langle \psi_k^0 | \boldsymbol{\mu} | \psi_l^0 \rangle,, \qquad k,l,=1,.1,,g\,, \quad .$$

If $g = 1$ (as is often the case for electronic states of molecules) the first-order energy becomes proportional to the expectation (average) value of the dipole operator $\hat{\imath}$,

$$E^{(1)} = -\mathbf{F} \cdot \langle \psi_1^0 | \boldsymbol{\mu} | \psi_1^0 \rangle = -\mathbf{F} \cdot \langle \boldsymbol{\mu} \rangle.$$

Because a dipole moment is a polar vector, the diagonal elements of the perturbation matrix V_{int} vanish for systems with an inversion center (such as atoms). Molecules with an inversion center in a non-degenerate electronic state do not have a (permanent) dipole and hence do not show a linear Stark effect.

In order to obtain a non-zero matrix V_{int} for systems with an inversion center it is necessary that some of the unperturbed functions ψ_i^0 have opposite parity (obtain plus and minus under inversion), because only functions of opposite parity give non-vanishing matrix elements. Degenerate zeroth-order states of opposite parity occur for excited hydrogen-like (one-electron) atoms. Such atoms have the principal quantum number n among their quantum numbers. The excited state of hydrogen-like atoms with principal quantum number n is n^2-fold degenerate and

$$n^2 = \sum_{\ell=0}^{n-1} (2\ell + 1),$$

where ℓ is the azimuthal (angular momentum) quantum number. For instance, the excited $n = 4$ state contains the following ℓ states,

$$16 = 1+3+5+7 \quad n = 4 \text{ contains } s \oplus p \oplus d \oplus f.$$

The one-electron states with even ℓ are even under parity, while those with odd ℓ are odd under parity. Hence hydrogen-like atoms with n>1 show first-order Stark effect.

The first-order Stark effect occurs in rotational transitions of symmetric top molecules (but not for linear and asymmetric molecules). In first approximation a molecule may be seen as a rigid rotor. A symmetric top rigid rotor has the unperturbed eigenstates

$$|JKM = (D^J_{MK})^* \quad \text{with} \quad M,K = -J,-J+1,\ldots,J$$

with 2(2J+1)-fold degenerate energy for |K| > 0 and (2J+1)-fold degenerate energy for K=0. Here D^J_{MK} is an element of the Wigner D-matrix. The first-order perturbation matrix on basis of the unperturbed rigid rotor function is non-zero and can be diagonalized. This gives shifts and splittings in the rotational spectrum. Quantitative analysis of these Stark shift yields the permanent electric dipole moment of the symmetric top molecule.

Second Order

As stated, the quadratic Stark effect is described by second-order perturbation theory. The zeroth-order problems

$$H^{(0)}\psi_k^0 = E_k^{(0)}\psi_k^0, \quad k = 0,1,\ldots, \quad E_0^{(0)} < E_1^{(0)} \le E_2^{(0)},\ldots$$

are assumed to be solved. It is usual to assume that the zeroth-order state to be perturbed is non-degenerate. If we take the ground state as the non-degenerate state under consideration (for hydrogen-like atoms: n = 1), perturbation theory gives

$$E^{(2)} = \sum_{k>0} \frac{\langle \psi_0^0 | V_{\text{int}} | \psi_k^0 \rangle \langle \psi_k^0 | V_{\text{int}} | \psi_0^0 \rangle}{E_0^{(0)} - E_k^{(0)}} = -\frac{1}{2} \sum_{i,j=1}^{3} F_i \alpha_{ij} F_j$$

with the components of the polarizability tensor α defined by

$$\alpha_{ij} \equiv -2 \sum_{k>0} \frac{\langle \psi_0^0 | \mu_i | \psi_k^0 \rangle \langle \psi_k^0 | \mu_j | \psi_0^0 \rangle}{E_0^{(0)} - E_k^{(0)}}.$$

The energy E gives the quadratic Stark effect.

Because of their often spherical symmetry the polarizability tensor of atoms is often isotropic,

$$\alpha_{ij} = \alpha_0 \delta_{ij} \Rightarrow E^{(2)} = -\frac{1}{2}\alpha_0 F^2,$$

which is the quadratic Stark shift for atoms. For many molecules this expression is not too bad an approximation, because molecular tensors are often reasonably isotropic.

Problems

The perturbative treatment of the Stark effect has some problems. In the presence of an electric field, states of atoms and molecules that were previously bound (square-integrable), become formally (non-square-integrable) resonances of finite width. These resonances may decay in finite time via field ionization. For low lying states and not too strong fields the decay times are so long, however, that for all practical purposes the system can be regarded as bound. For highly excited states and/or very strong fields ionization may have to be accounted for.

Quantum-confined Stark Effect

In a semiconductor heterostructure, where a small bandgap material is sandwiched between two layers of a larger bandgap material, the Stark effect can be dramatically enhanced by bound excitons. This is because the electron and hole which form the exciton are pulled in opposite directions by the applied electric field, but they remain confined in the smaller bandgap material, so the exciton is not merely pulled apart by the field. The quantum-confined Stark effect is widely used for semiconductor-based optical modulators, particularly for optical fiber communications.

Hydrogen Spectral Series

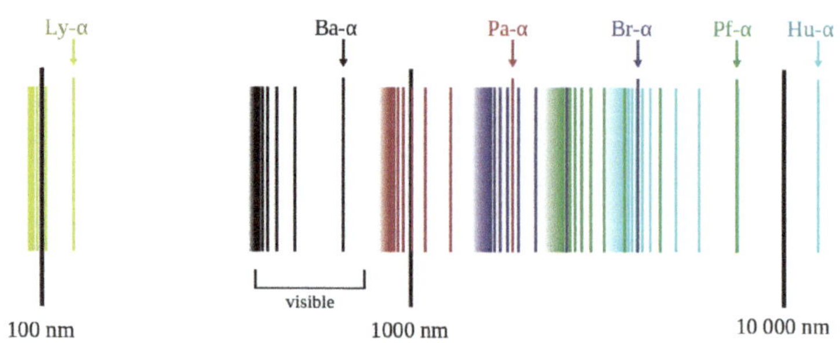

The spectral series of hydrogen, on a logarithmic scale.

The emission spectrum of atomic hydrogen is divided into a number of spectral series, with wavelengths given by the Rydberg formula. These observed spectral lines are due to the electron making transitions between two energy levels in the atom. The classification of the series by the Rydberg formula was important in the development of quantum mechanics. The spectral series are important in astronomical spectroscopy for detecting the presence of hydrogen and calculating red shifts.

Physics

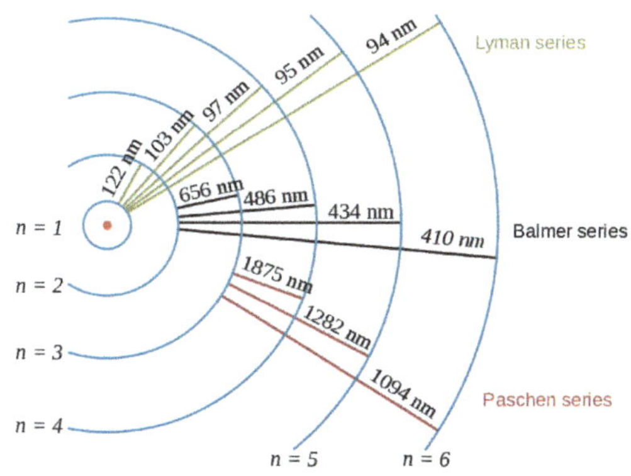

Electron transitions and their resulting wavelengths for hydrogen. Energy levels are not to scale.

A hydrogen atom consists of an electron orbiting its nucleus. The electromagnetic force between the electron and the nuclear proton leads to a set of quantum states for the electron, each with its own energy. These states were visualized by the Bohr model of the hydrogen atom as being distinct orbits around the nucleus. Each energy state, or orbit, is designated by an integer, n as shown in the figure.

Spectral emission occurs when an electron transitions, or jumps, from a higher energy state to a lower energy state. To distinguish the two states, the lower energy state is commonly designated as n', and the higher energy state is designated as n. The energy of an emitted photon corresponds to the energy difference between the two states. Because the energy of each state is fixed, the energy difference between them is fixed, and the transition will always produce a photon with the same energy.

The spectral lines are grouped into series according to n'. Lines are named sequentially starting from the longest wavelength/lowest frequency of the series, using Greek letters within each series. For example, the $2 \rightarrow 1$ line is called "Lyman-alpha" (Ly-α), while the $7 \rightarrow 3$ line is called "Paschen-delta" (Pa-δ).

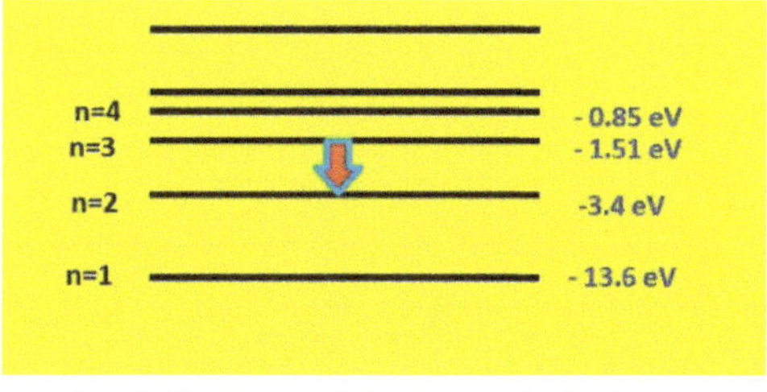

energy level diagram of electrons in hydrogen atom

There are emission lines from hydrogen that fall outside of these series, such as the 21 cm line. These emission lines correspond to much rarer atomic events such as hyperfine transitions. The fine structure also results in single spectral lines appearing as two or more closely grouped thinner lines, due to relativistic corrections.

Rydberg Formula

The energy differences between levels in the Bohr model, and hence the wavelengths of emitted/absorbed photons, is given by the Rydberg formula:

$$\frac{1}{\lambda} = RZ^2\left(\frac{1}{n'^2} - \frac{1}{n^2}\right)$$

where Z is the atomic number, i.e. the number of protons in the atomic nucleus of this element; n is the upper energy level; n' is the lower energy level; and R is the Rydberg constant (1.097373×10^7 m^{-1}). Meaningful values are returned only when n is greater than n'. Note that this equation is valid for all hydrogen-like species, i.e. atoms having only a single electron, and the particular case of hydrogen spectral lines are given by Z=1.

Series

Lyman Series (n′ = 1)

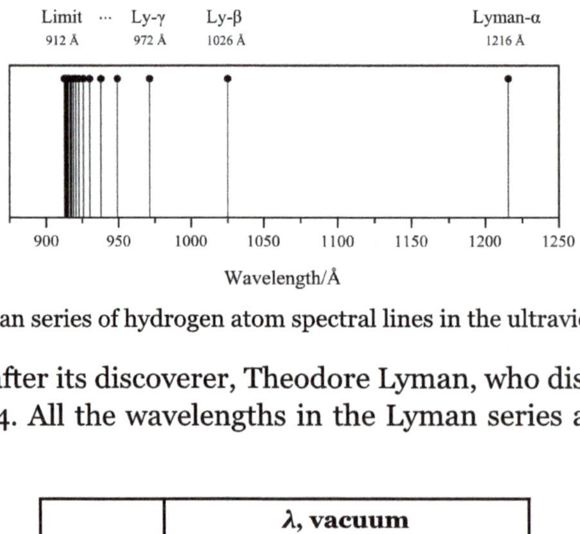

Lyman series of hydrogen atom spectral lines in the ultraviolet

The series is named after its discoverer, Theodore Lyman, who discovered the spectral lines from 1906–1914. All the wavelengths in the Lyman series are in the ultraviolet band.

n	λ, vacuum (nm)
2	121.57
3	102.57
4	97.254
5	94.974
6	93.780
∞	91.175

Balmer Series (n' = 2)

The four visible hydrogen emission spectrum lines in the Balmer series. H-alpha is the red line at the right.

Named after Johann Balmer, who discovered the Balmer formula, an empirical equation to predict the Balmer series, in 1885. Balmer lines are historically referred to as "H-alpha", "H-beta", "H-gamma" and so on, where H is the element hydrogen. Four of the Balmer lines are in the technically "visible" part of the spectrum, with wavelengths longer than 400 nm and shorter than 700 nm. Parts of the Balmer series can be seen in the solar spectrum. H-alpha is an important line used in astronomy to detect the presence of hydrogen.

n	λ, air (nm)
3	656.3
4	486.1
5	434.0
6	410.2
7	397.0
∞	364.6

Paschen Series (Bohr Series, n' = 3)

Named after the German physicist Friedrich Paschen who first observed them in 1908. The Paschen lines all lie in the infrared band. This series overlaps with the next (Brackett) series, i.e. the shortest line in the Brackett series has a wavelength that falls among the Paschen series. All subsequent series overlap.

n	λ, air (nm)
4	1875
5	1282
6	1094
7	1005
8	954.6
9	922.9
∞	820.4

Brackett Series (n′ = 4)

Named after the American physicist Frederick Sumner Brackett who first observed the spectral lines in 1922.

n	λ, air (nm)
5	4051
6	2625
7	2166
8	1944
9	1817
∞	1458

Pfund Series (n′ = 5)

Experimentally discovered in 1924 by August Herman Pfund.

n	λ, vacuum (nm)
6	7460
7	4654
8	3741
9	3297
10	3039
∞	2279

Humphreys Series (n′ = 6)

Discovered in 1953 by American physicist Curtis J. Humphreys.

n	λ, vacuum (μm)
7	12.37
8	7.503
9	5.908
10	5.129
11	4.673
∞	3.282

Further (n′ > 6)

Further series are unnamed, but follow exactly the same pattern as dictated by the Rydberg equation. Series are increasingly spread out and occur in increasing wavelengths. The lines are also increasingly faint, corresponding to increasingly rare atomic events. The seventh series of atomic hydrogen was first demonstrated experimentally at infrared wavelengths in 1972 by John Strong and Peter Hansen at the University of Massachusetts Amherst.

Extension to Other Systems

The concepts of the Rydberg formula can be applied to any system with a single particle orbiting a nucleus, for example a He^+ ion or a muonium exotic atom. The equation must be modified based on the system's Bohr radius; emissions will be of a similar character but at a different range of energies.

All other atoms possess at least two electrons in their neutral form and the interactions between these electrons makes analysis of the spectrum by such simple methods as described here impractical. The deduction of the Rydberg formula was a major step in physics, but it was long before an extension to the spectra of other elements could be accomplished.

References

- Carroll, Bradley W. (2007). An Introduction to Modern Astrophysics. CA, USA: Pearson Education. p. 256. ISBN 0-8053-0402-9.
- Philip R. Bunker; Per Jensen (2006). Molecular Symmetry and Spectroscopy. NRC Research Press. p. 414. ISBN 978-0-660-19628-2.
- Griffiths, David J. (2004). Introduction to Quantum Mechanics (2nd ed.). Prentice Hall. p. 247. ISBN 0-13-111892-7. OCLC 40251748.
- E. U. Condon & G. H. Shortley (1935). The Theory of Atomic Spectra. Cambridge University Press. ISBN 0-521-09209-4.

6

Understanding Angular Momentum Coupling

There are principally four quantum numbers for each electron orbital and the characteristic shape of each orbital reflects the electron's motion in that orbital. Angular momentum emulates the angular velocity at which an electron is moving in its orbit. This chapter investigates concepts like spin, spin quantum number, electron magnetic moment, spin-orbit interaction, J-coupling, fine structure and hyperfine structure.

Angular momentum coupling

In quantum mechanics, the procedure of constructing eigenstates of total angular momentum out of eigenstates of separate angular momenta is called angular momentum coupling. For instance, the orbit and spin of a single particle can interact through spin–orbit interaction, in which case the complete physical picture must include spin-orbit coupling. Or two charged particles, each with a well-defined angular momentum, may interact by Coulomb forces, in which case coupling of the two one-particle angular momenta to a total angular momentum is a useful step in the solution of the two-particle Schrödinger equation. In both cases the separate angular momenta are no longer constants of motion, but the sum of the two angular momenta usually still is. Angular momentum coupling in atoms is of importance in atomic spectroscopy. Angular momentum coupling of electron spins is of importance in quantum chemistry. Also in the nuclear shell model angular momentum coupling is ubiquitous.

In astronomy, spin-orbit coupling reflects the general law of conservation of angular momentum, which holds for celestial systems as well. In simple cases, the direction of the angular momentum vector is neglected, and the spin-orbit coupling is the ratio between the frequency with which a planet or other celestial body spins about its own axis to that with which it orbits another body. This is more commonly known as orbital resonance. Often, the underlying physical effects are tidal forces.

Angular Momentum Conservation

Conservation of angular momentum is the principle that the total angular momentum of a system has a constant magnitude and direction if the system is subjected to no external torque. Angular momentum is a property of a physical system that is a constant of motion (also referred to as a *conserved* property, time-independent and well-defined) in two situations:

1. The system experiences a spherically symmetric potential field.
2. The system moves (in quantum mechanical sense) in isotropic space.

In both cases the angular momentum operator commutes with the Hamiltonian of the system. By Heisenberg's uncertainty relation this means that the angular momentum and the energy (eigenvalue of the Hamiltonian) can be measured at the same time.

An example of the first situation is an atom whose electrons only experiences the Coulomb force of its atomic nucleus. If we ignore the electron-electron interaction (and other small interactions such as spin-orbit coupling), the *orbital angular momentum* l of each electron commutes with the total Hamiltonian. In this model the atomic Hamiltonian is a sum of kinetic energies of the electrons and the spherically symmetric electron-nucleus interactions. The individual electron angular momenta l_i commute with this Hamiltonian. That is, they are conserved properties of this approximate model of the atom.

An example of the second situation is a rigid rotor moving in field-free space. A rigid rotor has a well-defined, time-independent, angular momentum.

These two situations originate in classical mechanics. The third kind of conserved angular momentum, associated with spin, does not have a classical counterpart. However, all rules of angular momentum coupling apply to spin as well.

In general the conservation of angular momentum implies full rotational symmetry (described by the groups SO(3) and SU(2)) and, conversely, spherical symmetry implies conservation of angular momentum. If two or more physical systems have conserved angular momenta, it can be useful to combine these momenta to a total angular momentum of the combined system—a conserved property of the total system. The building of eigenstates of the total conserved angular momentum from the angular momentum eigenstates of the individual subsystems is referred to as *angular momentum coupling*.

Application of angular momentum coupling is useful when there is an interaction between subsystems that, without interaction, would have conserved angular momentum. By the very interaction the spherical symmetry of the subsystems is broken, but the angular momentum of the total system remains a constant of motion. Use of the latter fact is helpful in the solution of the Schrödinger equation.

Examples

As an example we consider two electrons, 1 and 2, in an atom (say the helium atom). If there is no electron-electron interaction, but only electron-nucleus interaction, the two electrons can be rotated around the nucleus independently of each other; nothing happens to their energy. Both operators, l_1 and l_2, are conserved. However, if we switch on the electron-electron interaction that depends on the distance $d(1,2)$ between the electrons, then only a simultaneous and equal rotation of the two electrons will leave $d(1,2)$ invariant. In such a case neither l_1 nor l_2 is a constant of motion in general, but the total orbital angular momentum $L = l_1 + l_2$ is. Given the eigenstates of l_1 and l_2, the construction of eigenstates of L (which still is conserved) is the *coupling of the angular momenta of electrons 1 and 2.*

The total orbital angular momentum quantum number L is restricted to integer values and must satisfy the triangular condition that $|l_1 - l_2| \leq L \leq l_1 + l_2$, such that the three nonnegative integer values could correspond to the three sides of a triangle.

In quantum mechanics, coupling also exists between angular momenta belonging to different Hilbert spaces of a single object, e.g. its spin and its orbital angular momentum. If the spin has half-integer values, such as 1/2 for an electron, then the total (orbital plus spin) angular momentum will also be restricted to half-integer values.

Reiterating slightly differently the above: one expands the quantum states of composed systems (i.e. made of subunits like two hydrogen atoms or two electrons) in basis sets which are made of tensor products of quantum states which in turn describe the subsystems individually. We assume that the states of the subsystems can be chosen as eigenstates of their angular momentum operators (and of their component along any arbitrary z axis).

The subsystems are therefore correctly described by a set of ℓ, m quantum numbers (see angular momentum for details). When there is interaction among the subsystems, the total Hamiltonian contains terms that do not commute with the angular operators acting on the subsystems only. However, these terms *do* commute with the *total* angular momentum operator. Sometimes one refers to the non-commuting interaction terms in the Hamiltonian as *angular momentum coupling terms*, because they necessitate the angular momentum coupling.

General Theory and Detailed Origin

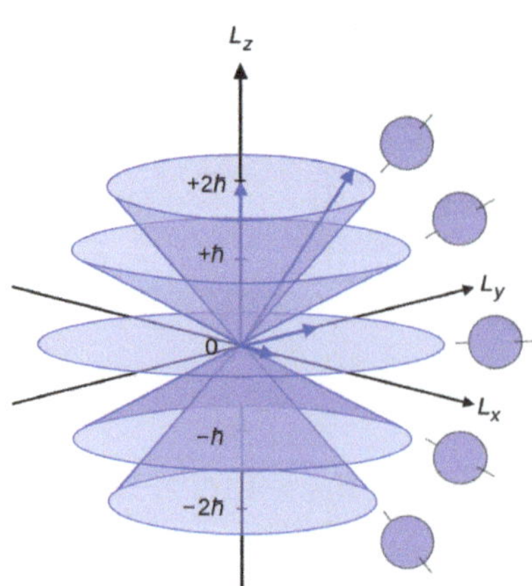

Orbital angular momentum (denoted l or L).

Spin-orbit Coupling

The behavior of atoms and smaller particles is well described by the theory of quantum mechanics, in which each particle has an intrinsic angular momentum called spin and specific configurations (of e.g. electrons in an atom) are described by a set of quantum numbers. Collections of particles also have angular momenta and corresponding quantum numbers, and under different circumstances the angular momenta of the parts couple in different ways to form the angular momentum of the whole. Angular momentum coupling is a category including some of the ways that subatomic particles can interact with each other.

In atomic physics, spin-orbit coupling, also known as spin-pairing, describes a weak magnetic interaction, or coupling, of the particle spin and the orbital motion of this particle, e.g. the electron spin and its motion around an atomic nucleus. One of its effects is to separate the energy of internal states of the atom, e.g. spin-aligned and spin-antialigned that would otherwise be identical in energy. This interaction is responsible for many of the details of atomic structure.

In the macroscopic world of orbital mechanics, the term *spin-orbit coupling* is sometimes used in the same sense as spin-orbital resonance.

LS Coupling

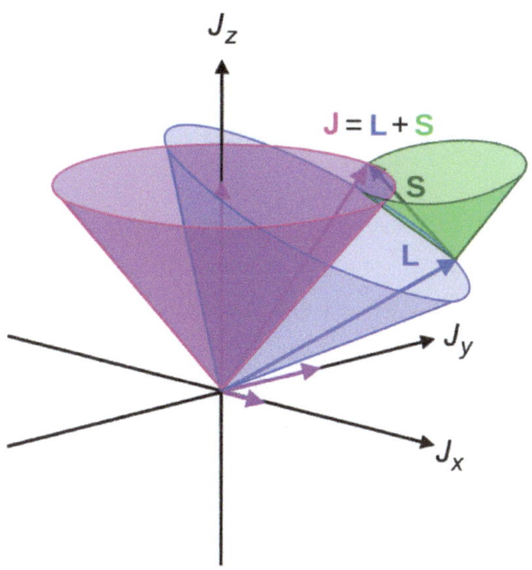

Illustration of L-S coupling. Total angular momentum J is purple, orbital L is blue, and spin S is green.

In light atoms (generally $Z \leq 30$), electron spins s_i interact among themselves so they combine to form a total spin angular momentum S. The same happens with orbital angular momenta ℓ_i, forming a total orbital angular momentum L. The interaction between the quantum numbers L and S is called *Russell–Saunders coupling* or *LS cou-*

pling. Then S and L couple together and form a total angular momentum J:

$$\mathbf{J} = \mathbf{L} + \mathbf{S},$$

where L and S are the totals:

$$\mathbf{L} = \sum_i \ell_i, \mathbf{S} = \sum_i \mathbf{s}_i.$$

This is an approximation which is good as long as any external magnetic fields are weak. In larger magnetic fields, these two momenta decouple, giving rise to a different splitting pattern in the energy levels (the Paschen–Back effect.), and the size of LS coupling term becomes small.

Jj Coupling

In heavier atoms the situation is different. In atoms with bigger nuclear charges, spin-orbit interactions are frequently as large as or larger than spin-spin interactions or orbit-orbit interactions. In this situation, each orbital angular momentum ℓ_i tends to combine with the corresponding individual spin angular momentum s_i, originating an individual total angular momentum j_i. These then couple up to form the total angular momentum J

$$\mathbf{J} = \sum_i \mathbf{j}_i = \sum_i (\ell_i + \mathbf{s}_i).$$

This description, facilitating calculation of this kind of interaction, is known as *jj coupling*.

Spin-spin Coupling

Spin-spin coupling is the coupling of the intrinsic angular momentum (spin) of different particles. Such coupling between pairs of nuclear spins is an important feature of nuclear magnetic resonance (NMR) spectroscopy as it can provide detailed information about the structure and conformation of molecules. Spin-spin coupling between nuclear spin and electronic spin is responsible for hyperfine structure in atomic spectra.

Term Symbols

Term symbols are used to represent the states and spectral transitions of atoms, they are found from coupling of angular momenta mentioned above. When the state of an atom has been specified with a term symbol, the allowed transitions can be found through selection rules by considering which transitions would conserve angular momentum. A photon has spin 1, and when there is a transition with emission or absorption of a photon the atom will need to change state to conserve angular momentum. The term symbol selection rules are. $\Delta S = 0$, $\Delta L = 0, \pm 1$, $\Delta l = \pm 1$, $\Delta J = 0, \pm 1$

The expression "term symbol" is derived from the "term series" associated with the Rydberg states of an atom and their energy levels. In the Rydberg formula the frequency or wave number of the light emitted by a hydrogen-like atom is proportional to the difference between the two terms of a transition. The series known to early spectroscopy were designated *sharp*, *principal*, *diffuse* and *fundamental* and consequently the letters S, P, D, and F were used to represent the orbital angular momentum states of an atom.

Relativistic Effects

In very heavy atoms, relativistic shifting of the energies of the electron energy levels accentuates spin-orbit coupling effect. Thus, for example, uranium molecular orbital diagrams must directly incorporate relativistic symbols when considering interactions with other atoms.

Nuclear Coupling

In atomic nuclei, the spin-orbit interaction is much stronger than for atomic electrons, and is incorporated directly into the nuclear shell model. In addition, unlike atomic-electron term symbols, the lowest energy state is not $L - S$, but rather, $\ell + s$. All nuclear levels whose ℓ value (orbital angular momentum) is greater than zero are thus split in the shell model to create states designated by $\ell + s$ and $\ell - s$. Due to the nature of the shell model, which assumes an average potential rather than a central Coulombic potential, the nucleons that go into the $\ell + s$ and $\ell - s$ nuclear states are considered degenerate within each orbital (e.g. The $2p3/2$ contains four nucleons, all of the same energy. Higher in energy is the $2p1/2$ which contains two equal-energy nucleons).

Spin (Physics)

In quantum mechanics and particle physics, spin is an intrinsic form of angular momentum carried by elementary particles, composite particles (hadrons), and atomic nuclei.

Spin is one of two types of angular momentum in quantum mechanics, the other being *orbital angular momentum*. The orbital angular momentum operator is the quantum-mechanical counterpart to the classical angular momentum of orbital revolution: it arises when a particle executes a rotating or twisting trajectory (such as when an electron orbits a nucleus). The existence of spin angular momentum is inferred from experiments, such as the Stern–Gerlach experiment, in which particles are observed to possess angular momentum that cannot be accounted for by orbital angular momentum alone.

In some ways, spin is like a vector quantity; it has a definite magnitude, and it has a "direction" (but quantization makes this "direction" different from the direction of an ordinary vector). All elementary particles of a given kind have the same magnitude of spin angular momentum, which is indicated by assigning the particle a *spin quantum number*.

The SI unit of spin is the (N·m·s) or ($kg \cdot m^2 \cdot s^{-1}$), just as with classical angular momentum. In practice, spin is given as a dimensionless *spin quantum number* by dividing the spin angular momentum by the reduced Planck constant $\hbar$, which has the same units of angular momentum. Very often, the "spin quantum number" is simply called "spin" leaving its meaning as the unitless "spin quantum number" to be inferred from context.

When combined with the spin-statistics theorem, the spin of electrons results in the Pauli exclusion principle, which in turn underlies the periodic table of chemical elements.

Wolfgang Pauli was the first to propose the concept of spin, but he did not name it. In 1925, Ralph Kronig, George Uhlenbeck and Samuel Goudsmit at Leiden University suggested an (erroneous) physical interpretation of particles spinning around their own axis. The mathematical theory was worked out in depth by Pauli in 1927. When Paul Dirac derived his relativistic quantum mechanics in 1928, electron spin was an essential part of it.

Quantum Number

As the name suggests, spin was originally conceived as the rotation of a particle around some axis. This picture is correct so far as spin obeys the same mathematical laws as quantized angular momenta do. On the other hand, spin has some peculiar properties that distinguish it from orbital angular momenta:

- Spin quantum numbers may take half-integer values.
- Although the direction of its spin can be changed, an elementary particle cannot be made to spin faster or slower.
- The spin of a charged particle is associated with a magnetic dipole moment with a g-factor differing from 1. This could only occur classically if the internal charge of the particle were distributed differently from its mass.

The conventional definition of the spin quantum number, s, is $s = n/2$, where n can be any non-negative integer. Hence the allowed values of s are 0, 1/2, 1, 3/2, 2, etc. The value of s for an elementary particle depends only on the type of particle, and cannot be altered in any known way (in contrast to the *spin direction* described below). The spin angular momentum, S, of any physical system is quantized. The allowed values of S are

$$S = \hbar\sqrt{s(s+1)} = \frac{h}{4\pi}\sqrt{n(n+2)},$$

where h is the Planck constant and $\hbar = h/2\pi$ is the reduced Planck constant. In contrast, orbital angular momentum can only take on integer values of s; i.e., even-numbered values of n.

Fermions and Bosons

Those particles with half-integer spins, such as 1/2, 3/2, 5/2, are known as fermions, while those particles with integer spins, such as 0, 1, 2, are known as bosons. The two families of particles obey different rules and *broadly* have different roles in the world around us. A key distinction between the two families is that fermions obey the Pauli exclusion principle; that is, there cannot be two identical fermions simultaneously having the same quantum numbers (meaning, roughly, having the same position, velocity and spin direction). In contrast, bosons obey the rules of Bose–Einstein statistics and have no such restriction, so they may "bunch together" even if in identical states. Also, composite particles can have spins different from their component particles. For example, a helium atom in the ground state has spin 0 and behaves like a boson, even though the quarks and electrons which make it up are all fermions.

This has profound practical applications:

- Quarks and leptons (including electrons and neutrinos), which make up what is classically known as matter, are all fermions with spin 1/2. The common idea that "matter takes up space" actually comes from the Pauli exclusion principle acting on these particles to prevent the fermions that make up matter from being in the same quantum state. Further compaction would require electrons to occupy the same energy states, and therefore a kind of pressure (sometimes known as degeneracy pressure of electrons) acts to resist the fermions being overly close. It is also this pressure which prevents stars collapsing inwardly, and which, when it finally gives way under immense gravitational pressure in a dying massive star, triggers inward collapse into a black hole.

  Elementary fermions with other spins (3/2, 5/2, etc.) are not known to exist, as of 2014.

- Elementary particles which are thought of as carrying forces are all bosons with spin 1. They include the photon which carries the electromagnetic force, the gluon (strong force), and the W and Z bosons (weak force). The ability of bosons to occupy the same quantum state is used in the laser, which aligns many photons having the same quantum number (the same direction and frequency), superfluid liquid helium resulting from helium-4 atoms being bosons, and superconductivity where pairs of electrons (which individually are fermions) act as single composite bosons.

Elementary bosons with other spins (0, 2, 3 etc.) were not historically known to exist, although they have received considerable theoretical treatment and are well established within their respective mainstream theories. In particular, theoreticians have proposed the graviton (predicted to exist by some quantum gravity theories) with spin 2, and the Higgs boson (explaining electroweak symmetry breaking) with spin 0. Since 2013, the Higgs boson with spin 0 has been considered proven to exist. It is the first scalar particle (spin 0) known to exist in nature.

Theoretical and experimental studies have shown that the spin possessed by elementary particles cannot be explained by postulating that they are made up of even smaller particles rotating about a common center of mass analogous to a classical electron radius; as far as can be determined at present, these elementary particles have no inner structure. The spin of an elementary particle is therefore seen as a truly intrinsic physical property, akin to the particle's electric charge and rest mass.

Spin-statistics Theorem

The proof that particles with half-integer spin (fermions) obey Fermi–Dirac statistics and the Pauli Exclusion Principle, and particles with integer spin (bosons) obey Bose–Einstein statistics, occupy "symmetric states", and thus can share quantum states, is known as the spin-statistics theorem. The theorem relies on both quantum mechanics and the theory of special relativity, and this connection between spin and statistics has been called "one of the most important applications of the special relativity theory".

Magnetic Moments

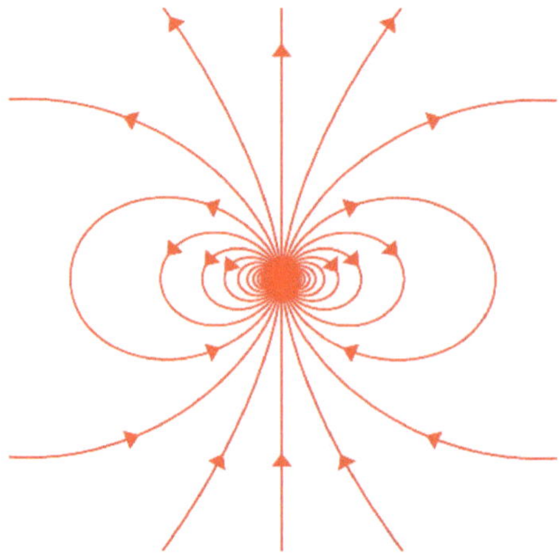

Schematic diagram depicting the spin of the neutron as the black arrow and magnetic field lines associated with the neutron magnetic moment. The neutron has a negative magnetic moment. While the spin of the neutron is upward in this diagram, the magnetic field lines at the center of the dipole are downward.

Particles with spin can possess a magnetic dipole moment, just like a rotating electrically charged body in classical electrodynamics. These magnetic moments can be experimentally observed in several ways, e.g. by the deflection of particles by inhomogeneous magnetic fields in a Stern–Gerlach experiment, or by measuring the magnetic fields generated by the particles themselves.

The intrinsic magnetic moment μ of a spin 1/2 particle with charge q, mass m, and spin angular momentum S, is

$$\text{ì} = \frac{g_s q}{2m}\mathbf{S}$$

where the dimensionless quantity g_s is called the spin g-factor. For exclusively orbital rotations it would be 1 (assuming that the mass and the charge occupy spheres of equal radius).

The electron, being a charged elementary particle, possesses a nonzero magnetic moment. One of the triumphs of the theory of quantum electrodynamics is its accurate prediction of the electron g-factor, which has been experimentally determined to have the value −2.0023193043622(15), with the digits in parentheses denoting measurement uncertainty in the last two digits at one standard deviation. The value of 2 arises from the Dirac equation, a fundamental equation connecting the electron's spin with its electromagnetic properties, and the correction of 0.002319304... arises from the electron's interaction with the surrounding electromagnetic field, including its own field. Composite particles also possess magnetic moments associated with their spin. In particular, the neutron possesses a non-zero magnetic moment despite being electrically neutral. This fact was an early indication that the neutron is not an elementary particle. In fact, it is made up of quarks, which are electrically charged particles. The magnetic moment of the neutron comes from the spins of the individual quarks and their orbital motions.

Neutrinos are both elementary and electrically neutral. The minimally extended Standard Model that takes into account non-zero neutrino masses predicts neutrino magnetic moments of:

$$\mu_\nu \approx 3\times10^{-19}\,\mu_\mathrm{B}\frac{m_\nu}{\mathrm{eV}}$$

where the μ_ν are the neutrino magnetic moments, m_ν are the neutrino masses, and μ_B is the Bohr magneton. New physics above the electroweak scale could, however, lead to significantly higher neutrino magnetic moments. It can be shown in a model independent way that neutrino magnetic moments larger than about $10^{-14}\,\mu_\mathrm{B}$ are unnatural, because they would also lead to large radiative contributions to the neutrino mass. Since the neutrino masses cannot exceed about 1 eV, these radiative corrections must then be assumed to be fine tuned to cancel out to a large degree.

The measurement of neutrino magnetic moments is an active area of research. As of

2001, the latest experimental results have put the neutrino magnetic moment at less than 1.2×10^{-10} times the electron's magnetic moment.

In ordinary materials, the magnetic dipole moments of individual atoms produce magnetic fields that cancel one another, because each dipole points in a random direction. Ferromagnetic materials below their Curie temperature, however, exhibit magnetic domains in which the atomic dipole moments are locally aligned, producing a macroscopic, non-zero magnetic field from the domain. These are the ordinary "magnets" with which we are all familiar.

In paramagnetic materials, the magnetic dipole moments of individual atoms spontaneously align with an externally applied magnetic field. In diamagnetic materials, on the other hand, the magnetic dipole moments of individual atoms spontaneously align oppositely to any externally applied magnetic field, even if it requires energy to do so.

The study of the behavior of such "spin models" is a thriving area of research in condensed matter physics. For instance, the Ising model describes spins (dipoles) that have only two possible states, up and down, whereas in the Heisenberg model the spin vector is allowed to point in any direction. These models have many interesting properties, which have led to interesting results in the theory of phase transitions.

Direction

Spin Projection Quantum Number and Multiplicity

In classical mechanics, the angular momentum of a particle possesses not only a magnitude (how fast the body is rotating), but also a direction (either up or down on the axis of rotation of the particle). Quantum mechanical spin also contains information about direction, but in a more subtle form. Quantum mechanics states that the component of angular momentum measured along any direction can only take on the values

$$S_i = \hbar s_i, \quad s_i \in \{-s, -(s-1), \ldots, s-1, s\}$$

where S_i is the spin component along the i-axis (either x, y, or z), s_i is the spin projection quantum number along the i-axis, and s is the principal spin quantum number (discussed in the previous section). Conventionally the direction chosen is the z-axis:

$$S_z = \hbar s_z, \quad s_z \in \{-s, -(s-1), \ldots, s-1, s\}$$

where S_z is the spin component along the z-axis, s_z is the spin projection quantum number along the z-axis.

One can see that there are $2s + 1$ possible values of s_z. The number "$2s + 1$" is the multiplicity of the spin system. For example, there are only two possible values for a spin-1/2 particle: $s_z = +1/2$ and $s_z = -1/2$. These correspond to quantum states in which the spin is pointing in the +z or −z directions respectively, and are often referred to as "spin up" and "spin down". For a spin-3/2 particle, like a delta baryon, the possible values are +3/2, +1/2, −1/2, −3/2.

Vector

A single point in space can spin continuously without becoming tangled. Notice that after a 360-degree rotation, the spiral flips between clockwise and counterclockwise orientations. It returns to its original configuration after spinning a full 720 degrees.

For a given quantum state, one could think of a spin vector S whose components are the expectation values of the spin components along each axis, i.e., $S=[S_x,S_y,S_z]$. This vector then would describe the "direction" in which the spin is pointing, corresponding to the classical concept of the axis of rotation. It turns out that the spin vector is not very useful in actual quantum mechanical calculations, because it cannot be measured directly: s_x, s_y and s_z cannot possess simultaneous definite values, because of a quantum uncertainty relation between them. However, for statistically large collections of particles that have been placed in the same pure quantum state, such as through the use of a Stern–Gerlach apparatus, the spin vector does have a well-defined experimental meaning: It specifies the direction in ordinary space in which a subsequent detector must be oriented in order to achieve the maximum possible probability (100%) of detecting every particle in the collection. For spin-1/2 particles, this maximum probability drops off smoothly as the angle between the spin vector and the detector increases, until at an angle of 180 degrees—that is, for detectors oriented in the opposite direction to the spin vector—the expectation of detecting particles from the collection reaches a minimum of 0%.

As a qualitative concept, the spin vector is often handy because it is easy to picture classically. For instance, quantum mechanical spin can exhibit phenomena analogous to classical gyroscopic effects. For example, one can exert a kind of "torque" on an electron by putting it in a magnetic field (the field acts upon the electron's intrinsic magnetic dipole moment—see the following section). The result is that the spin vector undergoes precession, just like a classical gyroscope. This phenomenon is known as electron spin resonance (ESR). The equivalent behaviour of protons in atomic nuclei is used in nuclear magnetic resonance (NMR) spectroscopy and imaging.

Mathematically, quantum mechanical spin states are described by vector-like objects known as spinors. There are subtle differences between the behavior of spinors and vectors under coordinate rotations. For example, rotating a spin-1/2 particle by 360 degrees does not bring it back to the same quantum state, but to the state with the opposite quantum phase; this is detectable, in principle, with interference experiments. To return the particle to its exact original state, one needs a 720-degree rotation. (The Plate trick and Mobius strip give non-quantum analogies.) A spin-zero particle can only have a single quantum state, even after torque is applied. Rotating a spin-2 particle 180 degrees can bring it back to the same quantum state and a spin-4 particle should be rotated 90 degrees to bring it back to the same quantum state. The spin 2 particle can be analogous to a straight stick that looks the same even after it is rotated 180 degrees and a spin 0 particle can be imagined as sphere which looks the same after whatever angle it is turned through.

Mathematical Formulation

Operator

Spin obeys commutation relations analogous to those of the orbital angular momentum:

$$[S_i, S_j] = i\hbar\varepsilon_{ijk}S_k$$

where ε_{ijk} is the Levi-Civita symbol, and the expression is summed over the kth index. It follows (as with angular momentum) that the eigenvectors of S^2 and S_z (expressed as kets in the total S basis) are:

$$S^2 \mid s, m_s\rangle = \hbar^2 s(s+1) \mid s, m_s\rangle S_z \mid s, m_s\rangle = \hbar m_s \mid s, m_s\rangle.$$

The spin raising and lowering operators acting on these eigenvectors give:

$$S_\pm \mid s, m_s\rangle = \hbar\sqrt{s(s+1) - m_s(m_s \pm 1)} \mid s, m_s \pm 1\rangle$$

where $S_\pm = S_x \pm i\, S_y$.

But unlike orbital angular momentum the eigenvectors are not spherical harmonics. They are not functions of θ and φ. There is also no reason to exclude half-integer values of s and m_s.

In addition to their other properties, all quantum mechanical particles possess an intrinsic spin (though it may have the intrinsic spin 0, too). The spin is quantized in units of the reduced Planck constant, such that the state function of the particle is, say, not $\psi = \psi(\mathrm{r})$, but $\psi = \psi(\mathrm{r},\sigma)$ where σ is out of the following discrete set of values:

$$\sigma \in \{-s\hbar, -(s-1)\hbar, \cdots, +(s-1)\hbar, +s\hbar\}.$$

One distinguishes bosons (integer spin) and fermions (half-integer spin). The total angular momentum conserved in interaction processes is then the sum of the orbital angular momentum and the spin.

Pauli Matrices

The quantum mechanical operators associated with spin-1/2 observables are:

$$S = \frac{\hbar}{2}\acute{\mathbf{o}}$$

where in Cartesian components:

$$S_x = \frac{\hbar}{2}\sigma_x, \quad S_y = \frac{\hbar}{2}\sigma_y, \quad S_z = \frac{\hbar}{2}\sigma_z.$$

For the special case of spin-1/2 particles, σ_x, σ_y and σ_z are the three Pauli matrices, given by:

$$\sigma_x = \begin{pmatrix} 0 & 1 \\ 1 & 0 \end{pmatrix} \quad \sigma_y = \begin{pmatrix} 0 & -i \\ i & 0 \end{pmatrix} \quad \sigma_z = \begin{pmatrix} 1 & 0 \\ 0 & -1 \end{pmatrix}.$$

Pauli Exclusion Principle

For systems of N identical particles this is related to the Pauli exclusion principle, which states that by interchanges of any two of the N particles one must have

$$\psi(\cdots \mathbf{r}_i, \sigma_i \cdots \mathbf{r}_j, \sigma_j \cdots) = (-1)^{2s}\psi(\cdots \mathbf{r}_j, \sigma_j \cdots \mathbf{r}_i, \sigma_i \cdots).$$

Thus, for bosons the prefactor $(-1)^{2s}$ will reduce to +1, for fermions to −1. In quantum mechanics all particles are either bosons or fermions. In some speculative relativistic quantum field theories "supersymmetric" particles also exist, where linear combinations of bosonic and fermionic components appear. In two dimensions, the prefactor $(-1)^{2s}$ can be replaced by any complex number of magnitude 1 such as in the anyon.

The above permutation postulate for N-particle state functions has most-important consequences in daily life, e.g. the periodic table of the chemical elements.

Rotations

As described above, quantum mechanics states that components of angular momentum measured along any direction can only take a number of discrete values. The most convenient quantum mechanical description of particle's spin is therefore with a set of complex numbers corresponding to amplitudes of finding a given value of projection of its intrinsic angular momentum on a given axis. For instance, for a spin-1/2 particle, we would need two numbers $a_{\pm 1/2}$, giving amplitudes of finding it with projection of angular momentum equal to $\hbar/2$ and $-\hbar/2$, satisfying the requirement

$$\left|a_{\frac{1}{2}}\right|^2 + \left|a_{-\frac{1}{2}}\right|^2 = 1.$$

For a generic particle with spin s, we would need $2s + 1$ such parameters. Since these numbers depend on the choice of the axis, they transform into each other non-trivially when this axis is rotated. It's clear that the transformation law must be linear, so we can represent it by associating a matrix with each rotation, and the product of two transformation matrices corresponding to rotations A and B must be equal (up to phase) to the matrix representing rotation AB. Further, rotations preserve the quantum mechanical inner product, and so should our transformation matrices:

$$\sum_{m=-j}^{j} a_m^* b_m = \sum_{m=-j}^{j}\left(\sum_{n=-j}^{j} U_{nm} a_n\right)^*\left(\sum_{k=-j}^{j} U_{km} b_k\right)$$

$$\sum_{n=-j}^{j}\sum_{k=-j}^{j} U_{np}^* U_{kq} = \delta_{pq}.$$

Mathematically speaking, these matrices furnish a unitary projective representation of the rotation group SO(3). Each such representation corresponds to a representation of the covering group of SO(3), which is SU(2). There is one n-dimensional irreducible representation of SU(2) for each dimension, though this representation is n-dimensional real for odd n and n-dimensional complex for even n (hence of real dimension $2n$). For a rotation by angle θ in the plane with normal vector $\hat{\theta}$, U can be written

$$U = e^{-\frac{i}{\hbar}\theta\cdot S},$$

where , $\theta = \theta\hat{\theta}$ and S is the vector of spin operators.

Working in the coordinate system where $\hat{\theta} = \hat{z}$, we would like to show that S_x and S_y are rotated into each other by the angle θ. Starting with S_x. Using units where $\hbar = 1$:

$$S_x \to U^\dagger S_x U \qquad = e^{i\theta S_z} S_x e^{-i\theta S_z}$$

$$= S_x + (i\theta)[S_z, S_x] + \left(\frac{1}{2!}\right)(i\theta)^2[S_z,[S_z,S_x]] + \left(\frac{1}{3!}\right)(i\theta)^3[S_z,[S_z,[S_z,S_x]]] + \cdots$$

Using the spin operator commutation relations, we see that the commutators evaluate to $i\,S_y$ for the odd terms in the series, and to S_x for all of the even terms. Thus:

$$U^{\dagger}S_xU \;= S_x\left[1-\frac{\theta^2}{2!}+\ldots\right]-S_y\left[\theta-\frac{\theta^3}{3!}\cdots\right]$$
$$= S_x\cos\theta - S_y\sin\theta$$

as expected. Note that since we only relied on the spin operator commutation relations, this proof holds for any dimension (i.e., for any principal spin quantum number s).

A generic rotation in 3-dimensional space can be built by compounding operators of this type using Euler angles:

$$\mathcal{R}(\alpha,\beta,\gamma)=e^{-i\alpha S_z}e^{-i\beta S_y}e^{-i\gamma S_z}$$

An irreducible representation of this group of operators is furnished by the Wigner D-matrix:

$$D^{s}_{m'm}(\alpha,\beta,\gamma)\equiv sm'\,|\,\mathcal{R}(\alpha,\beta,\gamma)\,|\,sm = e^{-im'\alpha}d^{s}_{m'm}(\beta)e^{-im\gamma},$$

where

$$d^{s}_{m'm}(\beta)=sm'\,|\,e^{-i\beta s_y}\,|\,sm$$

is Wigner's small d-matrix. Note that for $\gamma = 2\pi$ and $\alpha = \beta = 0$; i.e., a full rotation about the z-axis, the Wigner D-matrix elements become

$$D^{s}_{m'm}(0,0,2\pi)=d^{s}_{m'm}(0)e^{-im2\pi}=\delta_{m'm}(-1)^{2m}.$$

Recalling that a generic spin state can be written as a superposition of states with definite m, we see that if s is an integer, the values of m are all integers, and this matrix corresponds to the identity operator. However, if s is a half-integer, the values of m are also all half-integers, giving $(-1)^{2m} = -1$ for all m, and hence upon rotation by 2π the state picks up a minus sign. This fact is a crucial element of the proof of the spin-statistics theorem.

Lorentz Transformations

We could try the same approach to determine the behavior of spin under general Lorentz transformations, but we would immediately discover a major obstacle. Unlike SO(3), the group of Lorentz transformations SO(3,1) is non-compact and therefore does not have any faithful, unitary, finite-dimensional representations.

In case of spin-1/2 particles, it is possible to find a construction that includes both a finite-dimensional representation and a scalar product that is preserved by this representation. We associate a 4-component Dirac spinor ψ with each particle. These spinors transform under Lorentz transformations according to the law

$$\psi' = \exp\left(\tfrac{1}{8}\omega_{\mu\nu}[\gamma_\mu, \gamma_\nu]\right)\psi$$

where γ_ν are gamma matrices and $\omega_{\mu\nu}$ is an antisymmetric 4 × 4 matrix parametrizing the transformation. It can be shown that the scalar product

$$\langle\psi \mid \phi\rangle = \bar{\psi}\phi = \psi^\dagger\gamma_0\phi$$

is preserved. It is not, however, positive definite, so the representation is not unitary.

Measurement of Spin Along the x-, y-, or z-axes

Each of the (Hermitian) Pauli matrices has two eigenvalues, +1 and −1. The corresponding normalized eigenvectors are:

$$\psi_{x+} = \frac{1}{\sqrt{2}}\begin{pmatrix}1\\1\end{pmatrix}, \quad \psi_{x-} = \frac{1}{\sqrt{2}}\begin{pmatrix}1\\-1\end{pmatrix},$$

$$\psi_{y+} = \frac{1}{\sqrt{2}}\begin{pmatrix}1\\i\end{pmatrix}, \quad \psi_{y-} = \frac{1}{\sqrt{2}}\begin{pmatrix}1\\-i\end{pmatrix},$$

$$\psi_{z+} = \begin{pmatrix}1\\0\end{pmatrix}, \quad \psi_{z-} = \begin{pmatrix}0\\1\end{pmatrix}.$$

By the postulates of quantum mechanics, an experiment designed to measure the electron spin on the x-, y-, or z-axis can only yield an eigenvalue of the corresponding spin operator (S_x, S_y or S_z) on that axis, i.e. $\hbar/2$ or $-\hbar/2$. The quantum state of a particle (with respect to spin), can be represented by a two component spinor:

$$\psi = \begin{pmatrix}a+bi\\c+di\end{pmatrix}.$$

When the spin of this particle is measured with respect to a given axis (in this example, the x-axis), the probability that its spin will be measured as $\hbar/2$ is just $\left|\psi_{x+} \mid \psi\right|^2$. Correspondingly, the probability that its spin will be measured as $-\hbar/2$ is just $\left|\langle\psi_{x-} \mid \psi\rangle\right|^2$. Following the measurement, the spin state of the particle will collapse into the corresponding eigenstate. As a result, if the particle's spin along a given axis has been measured to have a given eigenvalue, all measurements will yield the same eigenvalue (since $\left|\langle\psi_{x+} \mid \psi_{x+}\rangle\right|^2 = 1$, , etc), provided that no measurements of the spin are made along other axes.

Measurement of Spin Along an Arbitrary Axis

The operator to measure spin along an arbitrary axis direction is easily obtained from the Pauli spin matrices. Let $u = (u_x, u_y, u_z)$ be an arbitrary unit vector. Then the operator for spin in this direction is simply

$$S_u = \frac{\hbar}{2}(u_x\sigma_x + u_y\sigma_y + u_z\sigma_z).$$

The operator S_u has eigenvalues of $\pm\hbar/2$, just like the usual spin matrices. This method of finding the operator for spin in an arbitrary direction generalizes to higher spin states, one takes the dot product of the direction with a vector of the three operators for the three *x*-, *y*-, *z*-axis directions.

A normalized spinor for spin-1/2 in the (u_x, u_y, u_z) direction (which works for all spin states except spin down where it will give 0/0), is:

$$\frac{1}{\sqrt{2+2u_z}}\begin{pmatrix} 1+u_z \\ u_x + iu_y \end{pmatrix}.$$

The above spinor is obtained in the usual way by diagonalizing the σ_u matrix and finding the eigenstates corresponding to the eigenvalues. In quantum mechanics, vectors are termed "normalized" when multiplied by a normalizing factor, which results in the vector having a length of unity.

Compatibility of Spin Measurements

Since the Pauli matrices do not commute, measurements of spin along the different axes are incompatible. This means that if, for example, we know the spin along the *x*-axis, and we then measure the spin along the *y*-axis, we have invalidated our previous knowledge of the *x*-axis spin. This can be seen from the property of the eigenvectors (i.e. eigenstates) of the Pauli matrices that:

$$\left|\psi_{x\pm} \mid \psi_{y\pm}\right|^2 = \left|\psi_{x\pm} \mid \psi_{z\pm}\right|^2 = \left|\psi_{y\pm} \mid \psi_{z\pm}\right|^2 = \frac{1}{2}.$$

So when physicists measure the spin of a particle along the *x*-axis as, for example, $\hbar/2$, the particle's spin state collapses into the eigenstate $\mid\psi_{x+}$. When we then subsequently measure the particle's spin along the *y*-axis, the spin state will now collapse into either $\mid\psi_{y+}$ or $\mid\psi_{y-}\rangle$,, each with probability 1/2. Let us say, in our example, that we measure $-\hbar/2$. When we now return to measure the particle's spin along the *x*-axis again, the probabilities that we will measure $\hbar/2$ or $-\hbar/2$ are each 1/2 (i.e. they are $\left|\langle\psi_{x+} \mid \psi_{y-}\rangle\right|^2$ and $\left|\langle\psi_{x-} \mid \psi_{y-}\rangle\right|^2$ respectively). This implies that the original measurement of the spin along the x-axis is no longer valid, since the spin along the *x*-axis will now be measured to have either eigenvalue with equal probability.

Higher Spins

The spin-1/2 operator $S = \hbar/2\sigma$ form the fundamental representation of SU(2). By taking Kronecker products of this representation with itself repeatedly, one may construct all higher irreducible representations. That is, the resulting spin operators for higher spin systems in three spatial dimensions, for arbitrarily large s, can be calculated using this spin operator and ladder operators.

The resulting spin matrices for spin 1 are:

$$S_x = \frac{\hbar}{\sqrt{2}}\begin{pmatrix} 0 & 1 & 0 \\ 1 & 0 & 1 \\ 0 & 1 & 0 \end{pmatrix} S_y = \frac{\hbar}{\sqrt{2}}\begin{pmatrix} 0 & -i & 0 \\ i & 0 & -i \\ 0 & i & 0 \end{pmatrix} S_z = \hbar\begin{pmatrix} 1 & 0 & 0 \\ 0 & 0 & 0 \\ 0 & 0 & -1 \end{pmatrix}$$

for spin 3/2 they are

$$S_x = \frac{\hbar}{2}\begin{pmatrix} 0 & \sqrt{3} & 0 & 0 \\ \sqrt{3} & 0 & 2 & 0 \\ 0 & 2 & 0 & \sqrt{3} \\ 0 & 0 & \sqrt{3} & 0 \end{pmatrix} S_y = \frac{\hbar}{2}\begin{pmatrix} 0 & -i\sqrt{3} & 0 & 0 \\ i\sqrt{3} & 0 & -2i & 0 \\ 0 & 2i & 0 & -i\sqrt{3} \\ 0 & 0 & i\sqrt{3} & 0 \end{pmatrix} S_z = \frac{\hbar}{2}\begin{pmatrix} 3 & 0 & 0 & 0 \\ 0 & 1 & 0 & 0 \\ 0 & 0 & -1 & 0 \\ 0 & 0 & 0 & -3 \end{pmatrix}$$

The generalization of these matrices for arbitrary *s* is

$$\begin{aligned} \left(S_x\right)_{ab} &= \frac{\hbar}{2}(\delta_{a,b+1} + \delta_{a+1,b})\sqrt{(s+1)(a+b-1)-ab} \\ \left(S_y\right)_{ab} &= \frac{\hbar}{2i}(\delta_{a,b+1} - \delta_{a+1,b})\sqrt{(s+1)(a+b-1)-ab} \quad 1 \le a,b \le 2s+1 \\ \left(S_z\right)_{ab} &= \hbar(s+1-a)\delta_{a,b} = \hbar(s+1-b)\delta_{a,b}. \end{aligned}$$

Also useful in the quantum mechanics of multiparticle systems, the general Pauli group G_n is defined to consist of all *n*-fold tensor products of Pauli matrices.

The analog formula of Euler's formula in terms of the Pauli matrices:

$$e^{i\theta(\hat{\mathbf{n}}\cdot\acute{\sigma})} = I\cos\theta + i(\hat{\mathbf{n}}\cdot\acute{\sigma})\sin\theta$$

for higher spins is tractable, but less simple.

Parity

In tables of the spin quantum number *s* for nuclei or particles, the spin is often followed by a "+" or "−". This refers to the parity with "+" for even parity (wave function unchanged by spatial inversion) and "−" for odd parity (wave function negated by spatial inversion). For example, see the isotopes of bismuth.

Applications

Spin has important theoretical implications and practical applications. Well-established *direct* applications of spin include:

- Nuclear magnetic resonance (NMR) spectroscopy in chemistry;
- Electron spin resonance spectroscopy in chemistry and physics;
- Magnetic resonance imaging (MRI) in medicine, a type of applied NMR, which relies on proton spin density;
- Giant magnetoresistive (GMR) drive head technology in modern hard disks.

Electron spin plays an important role in magnetism, with applications for instance in computer memories. The manipulation of *nuclear spin* by radiofrequency waves (nuclear magnetic resonance) is important in chemical spectroscopy and medical imaging.

Spin-orbit coupling leads to the fine structure of atomic spectra, which is used in atomic clocks and in the modern definition of the second. Precise measurements of the g-factor of the electron have played an important role in the development and verification of quantum electrodynamics. *Photon spin* is associated with the polarization of light.

An emerging application of spin is as a binary information carrier in spin transistors. The original concept, proposed in 1990, is known as Datta-Das spin transistor. Electronics based on spin transistors are referred to as spintronics. The manipulation of spin in dilute magnetic semiconductor materials, such as metal-doped ZnO or TiO_2 imparts a further degree of freedom and has the potential to facilitate the fabrication of more efficient electronics.

There are many *indirect* applications and manifestations of spin and the associated Pauli exclusion principle, starting with the periodic table of chemistry.

History

Spin was first discovered in the context of the emission spectrum of alkali metals. In 1924 Wolfgang Pauli introduced what he called a "two-valued quantum degree of freedom" associated with the electron in the outermost shell. This allowed him to formulate the Pauli exclusion principle, stating that no two electrons can share the same quantum state at the same time.

Wolfgang Pauli lecturing

The physical interpretation of Pauli's "degree of freedom" was initially unknown. Ralph Kronig, one of Landé's assistants, suggested in early 1925 that it was produced by the self-rotation of the electron. When Pauli heard about the idea, he criticized it severely, noting that the electron's hypothetical surface would have to be moving faster than the speed of light in order for it to rotate quickly enough to produce the necessary angular momentum. This would violate the theory of relativity. Largely due to Pauli's criticism, Kronig decided not to publish his idea.

In the autumn of 1925, the same thought came to two Dutch physicists, George Uhlenbeck and Samuel Goudsmit at Leiden University. Under the advice of Paul Ehrenfest, they published their results. It met a favorable response, especially after Llewellyn Thomas managed to resolve a factor-of-two discrepancy between experimental results and Uhlenbeck and Goudsmit's calculations (and Kronig's unpublished results). This discrepancy was due to the orientation of the electron's tangent frame, in addition to its position.

Mathematically speaking, a fiber bundle description is needed. The tangent bundle effect is additive and relativistic; that is, it vanishes if c goes to infinity. It is one half of the value obtained without regard for the tangent space orientation, but with opposite sign. Thus the combined effect differs from the latter by a factor two (Thomas precession).

Despite his initial objections, Pauli formalized the theory of spin in 1927, using the modern theory of quantum mechanics invented by Schrödinger and Heisenberg. He pioneered the use of Pauli matrices as a representation of the spin operators, and introduced a two-component spinor wave-function.

Pauli's theory of spin was non-relativistic. However, in 1928, Paul Dirac published the Dirac equation, which described the relativistic electron. In the Dirac equation, a four-component spinor (known as a "Dirac spinor") was used for the electron wave-function. In 1940, Pauli proved the *spin-statistics theorem*, which states that fermions have half-integer spin and bosons have integer spin.

In retrospect, the first direct experimental evidence of the electron spin was the Stern–Gerlach experiment of 1922. However, the correct explanation of this experiment was only given in 1927.

Spin Quantum Number

In atomic physics, the spin quantum number is a quantum number that parameterizes the intrinsic angular momentum (or spin angular momentum, or simply spin) of a given particle. The spin quantum number is the fourth of a set of quantum numbers (the principal quantum number, the azimuthal quantum number, the magnetic quantum number, and the spin quantum number), which completely describe the quantum state of an electron. It is designated by the letter *s*. It describes the energy, shape and orientation of orbitals.

Derivation

As a solution for a certain partial differential equation, the quantized angular momentum (see angular momentum quantum number) can be written as:

$$\mathbf{s} = \sqrt{s(s+1)}\hbar$$

where

$\mathbf{s}$ is the quantized spin vector

$\|\mathbf{s}\|$ is the norm of the spin vector

s is the spin quantum number associated with the spin angular momentum

$\hbar$ is the reduced Planck constant.

Given an arbitrary direction *z* (usually determined by an external magnetic field) the spin *z*-projection is given by

$$s_z = m_s \hbar$$

where m_s is the secondary spin quantum number, ranging from −*s* to +*s* in steps of one. This generates $2s + 1$ different values of m_s.

The allowed values for *s* are non-negative integers or half-integers. Fermions (such as the electron, proton or neutron) have half-integer values, whereas bosons (e.g., photon, mesons) have integer spin values.

Algebra

The algebraic theory of spin is a carbon copy of the Angular momentum in quantum mechanics theory. First of all, spin satisfies the fundamental commutation relation:

$$[S_i, S_j] = i\hbar \epsilon_{ijk} S_k,$$
$$\left[S_i, S^2\right] = 0$$

where ε_{lmn} is the (antisymmetric) Levi-Civita symbol. This means that it is impossible to know two coordinates of the spin at the same time because of the restriction of the uncertainty principle.

Next, the eigenvectors of S^2 and S_z satisfy:

$$S^2 \mid s, m_s = \hbar^2 s(s+1) \mid s, m_s$$

$$S_z \mid s, m_s \rangle = \hbar m_s \mid s, m_s \rangle$$

$$S_\pm \mid s, m_s = \hbar\sqrt{s(s+1) - m_s(m_s \pm 1)} \mid s, m_s \pm 1$$

where $S_\pm = S_x \pm iS_y$ are the creation and annihilation (or "raising" and "lowering" or "up" and "down") operators.

Electron Spin

Early attempts to explain the behavior of electrons in atoms focused on solving the Schrödinger wave equation for the hydrogen atom, the simplest possible case, with a single electron bound to the atomic nucleus. This was successful in explaining many features of atomic spectra.

The solutions required each possible state of the electron to be described by three "quantum numbers". These were identified as, respectively, the electron "shell" number n, the "orbital" number l, and the "orbital angular momentum" number m. Angular momentum is a so-called "classical" concept measuring the momentum of a mass in circular motion about a point. The shell numbers start at 1 and increase indefinitely. Each shell of number n contains n^2 orbitals. Each orbital is characterized by its number l, where l takes integer values from 0 to $n-1$, and its angular momentum number m, where m takes integer values from $+l$ to $-l$. By means of a variety of approximations and extensions, physicists were able to extend their work on hydrogen to more complex atoms containing many electrons.

Atomic spectra measure radiation absorbed or emitted by electrons "jumping" from one "state" to another, where a state is represented by values of n, l, and m. The so-called "Transition rule" limits what "jumps" are possible. In general, a jump or "transition" is allowed only if all three numbers change in the process. This is because a transition will be able to cause the emission or absorption of electromagnetic radiation only if it involves a change in the electromagnetic dipole of the atom.

However, it was recognized in the early years of quantum mechanics that atomic spectra measured in an external magnetic field (see Zeeman effect) cannot be predicted with just n, l, and m. A solution to this problem was suggested in early 1925 by George Uhlenbeck and Samuel Goudsmit, students of Paul Ehrenfest (who rejected the idea), and independently by Ralph Kronig, one of Landé's assistants. Uhlenbeck, Goudsmit, and Kronig introduced the idea of the self-rotation of the electron, which would naturally give rise to an angular momentum vector in addition to the one associated with orbital motion (quantum numbers l and m).

The spin angular momentum is characterized by a quantum number; s = 1/2 specifically for electrons. In a way analogous to other quantized angular momenta, L, it is possible to obtain an expression for the total spin angular momentum:

$$S = \hbar\sqrt{\frac{1}{2}\left(\frac{1}{2}+1\right)} = \frac{\sqrt{3}}{2}\hbar$$

where

$\hbar$ is the reduced Planck constant.

The hydrogen spectra fine structure is observed as a doublet corresponding to two possibilities for the *z*-component of the angular momentum, where for any given direction *z*:

$$\mathbf{S_z} = \pm\frac{1}{2}\hbar$$

whose solution has only two possible *z*-components for the electron. In the electron, the two different spin orientations are sometimes called "spin-up" or "spin-down".

The spin property of an electron would give rise to magnetic moment, which was a requisite for the fourth quantum number. The electron spin magnetic moment is given by the formula:

$$\boldsymbol{\mu}_{\mathbf{s}} = -\frac{e}{2m}gS$$

where

e is the charge of the electron

g is the Landé g-factor

and by the equation:

$$\mu_z = \pm\frac{1}{2}g\mu_B$$

where μ_B is the Bohr magneton.

When atoms have even numbers of electrons the spin of each electron in each orbital has opposing orientation to that of its immediate neighbor(s). However, many atoms have an odd number of electrons or an arrangement of electrons in which there is an unequal number of "spin-up" and "spin-down" orientations. These atoms or electrons are said to have unpaired spins that are detected in electron spin resonance.

Detection of Spin

When lines of the hydrogen spectrum are examined at very high resolution, they are found to be closely spaced doublets. This splitting is called fine structure, and was one of the first experimental evidences for electron spin. The direct observation of the electron's intrinsic angular momentum was achieved in the Stern–Gerlach experiment.

The Stern–Gerlach Experiment

The theory of spatial quantization of the spin moment of the momentum of electrons of atoms situated in the magnetic field needed to be proved experimentally. In 1920 (two years before the theoretical description of the spin was created) Otto Stern and Walter Gerlach observed it in the experiment they conducted.

Silver atoms were evaporated using an electric furnace in a vacuum. Using thin slits, the atoms were guided into a flat beam and the beam sent through an in-homogeneous magnetic field before colliding with a metallic plate. The laws of classical physics predict that the collection of condensed silver atoms on the plate should form a thin solid line in the same shape as the original beam. However, the in-homogeneous magnetic field caused the beam to split in two separate directions, creating two lines on the metallic plate.

The phenomenon can be explained with the spatial quantization of the spin moment of momentum. In atoms the electrons are paired such that one spins upward and one downward, neutralizing the effect of their spin on the action of the atom as a whole. But in the valence shell of silver atoms, there is a single electron whose spin remains unbalanced.

The unbalanced spin creates spin magnetic moment, making the electron act like a very small magnet. As the atoms pass through the in-homogeneous magnetic field, the force moment in the magnetic field influences the electron's dipole until its position matches the direction of the stronger field. The atom would then be pulled toward or away from the stronger magnetic field a specific amount, depending on the value of the valence electron's spin. When the spin of the electron is +1/2 the atom moves away from the stronger field, and when the spin is −1/2 the atom moves toward it. Thus the beam of silver atoms is split while traveling through the in-homogeneous magnetic field, according to the spin of each atom's valence electron.

In 1927 Phipps and Taylor conducted a similar experiment, using atoms of hydrogen with similar results. Later scientists conducted experiments using other atoms that have only one electron in their valence shell: (copper, gold, sodium, potassium). Every time there two lines formed on the metallic plate.

The atomic nucleus also may have spin, but protons and neutrons are much heavier than electrons (about 1836 times), and the magnetic dipole moment is inversely proportional to the mass. So the nuclear magnetic dipole momentum is much smaller than that of the whole atom. This small magnetic dipole was later measured by Stern, Frisch and Easterman.

Energy Levels from the Dirac Equation

In 1928, Paul Dirac developed a relativistic wave equation, now termed the Dirac equation, which predicted the spin magnetic moment correctly, and at the same time treated the electron as a point-like particle. Solving the Dirac equation for the energy levels of an electron in the hydrogen atom, all four quantum numbers including *s* occurred naturally and agreed well with experiment.

Total Spin of an Atom or Molecule

For some atoms the spins of several unpaired electrons (s_1, s_2, ...) are coupled to form a *total spin* quantum number S. This occurs especially in light atoms (or in molecules formed only of light atoms) when spin-orbit coupling is weak compared to the coupling between spins or the coupling between orbital angular momenta, a situation known as LS coupling because L and S are constants of motion. Here L is the total orbital angular momentum quantum number.

For atoms with a well-defined S, the multiplicity of a state is defined as (2S+1). This is equal to the number of different possible values of the total (orbital plus spin) angular momentum J for a given (L, S) combination, provided that S ≤ L (the typical case). For example, if S = 1, there are three states which form a triplet. The eigenvalues of S_z for these three states are +1ħ, 0 and -1ħ. The term symbol of an atomic state indicates its values of L, S, and J.

Electron Magnetic Moment

In atomic physics, the electron magnetic moment, or more specifically the electron magnetic dipole moment, is the magnetic moment of an electron caused by its intrinsic properties of spin and electric charge. The value of the electron magnetic moment is approximately $-9284.764 \times 10^{-27}$ J.T^{-1}. The electron magnetic moment has recently been measured to an accuracy of 7.6 parts in 10^{-13}

Magnetic Moment of an Electron

The electron is a charged particle of charge (−1*e*), where *e* is the unit of elementary charge. Its angular momentum comes from two types of rotation: spin and orbital motion. From classical electrodynamics, a rotating electrically charged body creates a magnetic dipole with magnetic poles of equal magnitude but opposite polarity. This analogy holds as an electron indeed behaves like a tiny bar magnet. One consequence is that an external magnetic field exerts a torque on the electron magnetic moment depending on its orientation with respect to the field.

If the electron is visualized as a classical charged particle literally rotating about an axis with angular momentum L, its magnetic dipole moment μ is given by:

$$\mu = \frac{-e}{2m_e}\mathrm{L}.$$

where m_e is the electron rest mass. Note that the angular momentum L in this equation may be the spin angular momentum, the orbital angular momentum, or the total angular momentum. It turns out the classical result is off by a proportional factor for the spin magnetic moment. As a result, the classical result is corrected by multiplying it with a dimensionless correction factor g, known as the g-factor;

It is usual to express the magnetic moment in terms of the reduced Planck constant $\hbar$ and the Bohr magneton μ_B:

$$\mu = - \quad \mu \quad -$$

Since the magnetic moment is quantized in units of μ_B, correspondingly the angular momentum is quantized in units of $\hbar$.

Spin Magnetic Dipole Moment

The spin magnetic moment is intrinsic for an electron. It is:

$$\mu_s = -g_s \mu_B \frac{\mathrm{S}}{\hbar}.$$

Here S is the electron spin angular momentum. The spin g-factor is approximately two: $g_s \approx 2$. The magnetic moment of an electron is approximately twice what it should be in classical mechanics. The factor of two implies that the electron appears to be twice as effective in producing a magnetic moment as the corresponding classical charged body.

The spin magnetic dipole moment is approximately one μ_B because $g \approx 2$ and the electron is a spin one-half particle: $S = \hbar/2$.

$$\mu_S \approx 2\frac{e\hbar}{2m_e}\frac{\frac{\hbar}{2}}{\hbar} = \mu_B.$$

The z-component of the electron magnetic moment is:

$$(\mu_s)_z = -g_s\mu_B m_s$$

where m_s is the spin quantum number. Note that μ is a *negative* constant multiplied by the spin, so the magnetic moment is *antiparallel* to the spin angular momentum.

The spin g-factor $g_s = 2$ comes from the Dirac equation, a fundamental equation connecting the electron's spin with its electromagnetic properties. Reduction of the Dirac equation for an electron in a magnetic field to its non-relativistic limit yields the Schrödinger equation with a correction term which takes account of the interaction of the electron's intrinsic magnetic moment with the magnetic field giving the correct energy.

For the electron spin, the most accurate value for the spin g-factor has been experimentally determined to have the value

$$2.002\,319\,304\,361\,82 \pm (2.6 \times 10^{-13}).$$

Note that it is only two thousandths larger than the value from Dirac equation. The small correction is known as the anomalous magnetic dipole moment of the electron; it arises from the electron's interaction with virtual photons in quantum electrodynamics. In fact, one famous triumph of the Quantum Electrodynamics theory is the accurate prediction of the electron g-factor. The most accurate value for the electron magnetic moment is

$$-928.4764620(57)\times 10^{-26}\ \text{J}\cdot\text{T}^{-1}.$$

The Classical Theory of the G-factor

The Dirac theory is not necessary to explain the g-factor for the electron. The deviation of the electron g-factor from that of the rigid sphere can be readily explained assuming that the charge distribution inside electron is different from the mass distribution. The electron can still be assumed a rigid body. Assuming for example the simplest and the most physical spherical Gaussian distributions for the charge and the mass separately:

$$\rho_e(r) = eN_e e^{-\frac{r^2}{r_e^2}}$$

and

$$\rho_m(r) = m_e N_m e^{-\frac{r^2}{r_m^2}}$$

where r_m is the mass radius of the electron and r_e is the charge radius (N_e and N_m are the distributions normalization factors to $1\times e$ and $1\times m_e$), the tunable g-factor can be obtained as the ratio

$$g = \left(\frac{r_e}{r_m}\right)^8.$$

For the electron $g = 2$ they differ therefore very slightly, namely

$$\left(\frac{r_e}{r_m}\right) \approx 1.09051.$$

Orbital Magnetic Dipole Moment

The revolution of an electron around an axis through another object, such as the nucleus, gives rise to the orbital magnetic dipole moment. Suppose that the angular momentum for the orbital motion is L. Then the orbital magnetic dipole moment is:

$$\mu_L = -g_L \mu_B \frac{\mathbf{L}}{\hbar}.$$

Here g_L is the electron orbital g-factor and μ_B is the Bohr magneton. The value of g_L is exactly equal to one, by a quantum-mechanical argument analogous to the derivation of the classical gyromagnetic ratio

Total Magnetic Dipole Moment

The total magnetic dipole moment resulting from both spin and orbital angular momenta of an electron is related to the total angular momentum J by a similar equation:

$$\mu_J = g_J \mu_B \frac{\mathbf{J}}{\hbar}.$$

The g-factor g_J is known as the Landé g-factor, which can be related to g_L and g_S by quantum mechanics. See Landé g-factor for details.

Example: Hydrogen Atom

For a hydrogen atom, an electron occupying the atomic orbital $\Psi_{n,\ell,m}$, the magnetic dipole moment is given by:

$$\mu_L = -g_L \frac{\mu_B}{\hbar} \langle \Psi_{n,\ell,m} | L | \Psi_{n,\ell,m} \rangle = -\mu_B \sqrt{\ell(\ell+1)}.$$

Here L is the orbital angular momentum, n, ℓ and m are the principal, azimuthal and magnetic quantum numbers respectively. The z-component of the orbital magnetic dipole moment for an electron with a magnetic quantum number m_ℓ is given by:

$$(\mu_{\mathrm{L}})_z = -\mu_{\mathrm{B}} m_\ell .$$

Electron Spin in the Pauli and Dirac Theories

The necessity of introducing half-integral spin goes back experimentally to the results of the Stern–Gerlach experiment. A beam of atoms is run through a strong non-uniform magnetic field, which then splits into N parts depending on the intrinsic angular momentum of the atoms. It was found that for silver atoms, the beam was split in two—the ground state therefore could not be integral, because even if the intrinsic angular momentum of the atoms were as small as possible, 1, the beam would be split into 3 parts, corresponding to atoms with $L_z = -1$, 0, and +1. The conclusion is that silver atoms have net intrinsic angular momentum of $\frac{1}{2}$. Pauli set up a theory which explained this splitting by introducing a two-component wave function and a corresponding correction term in the Hamiltonian, representing a semi-classical coupling of this wave function to an applied magnetic field, as so:

$$H = \frac{1}{2m}\left[\sigma \cdot \left(P - \frac{e}{c}A\right)\right]^2 + e\phi .$$

Here A is the magnetic potential and ϕ the electric potential representing the electromagnetic field, and $\sigma = (\sigma_x, \sigma_y, \sigma_z)$ are the Pauli matrices. On squaring out the first term, a residual interaction with the magnetic field is found, along with the usual classical Hamiltonian of a charged particle interacting with an applied field:

$$H = \frac{1}{2m}\left(P - \frac{e}{c}A\right)^2 + e\phi - \frac{e\hbar}{2mc}\sigma \cdot B .$$

This Hamiltonian is now a 2 × 2 matrix, so the Schrödinger equation based on it must use a two-component wave function. Pauli had introduced the 2 × 2 sigma matrices as pure *phenomenology*— Dirac now had a *theoretical argument* that implied that spin was somehow the consequence of incorporating relativity into quantum mechanics. On introducing the external electromagnetic 4-potential into the Dirac equation in a similar way, known as minimal coupling, it takes the form (in natural units $\hbar = c = 1$)

$$\left[-i\gamma^{\mu}\left(\partial_{\mu} + ieA_{\mu}\right) + m\right]\psi = 0$$

where γ^{μ} are the gamma matrices (known as Dirac matrices) and i is the imaginary unit. A second application of the Dirac operator will now reproduce the Pauli term exactly as before, because the spatial Dirac matrices multiplied by i, have the same squaring and commutation properties as the Pauli matrices. What is more, the value of the gyromagnetic ratio of the electron, standing in front of Pauli's new term, is explained from first principles. This was a major achievement of the Dirac equation and gave physicists great faith in its overall correctness. The Pauli theory may be seen as the low energy limit of the Dirac theory in the following manner. First the equation is written in the form of coupled equations for 2-spinors with the units restored:

$$\begin{pmatrix} (mc^2 - E + e\phi) & c\sigma\cdot\left(P - \frac{e}{c}A\right) \\ -c\sigma\cdot\left(P - \frac{e}{c}A\right) & \left(mc^2 + E - e\phi\right) \end{pmatrix}\begin{pmatrix} \psi_+ \\ \psi_- \end{pmatrix} = \begin{pmatrix} 0 \\ 0 \end{pmatrix}.$$

so

$$(E - e\phi)\psi_+ - c\sigma\cdot\left(p - \frac{e}{c}A\right)\psi_- = mc^2\psi_+$$

$\phi \; \psi \quad \sigma \quad \psi$

$$-(E - e\phi)\psi_- + c\sigma\cdot\left(p - \frac{e}{c}A\right)\psi_+ = mc^2\psi_-$$

Assuming the field is weak and the motion of the electron non-relativistic, we have the total energy of the electron approximately equal to its rest energy, and the momentum reducing to the classical value,

$$E - e\phi \;\approx mc^2 p \approx mv$$

and so the second equation may be written

$$\psi_- \approx \frac{1}{2mc}\sigma\cdot\left(p - \frac{e}{c}A\right)\psi_+$$

which is of order v/c - thus at typical energies and velocities, the bottom components of the Dirac spinor in the standard representation are much suppressed in comparison to the top components. Substituting this expression into the first equation gives after some rearrangement

$$\left(E - mc^2\right)\psi_+ = \frac{1}{2m}\left[\sigma\cdot\left(p - \frac{e}{c}A\right)\right]^2\psi_+ + e\phi\psi_+$$

The operator on the left represents the particle energy reduced by its rest energy, which is just the classical energy, so we recover Pauli's theory if we identify his 2-spinor with the top components of the Dirac spinor in the non-relativistic approximation. A further approximation gives the Schrödinger equation as the limit of the Pauli theory. Thus the Schrödinger equation may be seen as the far non-relativistic approximation of the Dirac equation when one may neglect spin and work only at low energies and velocities. This also was a great triumph for the new equation, as it traced the mysterious *i* that appears in it, and the necessity of a complex wave function, back to the geometry of space-time through the Dirac algebra. It also highlights why the Schrödinger equation, although superficially in the form of a diffusion equation, actually represents the propagation of waves.

It should be strongly emphasized that this separation of the Dirac spinor into large and small components depends explicitly on a low-energy approximation. The entire Dirac spinor represents an *irreducible* whole, and the components we have just neglected to arrive at the Pauli theory will bring in new phenomena in the relativistic regime - antimatter and the idea of creation and annihilation of particles.

In a general case (if a certain linear function of electromagnetic field does not vanish identically), three out of four components of the spinor function in the Dirac equation can be algebraically eliminated, yielding an equivalent fourth-order partial differential equation for just one component. Furthermore, this remaining component can be made real by a gauge transform.

Measurement

The existence of the anomalous magnetic moment of the electron has been detected experimentally by magnetic resonance method. This allows the determination of hyperfine splitting of electron shell energy levels in atoms of protium and deuterium using the measured resonance frequency for several transitions.

The magnetic moment of the electron has been measured using a one-electron quantum cyclotron and quantum nondemolition spectroscopy.

Spin–orbit Interaction

In quantum physics, the spin–orbit interaction (also called spin–orbit effect or spin–orbit coupling) is an interaction of a particle's spin with its motion. The first and best known example of this is that spin–orbit interaction causes shifts in an electron's atomic energy levels due to electromagnetic interaction between the electron's spin and the magnetic field generated by the electron's orbit around the nucleus. This is detectable as a splitting of spectral lines, which can be thought of as a Zeeman Effect due to the internal field. A similar effect, due to the relationship between angular momentum and the strong nuclear force, occurs for protons and neutrons moving inside the nucleus, leading to a shift in their energy levels in the nucleus shell model. In the field of spintronics, spin–orbit effects for electrons in semiconductors and other materials are explored for technological applications. The spin–orbit interaction is one cause of magnetocrystalline anisotropy.

Spin–orbit Interaction in Atomic Energy Levels

This section presents a relatively simple and quantitative description of the spin–orbit interaction for an electron bound to an atom, up to first order in perturbation theory, using some semiclassical electrodynamics and non-relativistic quantum mechanics. This gives results that agree reasonably well with observations. A more rigorous derivation of the same result would start with the Dirac equation, and achieving a more precise result would involve calculating small corrections from quantum electrodynamics.

Energy of a Magnetic Moment

The energy of a magnetic moment in a magnetic field is given by:

$$\Delta H = -\mu \cdot B,$$

where μ is the magnetic moment of the particle and B is the magnetic field it experiences.

Magnetic Field

We shall deal with the magnetic field first. Although in the rest frame of the nucleus, there is no magnetic field acting on the electron, there *is* one in the rest frame of the electron (see Classical electromagnetism and special relativity). Ignoring for now that this frame is not inertial, in SI units we end up with the equation

$$\mathbf{B} = -\frac{\mathbf{v} \times \mathbf{E}}{c^2},$$

where v is the velocity of the electron and E the electric field it travels through. Here weak relativistic is assumed that Lorentz factor $\gamma 1$. Now we know that E is radial so we can rewrite $\boldsymbol{E} = |E/r|\boldsymbol{r}$. Also we know that the momentum of the electron $\boldsymbol{p} = m_e \boldsymbol{v}$. Substituting this in and changing the order of the cross product gives:

$$\mathbf{B} = \frac{\mathbf{r} \times \mathbf{p}}{m_e c^2} \left| \frac{E}{r} \right|.$$

Next, we express the electric field as the gradient of the electric potential $\mathbf{E} = -\nabla V$. Here we make the central field approximation, that is, that the electrostatic potential is spherically symmetric, so is only a function of radius. This approximation is exact for hydrogen and hydrogen-like systems. Now we can say

$$|E| = \frac{\partial V}{\partial r} = \frac{1}{e} \frac{\partial U(r)}{\partial r},$$

where $U = eV$ is the potential energy of the electron in the central field, and e is the elementary charge. Now we remember from classical mechanics that the angular momentum of a particle $\boldsymbol{L} = \boldsymbol{r} \times \boldsymbol{p}$. Putting it all together we get

$$\boldsymbol{B} = \frac{1}{m_e e c^2} \frac{1}{r} \frac{\partial U(r)}{\partial r} \boldsymbol{L}.$$

It is important to note at this point that B is a positive number multiplied by L, meaning that the magnetic field is parallel to the orbital angular momentum of the particle, which is itself perpendicular to the particle's velocity.

Magnetic Moment of the Electron

The magnetic moment of the electron is

$$\mu_S = -g_S \mu_B \frac{\mathbf{S}}{\hbar}.$$

where $\boldsymbol{S}$ is the spin angular momentum vector, μ_B is the Bohr magneton and $g_\mathrm{s} \approx 2$ is the electron spin g-factor. Here, ***ì*** is a negative constant multiplied by the spin, so the magnetic moment is antiparallel to the spin angular momentum.

The spin–orbit potential consists of two parts. The Larmor part is connected to the interaction of the magnetic moment of the electron with the magnetic field of the nucleus in the co-moving frame of the electron. The second contribution is related to Thomas precession.

Larmor Interaction Energy

The Larmor interaction energy is

$$\Delta H_\mathrm{L} = -\mu \cdot \boldsymbol{B}.$$

Substituting in this equation expressions for the magnetic moment and the magnetic field, one gets

$$\Delta H_\mathrm{L} = \frac{2\mu_\mathrm{B}}{\hbar m_\mathrm{e} e c^2} \frac{1}{r} \frac{\partial U(r)}{\partial r} \mathbf{L} \cdot \mathbf{S}.$$

Now, we have to take into account Thomas precession correction for the electron's curved trajectory.

Thomas Interaction Energy

In 1926 Llewellyn Thomas relativistically recomputed the doublet separation in the fine structure of the atom. Thomas precession rate, **Ù**$_\mathrm{T}$, is related to the angular frequency of the orbital motion, ***ù*** , of a spinning particle as follows

$$\Omega_\mathrm{T} = \omega(\gamma - 1),$$

where γ is the Lorentz factor of the moving particle. The Hamiltonian producing the spin precession **Ù**$_\mathrm{T}$ is given by

$$\Delta H_\mathrm{T} = \Omega_\mathrm{T} \cdot \mathbf{S}.$$

To the first order in $(v/c)^2$, we obtain

$$\Delta H_T = -\frac{\mu_B}{\hbar m_e e c^2}\frac{1}{r}\frac{\partial U(r)}{\partial r} L \cdot S.$$

Total Interaction Energy

The total spin–orbit potential in an external electrostatic potential takes the form

$$\Delta H \equiv \Delta H_L + \Delta H_T = \frac{\mu_B}{\hbar m_e e c^2}\frac{1}{r}\frac{\partial U(r)}{\partial r} \mathbf{L} \cdot \mathbf{S}.$$

The net effect of Thomas precession is the reduction of the Larmor interaction energy by factor 1/2 which came to be known as the *Thomas half.*

Evaluating the Energy Shift

Thanks to all the above approximations, we can now evaluate the detailed energy shift in this model. In particular, we wish to find a basis that diagonalizes both H_o (the non-perturbed Hamiltonian) and ΔH. To find out what basis this is, we first define the total angular momentum operator

$$\mathbf{J} = \mathbf{L} + \mathbf{S}.$$

Taking the dot product of this with itself, we get

$$\mathbf{J}^2 = \mathbf{L}^2 + \mathbf{S}^2 + 2\mathbf{L} \cdot \mathbf{S}$$

(since L and S commute), and therefore

$$\boldsymbol{L} \cdot \boldsymbol{S} = \frac{1}{2}(\boldsymbol{J}^2 - \boldsymbol{L}^2 - \boldsymbol{S}^2) = \frac{\hbar^2}{2}\begin{pmatrix} L_z & L_- \\ L_+ & -L_z \end{pmatrix}$$

It can be shown that the five operators H_0, J^2, L^2, S^2, and J_z all commute with each other and with ΔH. Therefore, the basis we were looking for is the simultaneous eigenbasis of these five operators (i.e., the basis where all five are diagonal). Elements of this basis have the five quantum numbers: n (the "principal quantum number") j (the "total angular momentum quantum number"), l (the "orbital angular momentum quantum number"), s (the "spin quantum number"), and j_z (the "z-component of total angular momentum").

To evaluate the energies, we note that

$$\left\langle \frac{1}{r^3} \right\rangle = \frac{2}{a^3 n^3 l(l+1)(2l+1)}$$

for hydrogenic wavefunctions (here $a = \hbar / Z\alpha m_e c$ is the Bohr radius divided by the nuclear charge Z); and

$$\boldsymbol{L}\cdot\boldsymbol{S} = \frac{1}{2}(\boldsymbol{J}^2 - \boldsymbol{L}^2 - \boldsymbol{S}^2)$$

$$= \frac{\hbar^2}{2}(j(j+1) - l(l+1) - s(s+1))$$

Final Energy Shift

We can now say

$$\Delta E = \frac{\beta}{2}(j(j+1) - l(l+1) - s(s+1))$$

where

$$\beta = \beta(n,l) = Z^4 \frac{\mu_0}{4\pi} g_s \mu_B^2 \frac{1}{n^3 a_0^3 l(l+1/2)(l+1)}$$

Spin–orbit Interaction in Solids

A crystalline solid (semiconductor, metal etc.) is characterized by its band structure. While on the overall scale (including the core levels) the spin–orbit interaction is still a small perturbation, it may play a relatively more important role if we zoom in to bands close to the Fermi level (E_F). The atomic $\boldsymbol{L}\cdot\boldsymbol{S}$ interaction for example splits bands which would be otherwise degenerate and the particular form of this spin–orbit splitting (typically of the order of few to few hundred millielectronvolts) depends on the particular system. The bands of interest can be then described by various effective models, usually based on some perturbative approach. An example of how the atomic spin–orbit interaction influences the band structure of a crystal is explained in the article about Rashba interaction.

In crystalline solid contained paramagnetic ions e.g. ions with unclosed d or f atomic subshell, localized electronic states exist. In this case, atomic-like electronic levels structure is shaped by intrinsic magnetic spin-orbit interactions and interactions with crystalline electric fields. Such structure is named the fine electronic structure. For rare-earth ions the spin-orbit interactions are much stronger than the CEF interactions. The strong spin-orbit coupling makes J become a relatively good quantum number, because the first excited multiplet is at least ~130 meV (1500 K) above the primary multiplet. The result is that filling it at room temperature (300K) is negligibly small. In this case, a 2J+1-fold degenerated primary multiplet split by an external crystal electric field can be treated as the basic contribution to the analysis of such systems' properties. In the case of approximate calculations for basis $|J, J_z>$, to determine which is the primary multiplet, the Hund principles, known from atomic physics, are applied:

- The ground state of the terms' structure has the maximum value S allowed by the Pauli exclusion principle.
- The ground state has a maximum allowed L value, with maximum S.
- The primary multiplet has a corresponding J=|L-S| when the shell is less than half full, and J=L+S, where the fill is greater.

The S, L and J of the ground multiplet are determined by Hund's rules. The ground multiplet is 2J+1 degenerated – its degeneracy is removed by CEF interactions and magnetic interactions. CEF interactions and magnetic interactions resemble, somehow, Stark and Zeeman effect known from atomic physics. The energies and eigenfunctions of the discrete fine electronic structure are obtained by diagonalization of the (2J+1)-dimensional matrix. The fine electronic structure can be directly detected by many different spectroscopic methods. The one of the best of them is Inelastic Neutron Scattering (INS) experiments. The case of strong cubic CEF (for 3*d* transition -metal ions) interactions form group of levels (e.g. T2g, E.g., A2g) which are partially split by spin-orbit interactions and (if occur) lower symmetry CEF interactions. The energies and eigenfunctions of the discrete fine electronic structure (for the lowest term) are obtained by diagonalization of the (2L+1)(2S+1)-dimensional matrix. In the temperature T = 0[K] (absolute zero) only the lowest state is occupied. The magnetic moment at T=0[K] is equal to the moment of the ground state. It allows evaluate the total, spin and orbital moments. The eigenstates and corresponding eigenfunctions $|\Gamma_n>$ can be found from direct diagonalization of hamiltonian matrix containing Crystal field and Spin–orbit interactions. Taking into consideration the thermal population of states, the thermal evolution of the single ion properties of the compound is established. This technique is based on the equivalent operator theory defined as the CEF widened by thermodynamic and analytical calculations defined as the supplement of the CEF theory by including thermodynamic and analytical calculations.

Examples of Effective Hamiltonians

Hole bands of a bulk (3D) zinc-blende semiconductor will be split by $Ä_0$ into heavy and light holes (which form a Ã quadruplet in the Ã-point of the Brillouin zone) and a split-off band (Ã doublet). Including two conduction bands (Ã doublet in the Ã-point), the system is described by the effective eight-band model of Kohn and Luttinger. If only top of the valence band is of interest (for example when $E_F Ä_0$, Fermi level measured from the top of the valence band), the proper four-band effective model is

$$H_{\mathrm{KL}}(k_x,k_y,k_z) = -\frac{\hbar^2}{2m}\left[(\gamma_1+\frac{5}{2}\gamma_2)k^2 - 2\gamma_2(J_x^2k_x^2+J_y^2k_y^2+J_z^2k_z^2) - 2\gamma_3\sum_{m\neq n}J_mJ_nk_mk_n\right]$$

where $\gamma_{1,2,3}$ are the Luttinger parameters (analogous to the single effective mass of a one-band model of electrons) and $J_{x,y,z}$ are angular momentum 3/2 matrices (m is the free electron mass). In combination with magnetization, this type of spin–orbit interaction will distort the electronic bands depending on the magnetization direction, thereby causing Magnetocrystalline anisotropy (a special type of Magnetic anisotropy). If the semiconductor moreover lacks the inversion symmetry, the hole bands will exhibit cubic Dresselhaus splitting. Within the four bands (light and heavy holes), the dominant term is

$$H_{\mathrm{D3}} = b_{41}^{8v8v}[(k_xk_y^2 - k_xk_z^2)J_x + (k_yk_z^2 - k_yk_x^2)J_y + (k_zk_x^2 - k_zk_y^2)J_z]$$

where the material parameter $b_{41}^{8v8v} = -81.93\,\mathrm{meV\cdot nm^3}$ for GaAs (see pp. 72 in Winkler's book, according to more recent data the Dresselhaus constant in GaAs is 9 eVÅ³; the total Hamiltonian will be $H_{\mathrm{KL}} + H_{\mathrm{D3}}$). Two-dimensional electron gas in an asymmetric quantum well (or heterostructure) will feel the Rashba interaction. The appropriate two-band effective Hamiltonian is

$$H_0 + H_{\mathrm{R}} = \frac{\hbar^2k^2}{2m^*}\sigma_0 + \alpha(k_y\sigma_x - k_x\sigma_y)$$

where σ_0 is the 2 × 2 identity matrix, $\sigma_{x,y}$ the Pauli matrices and m^* the electron effective mass. The spin–orbit part of the Hamiltonian, H_{R} is parametrized by α, sometimes called the Rashba parameter (its definition somewhat varies), which is related to the structure asymmetry.

Above expressions for spin–orbit interaction couple spin matrices $\boldsymbol{J}$ and $\boldsymbol{\acute{o}}$ to the quasi-momentum $\boldsymbol{k}$, and to the vector potential $\boldsymbol{A}$ of an AC electric field through the Peierls substitution $\boldsymbol{k} = -i\nabla - (\frac{e}{\hbar c})\boldsymbol{A}$. They are lower order terms of the Luttinger–Kohn $\boldsymbol{k}\cdot\boldsymbol{p}$ expansion in powers of k. Next terms of this expansion also produce terms that couple spin operators of the electron coordinate $\boldsymbol{r}$. Indeed, a cross product $(\boldsymbol{\acute{o}}\times\boldsymbol{k})$ is invariant with respect to time inversion. In cubic crystals, it has a symmetry of a vector and acquires a meaning of a spin–orbit contribution $\mathbf{r}_{\mathrm{SO}}$ to the operator of coordinate. For electrons in semiconductors with a narrow gap E_G between the conduction and heavy hole bands, Yafet derived the equation

$$r_{\mathrm{SO}} = \frac{\hbar^2 g}{4m_0}\left(\frac{1}{E_G} + \frac{1}{E_G+\Delta_0}\right)(\sigma\times k)$$

where m_0 is a free electron mass, and g is a g-factor properly renormalized for spin–orbit interaction. This operator couples electron spin $S = \frac{1}{2}\sigma$ directly to the electric field $\mathbf{E}$ through the interaction energy $-e(\mathbf{r}_{\mathrm{SO}}\cdot\mathbf{E})$.

J-coupling

Scalar or *J*-couplings (also called *indirect dipole–dipole coupling*) are mediated through chemical bonds connecting two spins. It is an indirect interaction between two nuclear spins which arises from hyperfine interactions between the nuclei and local electrons. *J*-coupling contains information about bond distance and angles. Most importantly, *J*-coupling provides information on the connectivity of molecules. In NMR spectroscopy, it is responsible for the appearance of many signals in the NMR spectra of fairly simple molecules.

Vector Model and Manifestations for Chemical Structure Assignments

The origin of *J*-coupling can be visualized by a vector model for a simple molecule such as hydrogen fluoride (HF). In HF, the two nuclei have spin 1/2. Four states are possible, depending on the relative alignment of the H and F nuclear spins with the external magnetic field. The selection rules of NMR spectroscopy dictate that $\Delta I = 1$, which means that a given photon (in the radio frequency range) can affect ("flip") only one of the two nuclear spins.

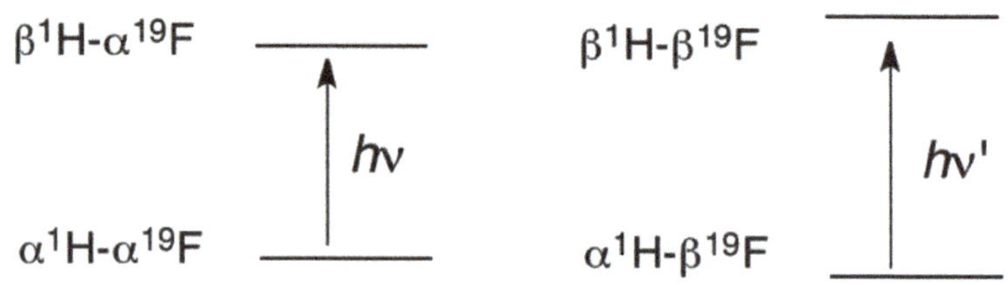

Energy diagram showing the effects of *J*-coupling for the molecule hydrogen fluoride.

J-coupling provides three parameters: the multiplicity (the "number of lines"), the magnitude of the coupling (strong, medium, weak), and the sign of the coupling.

Multiplicity

The multiplicity provides information on the number of centers coupled to the signal of interest, and their nuclear spin. For simple systems, as in ^1H-^1H coupling in NMR spectroscopy, the multiplicity reflects the number of adjacent, magnetically nonequivalent protons. Nuclei with spins greater than 1/2, which are called quadrupolar, can give rise to greater splitting, although in many cases coupling to quadrupolar nuclei is not observed. Many elements consist of nuclei with nuclear spin and without. In these cases the observed spectrum is the sum of spectra for each isotopomer. One of the great conveniences of NMR spectroscopy for organic molecules is that the many lighter elements are nearly monoisotopic: ^1H, ^{19}F, and ^{31}P each have spin 1/2. ^{12}C and ^{16}O have no nuclear spin.

Magnitude of J-coupling

For ^{1}H–^{1}H coupling, the magnitude of J provides information on the proximity of the coupling partners. Generally speaking 2-bond coupling (i.e. ^{1}H–C–^{1}H) is stronger than three-bond coupling (^{1}H–C–C–^{1}H). The magnitude of the coupling also provides information on the dihedral angles relating the coupling partners, as described by the Karplus relationship.

For heteronuclear coupling, the magnitude of J is related to the nuclear magnetic moments of the coupling partners. ^{19}F, with a high nuclear magnetic moment, gives rise to large coupling to protons. ^{103}Rh, with a very small nuclear magnetic moment, gives only small couplings to ^{1}H. To correct for the effect of the nuclear magnetic moment (or equivalently the gyromagnetic ratio γ), the "reduced coupling constant" K is often discussed, where

$$K = 4\pi^2 J / h\gamma_x \gamma_y.$$

The value of J also has a sign, and couplings constants of comparable magnitude often have opposite signs.

J-Coupling Hamiltonian

The Hamiltonian of a molecular system may be taken as:

$$H = D_1 + D_2 + D_3,$$

- D_1 = electron orbital–orbital, spin–orbital, spin–spin and electron-spin–external-field interactions
- D_2 = magnetic interactions between nuclear spin and electron spin
- D_3 = direct interaction of nuclei with each other

For a singlet molecular state and frequent molecular collisions, D_1 and D_3 are almost zero. The full form of the J-coupling interaction between spins 'I_j and I_k on the same molecule is:

$$H = 2\pi\, I_j \cdot J_{jk} \cdot I_k$$

where J_{jk} is the J-coupling tensor, a real 3 × 3 matrix. It depends on molecular orientation, but in an isotropic liquid it reduces to a number, the so-called scalar coupling. In 1D NMR, the scalar coupling leads to oscillations in the free induction decay as well as splittings of lines in the spectrum.

Decoupling

By selective radio frequency irradiation, NMR spectra can be fully or partially decoupled, eliminating or selectively reducing the coupling effect. Carbon-13 NMR spectra are often recorded with proton decoupling.

History

In October 1951, E. L. Hahn and D. E. Maxwell reported a *spin echo experiment* which indicates the existence of an interaction between two protons in dichloroacetaldehyde. In the echo experiment, two short, intense pulses of radiofrequency magnetic field are applied to the spin ensemble at the nuclear resonance condition and are separated by a time interval of τ. The echo appears with a given amplitude at time 2τ. For each setting of τ, the maximum value of the echo signal is measured and plotted as a function of τ. If the spin ensemble consists of a magnetic moment, a monotonic decay in the echo envelope is obtained. In the Hahn-Maxwell experiment, the decay was modulated by two frequencies: one frequency corresponded with the difference in chemical shift between the two non-equivalent spins and a second frequency, J, that was smaller and independent of magnetic field strength ($J/2\pi = 0.7$ Hz).

Such interaction came as a great surprise. The direct interaction between two magnetic dipoles depends on the relative position of two nuclei in such a way that when averaged over all possible orientations of the molecule it equals to zero.

In November 1951, N. F. Ramsey and E. M. Purcell proposed a mechanism that explained the observation and gave rise to an interaction of the form $I_1 \cdot I_2$. The mechanism is the magnetic interaction between each nucleus and the electron spin of its own atom together with the exchange coupling of the electron spins with each other.

In the 1990s, direct evidence was found for the presence of J-couplings between magnetically active nuclei on both sides of the hydrogen bond. Initially, it was surprising to observe such couplings across hydrogen bonds since J-couplings are usually associated with the presence of purely covalent bonds. However, it is now well established that the H-bond J-couplings follow the same electron-mediated polarization mechanism as their covalent counterparts.

The spin–spin coupling between nonbonded atoms in close proximity has sometimes been observed between fluorine, nitrogen, carbon, silicon and phosphorus atoms.

Fine Structure

In atomic physics, the fine structure describes the splitting of the spectral lines of atoms due to electron spin and relativistic corrections to the non-relativistic Schrödinger equation.

The gross structure of line spectra is the line spectra predicted by the quantum mechanics of non-relativistic electrons with no spin. For a hydrogenic atom, the gross structure energy levels only depend on the principal quantum number n. However, a more accurate model takes into account relativistic and spin effects, which break the degeneracy of the energy levels and split the spectral lines. The scale of the fine structure splitting relative to the gross structure energies is on the order of $(Z\alpha)^2$, where Z is the atomic number and α is the fine-structure constant, a dimensionless number equal to approximately $1/137$.

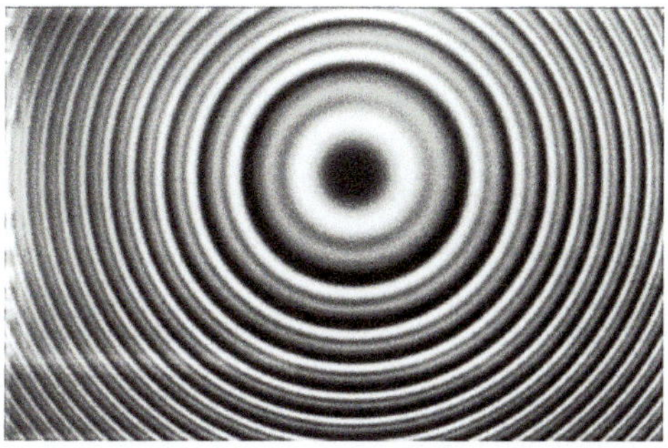

Interference fringes, showing fine structure (splitting) of a cooled deuterium source, viewed through a Fabry-Pérot étalon.

The fine structure energy corrections can be obtained by using perturbation theory. To do this one adds three corrective terms to the Hamiltonian: the leading order relativistic correction to the kinetic energy, the correction due to the spin-orbit coupling, and the Darwinian term. The full Hamiltonian is given by

$$H = H_0 + H_{\text{kinetic}} + H_{\text{so}} + H_{\text{Darwinian}},$$

where H_0 is the Hamiltonian from the Coulomb interaction. These corrections can also be obtained from the non-relativistic limit of the Dirac equation, since Dirac's theory naturally incorporates relativity and spin interactions.

Kinetic Energy Relativistic Correction

Classically, the kinetic energy term of the Hamiltonian is

$$T = \frac{p^2}{2m}$$

where p is the momentum and m is the mass of the electron.

However, when considering a more accurate theory of nature via. special relativity, we must use a relativistic form of the kinetic energy,

$$T = \sqrt{p^2c^2 + m^2c^4} - mc^2$$

where the first term is the total relativistic energy, and the second term is the rest energy of the electron. (c is the speed of light) Expanding this in a Taylor series (specifically a binomial series), we find

$$T = \frac{p^2}{2m} - \frac{p^4}{8m^3c^2} + \cdots$$

Then, the first order correction to the Hamiltonian is

$$H_{\text{kinetic}} = -\frac{p^4}{8m^3c^2}$$

Using this as a perturbation, we can calculate the first order energy corrections due to relativistic effects.

$$E_n^{(1)} = \langle \psi^0 | H' | \psi^0 \rangle = -\frac{1}{8m^3c^2}\langle \psi^0 | p^4 | \psi^0 \rangle = -\frac{1}{8m^3c^2}\langle \psi^0 | p^2 p^2 | \psi^0 \rangle$$

where ψ^0 is the unperturbed wave function. Recalling the unperturbed Hamiltonian, we see

$$\begin{aligned} H^0 |\psi^0\rangle &= E_n |\psi^0\rangle \\ \left(\frac{p^2}{2m} + V\right)|\psi^0\rangle &= E_n |\psi^0\rangle \\ p^2 |\psi^0\rangle &= 2m(E_n - V)|\psi^0\rangle \end{aligned}$$

We can use this result to further calculate the relativistic correction:

$$\begin{aligned} E_n^{(1)} &= -\frac{1}{8m^3c^2}\langle \psi^0 | p^2 p^2 |\rangle \\ E_n^{(1)} &= -\frac{1}{8m^3c^2}\langle \psi^0 | (2m)^2 (E_n - V)^2 | \psi^0 \rangle \\ E_n^{(1)} &= -\frac{1}{2mc^2}\left(E_n^2 - 2E_n\langle V \rangle + \langle V^2 \rangle\right) \end{aligned}$$

For the hydrogen atom, $V(r) = \frac{-e^2}{4\pi\varepsilon_0 r}$, $\left\langle \frac{1}{r} \right\rangle = \frac{1}{a_0 n^2}$,, and $\frac{1}{r^2} = \frac{1}{(l+1/2)n^3 a_0^2}$ where a_0 is the Bohr Radius, n is the principal quantum number and l is the azimuthal quantum number. Therefore, the first order relativistic correction for the hydrogen atom is

$$E_n^{(1)} = -\frac{1}{2mc^2}\left(E_n^2 + 2E_n \frac{e^2}{4\pi\varepsilon_0}\frac{1}{a_0 n^2} + \frac{1}{16\pi^2\varepsilon_0^2}\frac{e^4}{(l+\frac{1}{2})n^3 a_0^2}\right) = -\frac{E_n^2}{2mc^2}\left(\frac{4n}{l+\frac{1}{2}} - 3\right)$$

where we have used:

$$E_n = -\frac{e^2}{8\pi\varepsilon_0 a_0 n^2}$$

On final calculation, the order of magnitude for the relativistic correction to the ground state is -9.056×10^{-4} eV..

Spin-orbit Coupling

For a hydrogen-like atom with Z protons, orbital momentum $\vec{L}$ and electron spin $\vec{S}$, the spin-orbit term is given by:

$$H_{so} = \frac{1}{2}\left(\frac{Ze^2}{4\pi\varepsilon_0}\right)\left(\frac{g_s}{2m_e^2c^2}\right)\frac{\vec{L}\cdot\vec{S}}{r^3}$$

m_e is the electron mass, ε_0 is the vacuum permittivity and $\vec{L}$ is the spin g-factor. r is the distance of the electron from the nucleus.

The spin-orbit correction can be understood by shifting from the standard frame of reference (where the electron orbits the nucleus) into one where the electron is stationary and the nucleus instead orbits it. In this case the orbiting nucleus functions as an effective current loop, which in turn will generate a magnetic field. However, the electron itself has a magnetic moment due to its intrinsic angular momentum. The two magnetic vectors, $\vec{B}$ and $\vec{\mu}_s$ couple together so that there is a certain energy cost depending on their relative orientation. This gives rise to the energy correction of the form

$$\Delta E_{SO} = \xi(r)\vec{L}\cdot\vec{S}$$

Notice that there is a factor of 2, called the Thomas precession, which comes from the relativistic calculation that changes back to the electron's frame from the nucleus frame.

Since

$$\left\langle \frac{1}{r^3} \right\rangle = \frac{Z^3}{n^3 a_0^3}\frac{1}{l\left(l+\frac{1}{2}\right)(l+1)}$$

$$\left\langle \vec{L}\cdot\vec{S} \right\rangle = \frac{\hbar^2}{2}[j(j+1)-l(l+1)-s(s+1)]$$

the expectation value for the Hamiltonian is:

$$H_{SO} = \frac{E_n^{\,2}}{m_e c^2}\, n\, \frac{j(j+1)-l(l+1)-\frac{3}{4}}{l\left(l+\frac{1}{2}\right)(l+1)}$$

Thus the order of magnitude for the spin-orbital coupling is .

Remark: On the (n,l,s)=(n,0,1/2) and (n,l,s)=(n,1,-1/2) energy level, which the fine structure said their level are the same. If we take the g-factor to be 2.0031904622, then, the calculated energy level will be different by using 2 as g-factor. Only using 2 as the g-factor, we can match the energy level in the 1st order approximation of the relativistic correction. When using the higher order approximation for the relativistic term, the 2.0031904622 g-factor may agree with each other. However, if we use the g-factor as 2.0031904622, the result does not agree with the formula, which included every effect.

Darwin Term

There is one last term in the non-relativistic expansion of the Dirac equation. It is referred to as the Darwin term, and is given by:

$$
\begin{aligned}
H_{\text{Darwinian}} &= \frac{\hbar^2}{8m_e^2c^2} 4\pi \left(\frac{Ze^2}{4\pi\varepsilon_0} \right) \delta^3(\vec{r}) \\
\langle H_{\text{Darwinian}} \rangle &= \frac{\hbar^2}{8m_e^2c^2} 4\pi \left(\frac{Ze^2}{4\pi\varepsilon_0} \right) |\psi(0)|^2 \\
\psi(0) &= 0 \text{ for } l > 0 \\
\psi(0) &= \frac{1}{\sqrt{4\pi}} 2 \left(\frac{Z}{na_0} \right)^{\frac{3}{2}} \text{ for } l = 0 \\
H_{\text{Darwinian}} &= \frac{2n}{m_e c^2} E_n^2
\end{aligned}
$$

The Darwin term affects only the s-orbit. This is because the wave function of an electron with $l > 0$ vanishes at the origin, hence the delta function has no effect. For example, it gives the 2s-orbit the same energy as the 2p-orbit by raising the 2s-state by 9.057×10^{-5} eV.

The Darwin term changes the effective potential at the nucleus. It can be interpreted as a smearing out of the electrostatic interaction between the electron and nucleus due to zitterbewegung, or rapid quantum oscillations, of the electron. This can be demonstrated by a short calculation

Quantum fluctuations allow for the creation of virtual electron-positron pairs with a lifetime estimated by the uncertainty principle $\text{Ä}t \approx \hbar / \text{Ä}E \approx \hbar / mc^2$. The distance the particles can move during this time is $\xi \approx c\text{Ä}t \approx \hbar / mc = \lambda_c$, the Compton wavelength. The electrons of the atom interact with those pairs. This yields a fluctuating electron position $\vec{r} + \vec{\xi}$. Using a Taylor expansion, the effect on the potential U can be estimated:

$$
U(\vec{r} + \vec{\xi}) \approx U(\vec{r}) + \xi \cdot \nabla U(\vec{r}) + \frac{1}{2} \sum_{ij} \xi_i \xi_j \partial_i \partial_j U(\vec{r})
$$

Averaging over the fluctuations $\overline{\vec{\xi}}\ \overline{\vec{\xi}} = 0, \quad \overline{\xi_i \xi_j} = \frac{1}{3}\overline{\vec{\xi}^2}\delta_{ij}$,

gives the average potential

$$\overline{U\left(\vec{r}+\vec{\xi}\right)} = U\left(\vec{r}\right) + \frac{1}{6}\overline{\vec{\xi}^2}\nabla^2 U\left(\vec{r}\right).$$

Approximating $\overline{\vec{\xi}^2} \approx \lambda_c^2$, this yields the perturbation of the potential due to fluctuations:

$$\delta U \approx \frac{1}{6}\lambda_c^2 \nabla^2 U = \frac{\hbar^2}{6m^2c^2}\nabla^2 U$$

To compare with the expression above, plug in the Coulomb potential:

$$\nabla^2 U = -\nabla^2 \frac{Ze^2}{4\pi\varepsilon_0 r} = 4\pi\left(\frac{Ze^2}{4\pi\varepsilon_0}\right)\delta(\vec{r}) \quad \Rightarrow \quad \delta U \approx \frac{\hbar^2}{6m^2c^2}4\pi\left(\frac{Ze^2}{4\pi\varepsilon_0}\right)\delta(\vec{r})$$

This is only slightly different.

Another mechanism that affects only the s-state is the Lamb shift, a further, smaller correction that arises in quantum electrodynamics that should not be confused with the Darwin term. The Darwin term gives the s-state and p-state the same energy, but the Lamb shift makes the s-state higher in energy than the p-state.

Total Effect

The total effect, obtained by summing the three components up, is given by the following expression:

$$\Delta E = \frac{E_n (Z\alpha)^2}{n}\left(\frac{1}{j+\frac{1}{2}} - \frac{3}{4n}\right),$$

where j is the total angular momentum ($j = 1/2$ if $l = 0$ and $j = l \pm 1/2$ otherwise). It is worth noting that this expression was first obtained by A. Sommerfeld based on the old Bohr theory; i.e., before the modern quantum mechanics was formulated.

The total effect can also be obtained by using the Dirac equation. In this case, it is not necessary to treat the electron as non-relativistic or make other approximations. The exact energies are given by

$$E_{jn} = -m_e c^2 \left[1 - \left(1 + \left[\frac{\alpha}{n - j - \frac{1}{2} + \sqrt{\left(j + \frac{1}{2}\right)^2 - \alpha^2}} \right]^2 \right)^{-1/2} \right].$$

Hyperfine Structure

In atomic physics, hyperfine structure is the different effects leading to small shifts and splittings in the energy levels of atoms, molecules and ions. The name is a reference to the *fine structure*, which results from the interaction between the magnetic moments associated with electron spin and the electrons' orbital angular momentum. Hyperfine structure, with energy shifts typically orders of magnitudes smaller than those of a fine-structure shift, results from the interactions of the nucleus (or nuclei, in molecules) with internally generated electric and magnetic fields.

In atoms, hyperfine structure occurs due to the energy of the nuclear magnetic dipole moment in the magnetic field generated by the electrons, and the energy of the nuclear electric quadrupole moment in the electric field gradient due to the distribution of charge within the atom. Molecular hyperfine structure is generally dominated by these two effects, but also includes the energy associated with the interaction between the magnetic moments associated with different magnetic nuclei in a molecule, as well as between the nuclear magnetic moments and the magnetic field generated by the rotation of the molecule.

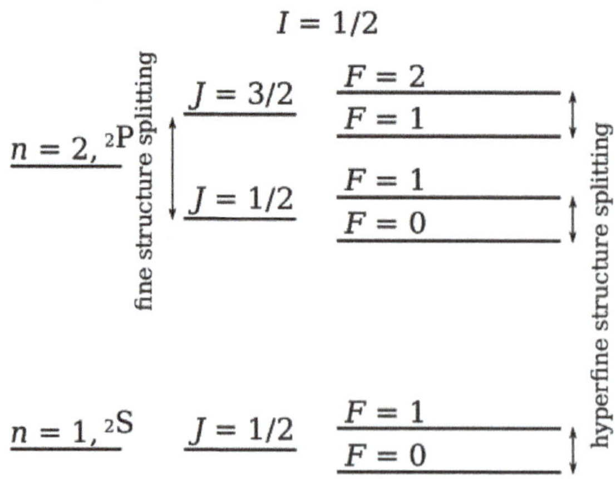

Schematic illustration of fine and hyperfine structure in hydrogen

History

The optical hyperfine structure was already observed in 1881 by Albert Abraham Michelson. It could, however, only be explained in terms of quantum mechanics when Wolfgang Pauli proposed the existence of a small nuclear magnetic moment in 1924.

In 1935, H. Schüler and Theodor Schmidt proposed the existence of a nuclear quadrupole moment in order to explain anomalies in the hyperfine structure.

Theory

The theory of hyperfine structure comes directly from electromagnetism, consisting of the interaction of the nuclear multipole moments (excluding the electric monopole) with internally generated fields. The theory is derived first for the atomic case, but can be applied to *each nucleus* in a molecule. Following this there is a discussion of the additional effects unique to the molecular case.

Atomic Hyperfine Structure

Magnetic Dipole

The dominant term in the hyperfine Hamiltonian is typically the magnetic dipole term. Atomic nuclei with a non-zero nuclear spin **I** have a magnetic dipole moment, given by:

$$\mu_{\mathrm{I}} = g_{\mathrm{I}} \mu_{\mathrm{N}} I,$$

where g_{I} is the *g*-factor and μ_{N} is the nuclear magneton.

There is an energy associated with a magnetic dipole moment in the presence of a magnetic field. For a nuclear magnetic dipole moment, μ_{I}, placed in a magnetic field, B, the relevant term in the Hamiltonian is given by:

$$\hat{H}_{\mathrm{D}} = -\mu_{\mathrm{I}} \cdot B.$$

In the absence of an externally applied field, the magnetic field experienced by the nucleus is that associated with the orbital (l) and spin (s) angular momentum of the electrons:

$$\mathbf{B} \equiv \mathbf{B}_{\mathrm{el}} = \mathbf{B}_{\mathrm{el}}^{l} + \mathbf{B}_{\mathrm{el}}^{s}.$$

Electron orbital angular momentum results from the motion of the electron about some fixed external point that we shall take to be the location of the nucleus. The magnetic field at the nucleus due to the motion of a single electron, with charge *-e* at a position r relative to the nucleus, is given by:

$$\mathbf{B}_{\mathrm{el}}^{l} = \frac{\mu_0}{4\pi} \frac{-e\mathbf{v} \times -\mathbf{r}}{r^3},$$

where −r gives the position of the nucleus relative to the electron. Written in terms of the Bohr magneton, this gives:

$$\mathbf{B}_{\text{el}}^{l} = -2\mu_{\text{B}} \frac{\mu_0}{4\pi} \frac{1}{r^3} \frac{\mathbf{r} \times m_{\text{e}} \mathbf{v}}{\hbar}.$$

Recognizing that m_ev is the electron momentum, p, and that r×p/ħ is the orbital angular momentum in units of ħ, l, we can write:

$$\mathbf{B}_{\text{el}}^{l} = -2\mu_{\text{B}} \frac{\mu_0}{4\pi} \frac{1}{r^3} \mathbf{l}.$$

For a many electron atom this expression is generally written in terms of the total orbital angular momentum, $\mathbf{L}$, by summing over the electrons and using the projection operator, ϕ_i^l, where $\sum_i \mathbf{l}_i \quad \sum_i {}_i\mathbf{L}$. For states with a well defined projection of the orbital angular momentum, L_z, we can write $\phi_i^l = \hat{l}_{z_i} / L_z$, giving:

$$\mathbf{B}_{\text{el}}^{l} = -2\mu_{\text{B}} \frac{\mu_0}{4\pi} \frac{1}{L_z} \sum_i \frac{\hat{l}_{zi}}{r_i^3} \mathbf{L}.$$

The electron spin angular momentum is a fundamentally different property that is intrinsic to the particle and therefore does not depend on the motion of the electron. Nonetheless it is angular momentum and any angular momentum associated with a charged particle results in a magnetic dipole moment, which is the source of a magnetic field. An electron with spin angular momentum, s, has a magnetic moment, μ_s, given by:

$$\mu_{\text{s}} = -g_s \mu_{\text{B}} \mathbf{s},$$

where g_s is the electron spin *g*-factor and the negative sign is because the electron is negatively charged (consider that negatively and positively charged particles with identical mass, travelling on equivalent paths, would have the same angular momentum, but would result in currents in the opposite direction).

The magnetic field of a dipole moment, μ_s, is given by:

$$B_{el}^{s} = \frac{\mu_0}{4\pi r^3}\left(3(\mu_{\text{s}} \cdot \hat{\mathbf{r}})\hat{\mathbf{r}} - \mu_{\text{s}}\right) + \frac{2\mu_0}{3} \mu_{\text{s}} \delta^3(r).$$

The complete magnetic dipole contribution to the hyperfine Hamiltonian is thus given by:

$$\begin{aligned} \hat{H}_D = \quad & 2 g_{\text{I}} \mu_{\text{N}} \mu_{\text{B}} \frac{\mu_0}{4\pi} \frac{1}{L_z} \sum_i \frac{\hat{l}_{zi}}{r_i^3} \mathbf{I} \cdot \mathbf{L} \\ & + g_{\text{I}} \mu_{\text{N}} g_{\text{s}} \mu_{\text{B}} \frac{\mu_0}{4\pi} \frac{1}{S_z} \sum_i \frac{\hat{s}_{zi}}{r_i^3} \left\{ 3(\mathbf{I} \cdot \hat{\mathbf{r}})(\mathbf{S} \cdot \hat{\mathbf{r}}) - \mathbf{I} \cdot \mathbf{S} \right\} \\ & + \frac{2}{3} g_{\text{I}} \mu_{\text{N}} g_{\text{s}} \mu_{\text{B}} \mu_0 \frac{1}{S_z} \sum_i \hat{s}_{zi} \delta^3(\mathbf{r}_i) \mathbf{I} \cdot \mathbf{S}. \end{aligned}$$

The first term gives the energy of the nuclear dipole in the field due to the electronic orbital angular momentum. The second term gives the energy of the "finite distance" interaction of the nuclear dipole with the field due to the electron spin magnetic moments. The final term, often known as the "Fermi contact" term relates to the direct interaction of the nuclear dipole with the spin dipoles and is only non-zero for states with a finite electron spin density at the position of the nucleus (those with unpaired electrons in *s*-subshells). It has been argued that one may get a different expression when taking into account the detailed nuclear magnetic moment distribution.

For states with $l \neq 0$ this can be expressed in the form

$$\hat{H}_D = 2g_I \mu_B \mu_N \frac{\mu_0}{4\pi} \frac{\mathbf{I}\cdot\mathbf{N}}{r^3},$$

where:

$$\mathbf{N} = \mathbf{l} - (g_s/2)\mathbf{s} + 3(\mathbf{s}\cdot\hat{\mathbf{r}})\hat{\mathbf{r}}.$$

If hyperfine structure is small compared with the fine structure (sometimes called *IJ*-coupling by analogy with *LS*-coupling), *I* and *J* are good quantum numbers and matrix elements of $\hat{H}_D$ can be approximated as diagonal in *I* and *J*. In this case (generally true for light elements), we can project N onto J (where J = L + S is the total electronic angular momentum) and we have:

$$\hat{H}_D = 2g_I \mu_B \mu_N \frac{\mu_0}{4\pi} \frac{\mathbf{N}\cdot\mathbf{J}}{\mathbf{J}\cdot\mathbf{J}} \frac{\mathbf{I}\cdot\mathbf{J}}{r^3}.$$

This is commonly written as

$$\hat{H}_D = \hat{A}\mathbf{I}\cdot\mathbf{J},$$

with $\hat{A}$ being the hyperfine structure constant which is determined by experiment. Since I.J = ½{F.F - I.I - J.J} (where F = I + J is the total angular momentum), this gives an energy of

$$\Delta E_D = \frac{1}{2}\langle\hat{A}\rangle[F(F+1) - I(I+1) - J(J+1)].$$

In this case the hyperfine interaction satisfies the Landé interval rule.

Electric Quadrupole

Atomic nuclei with spin $I \geq 1$ have an electric quadrupole moment. In the general case this is represented by a rank-2 tensor, $\underline{\underline{Q}}$, with components given by:

$$Q_{ij} = \frac{1}{e}\int\left(3x_i'x_j' - (r')^2\delta_{ij}\right)\rho(\mathbf{r}')d^3r',$$

where i and j are the tensor indices running from 1 to 3, x_i and x_j are the spatial variables x, y and z depending on the values of i and j respectively, δ_{ij} is the Kronecker delta and $\rho(\mathrm{r})$ is the charge density. Being a 3-dimensional rank-2 tensor, the quadrupole moment has $3^2 = 9$ components. From the definition of the components it is clear that the quadrupole tensor is a symmetric matrix ($Q_{ij} = Q_{ji}$) that is also traceless ($\Sigma_i Q_{ii} = 0$), giving only five components in the irreducible representation. Expressed using the notation of irreducible spherical tensors we have:

$$T_m^2(Q) = \sqrt{\frac{4\pi}{5}}\int\rho(\mathbf{r}')(r')^2 Y_m^2(\theta',\phi')d^3r'.$$

The energy associated with an electric quadrupole moment in an electric field depends not on the field strength, but on the electric field gradient, confusingly labelled $\underline{\underline{q}}$, another rank-2 tensor given by the outer product of the del operator with the electric field vector:

$$\underline{\underline{q}} = \nabla \otimes \mathbf{E},$$

with components given by:

$$q_{ij} = \frac{\partial^2 V}{\partial x_i \partial x_j}.$$

Again it is clear this is a symmetric matrix and, because the source of the electric field at the nucleus is a charge distribution entirely outside the nucleus, this can be expressed as a 5-component spherical tensor, $T^2(q)$, with:

$$T\ (q)\quad \frac{\sqrt{\ }}{}\, q_{zz}$$

$$T_{+1}^2(q) = -q_{xz} - iq_{yz}$$

$$T_{+2}^2(q) = \frac{1}{2}(q_{xx} - q_{yy}) + iq_{xy},$$

where:

$$T_{-m}^2(q) = (-1)^m T_{+m}^2(q)^*.$$

The quadrupolar term in the Hamiltonian is thus given by:

$$\hat{H}_Q = -eT^2(Q)\cdot T^2(q) = -e\sum_m (-1)^m T_m^2(Q)T_{-m}^2(q).$$

A typical atomic nucleus closely approximates cylindrical symmetry and therefore all off-diagonal elements are close to zero. For this reason the nuclear electric quadrupole moment is often represented by Q_{zz}.

Molecular Hyperfine Structure

The molecular hyperfine Hamiltonian includes those terms already derived for the atomic case with a magnetic dipole term for each nucleus with $I > 0$ and an electric quadrupole term for each nucleus with $I > 0$. The magnetic dipole terms were first derived for diatomic molecules by Frosch and Foley, and the resulting hyperfine parameters are often called the Frosch and Foley parameters.

In addition to the effects described above, there are a number of effects specific to the molecular case.

Direct Nuclear Spin–spin

Each nucleus with $I > 0$ has a non-zero magnetic moment that is both the source of a magnetic field and has an associated energy due to the presence of the combined field of all of the other nuclear magnetic moments. A summation over each magnetic moment dotted with the field due to each *other* magnetic moment gives the direct nuclear spin–spin term in the hyperfine Hamiltonian, $\hat{H}_{II}$.

$$\hat{H}_{II} = -\sum_{\alpha\neq\alpha'} \mathbf{i}_{\alpha} \cdot \mathbf{B}_{\alpha'},$$

where α and α› are indices representing the nucleus contributing to the energy and the nucleus that is the source of the field respectively. Substituting in the expressions for the dipole moment in terms of the nuclear angular momentum and the magnetic field of a dipole, both given above, we have

$$\hat{H}_{II} = \frac{\mu_0 \mu_{\mathrm{N}}^2}{4\pi} \sum_{\alpha\neq\alpha'} \frac{g_\alpha g_{\alpha'}}{R_{\alpha\alpha'}^3} \left\{ \mathbf{I}_\alpha \cdot \mathbf{I}_{\alpha'} - 3(\mathbf{I}_\alpha \cdot \hat{\mathbf{R}}_{\alpha\alpha'})(\mathbf{I}_{\alpha'} \cdot \hat{\mathbf{R}}_{\alpha\alpha'}) \right\}.$$

Nuclear Spin–rotation

The nuclear magnetic moments in a molecule exist in a magnetic field due to the angular momentum, T (**R** is the internuclear displacement vector), associated with the bulk rotation of the molecule, thus

$$\hat{H}_{\mathrm{IR}} = \frac{e\mu_0 \mu_{\mathrm{N}} \hbar}{4\pi} \sum_{\alpha\neq\alpha'} \frac{1}{R_{\alpha\alpha'}^3} \left\{ \frac{Z_\alpha g_{\alpha'}}{M_\alpha} \mathbf{I}_{\alpha'} + \frac{Z_{\alpha'} g_\alpha}{M_{\alpha'}} \mathbf{I}_\alpha \right\} \cdot \mathbf{T}.$$

Measurements

Hyperfine interactions can be measured, among other ways, in atomic and molecular spectra and in electron paramagnetic resonance spectra of free radicals and transition-metal ions.

Applications

Astrophysics

As the hyperfine splitting is very small, the transition frequencies usually are not optical, but in the range of radio- or microwave frequencies.

Hyperfine structure gives the 21 cm line observed in H I regions in interstellar medium.

Carl Sagan and Frank Drake considered the hyperfine transition of hydrogen to be a sufficiently universal phenomenon so as to be used as a base unit of time and length on the Pioneer plaque and later Voyager Golden Record.

In radio astronomy, heterodyne receivers are widely used in detection of the electromagnetic signals from celestial objects. The separations among various components of a hyperfine structure are usually small enough to fit into the receiver's IF band. Because optical depth varies with frequency, strength ratios among the hyperfine components differ from that of their intrinsic intensities. From this we can derive the object's physical parameters.

Nuclear Technology

The atomic vapor laser isotope separation (AVLIS) process uses the hyperfine splitting between optical transitions in uranium-235 and uranium-238 to selectively photo-ionize only the uranium-235 atoms and then separate the ionized particles from the non-ionized ones. Precisely tuned dye lasers are used as the sources of the necessary exact wavelength radiation.

Use in Defining the SI Second and Meter

The hyperfine structure transition can be used to make a microwave notch filter with very high stability, repeatability and Q factor, which can thus be used as a basis for very precise atomic clocks. Typically, the hyperfine structure transition frequency of a particular isotope of caesium or rubidium atoms is used as a basis for these clocks.

Due to the accuracy of hyperfine structure transition-based atomic clocks, they are now used as the basis for the definition of the second. One second is now *defined* to be *exactly* 9,192,631,770 cycles of the hyperfine structure transition frequency of caesium-133 atoms.

On October 21, 1983, the 17th CGPM defined the metre as the length of the path travelled by light in a vacuum during a time interval of 1/299,792,458 of a second.

Precision Tests of Quantum Electrodynamics

The hyperfine splitting in hydrogen and in muonium have been used to measure the value of the fine structure constant α. Comparison with measurements of α in other physical systems provides a stringent test of QED.

Qubit in Ion-trap Quantum Computing

The hyperfine states of a trapped ion are commonly used for storing qubits in ion-trap quantum computing. They have the advantage of having very long lifetimes, experimentally exceeding ~10 min (compared to ~1 s for metastable electronic levels).

The frequency associated with the states' energy separation is in the microwave region, making it possible to drive hyperfine transitions using microwave radiation. However, at present no emitter is available that can be focused to address a particular ion from a sequence. Instead, a pair of laser pulses can be used to drive the transition, by having their frequency difference (*detuning*) equal to the required transition's frequency. This is essentially a stimulated Raman transition. In addition, near-field gradients have been exploited to individually address two ions separated by approximately 4.3 micrometers directly with microwave radiation.

References

- R. Resnick, R. Eisberg (1985). Quantum Physics of Atoms, Molecules, Solids, Nuclei and Particles (2nd ed.). John Wiley & Sons. ISBN 978-0-471-87373-0.
- R.P. Feynman (1985). "Electrons and Their Interactions". QED: The Strange Theory of Light and Matter. Princeton, New Jersey: Princeton University Press. p. 115. ISBN 0-691-08388-6.
- Zelevinsky, Vladimir (2011), Quantum Physics Volume 1: From Basics to Symmetries and Perturbations, WILEY-VCH, ISBN 978-3-527-40979-2 p. 551
- Berestetskii, V. B.; E. M. Lifshitz; L. P. Pitaevskii (1982). Quantum electrodynamics. Butterworth-Heinemann. ISBN 978-0-7506-3371-0.
- Sommerfeld, Arnold (1919). Atombau und Spektrallinien'. Braunschweig: Friedrich Vieweg und Sohn. ISBN 3-87144-484-7.
- Brown, John; Alan Carrington (2003). Rotational Spectroscopy of Diatomic Molecules. Cambridge University Press. ISBN 978-0-521-53078-1.

7

Scattering: An Overview

The deviation or reflection of radiation from its original path due to any deviation in the medium like the introduction of particles, density fluctuations, bubbles etc. The chapter examines scattering and its types like Raman scattering, Rayleigh scattering, Rutherford scattering, Compton scattering, small-angle scattering etc. The topics discussed in the chapter are of great importance to broaden the existing knowledge on atomic physics.

Scattering

Scattering is a general physical process where some forms of radiation, such as light, sound, or moving particles, are forced to deviate from a straight trajectory by one or more paths due to localized non-uniformities in the medium through which they pass. In conventional use, this also includes deviation of reflected radiation from the angle predicted by the law of reflection. Reflections that undergo scattering are often called *diffuse reflections* and unscattered reflections are called *specular* (mirror-like) reflections.

Scattering may also refer to particle-particle collisions between molecules, atoms, electrons, photons and other particles. Examples include: cosmic ray scattering in the Earth's upper atmosphere; particle collisions inside particle accelerators; electron scattering by gas atoms in fluorescent lamps; and neutron scattering inside nuclear reactors.

The types of non-uniformities which can cause scattering, sometimes known as *scatterers* or *scattering centers*, are too numerous to list, but a small sample includes particles, bubbles, droplets, density fluctuations in fluids, crystallites in polycrystalline solids, defects in monocrystalline solids, surface roughness, cells in organisms, and textile fibers in clothing. The effects of such features on the path of almost any type of propagating wave or moving particle can be described in the framework of scattering theory.

Some areas where scattering and scattering theory are significant include radar sensing, medical ultrasound, semiconductor wafer inspection, polymerization process monitoring, acoustic tiling, free-space communications and computer-generated imagery. Particle-particle scattering theory is important in areas such as particle physics, atomic, molecular, and optical physics, nuclear physics and astrophysics.

Single and Multiple Scattering

When radiation is only scattered by one localized scattering center, this is called *single scattering*. It is very common that scattering centers are grouped together; in such cases, radiation may scatter many times, in what is known as *multiple scattering*. The main difference between the effects of single and multiple scattering is that single scattering can usually be treated as a random phenomenon, whereas multiple scattering is usually more stochastic. Because the location of a single scattering center is not usually well known relative to the path of the radiation, the outcome, which tends to depend strongly on the exact incoming trajectory, appears random to an observer. This type of scattering would be exemplified by an electron being fired at an atomic nucleus. In this case, the atom's exact position relative to the path of the electron is unknown and would be unmeasurable, so the exact trajectory of the electron after the collision cannot be predicted. Single scattering is therefore often described by probability distributions.

Zodiacal light is a glow originated in the scattering of sunlight by dust located between the planets that are spread through the plane of the Solar System.

With multiple scattering, the randomness of the interaction tends to be averaged out by the large number of scattering events, so that the final path of the radiation appears to be a deterministic distribution of intensity. This is exemplified by a light beam passing through thick fog. Multiple scattering is highly analogous to diffusion, and the terms *multiple scattering* and *diffusion* are interchangeable in many contexts. Optical elements designed to produce multiple scattering are thus known as *diffusers*. Coherent backscattering, an enhancement of backscattering that occurs when coherent radiation is multiply scattered by a random medium, is usually attributed to weak localization.

Not all single scattering is random, however. A well-controlled laser beam can be exactly positioned to scatter off a microscopic particle with a deterministic outcome, for instance. Such situations are encountered in radar scattering as well, where the targets tend to be macroscopic objects such as people or aircraft.

Similarly, multiple scattering can sometimes have somewhat random outcomes, particularly with coherent radiation. The random fluctuations in the multiply scattered intensity of coherent radiation are called speckles. Speckle also occurs if multiple parts of a coherent wave scatter from different centers. In certain rare circumstances, multiple scattering may only involve a small number of interactions such that the randomness is not completely averaged out. These systems are considered to be some of the most difficult to model accurately.

The description of scattering and the distinction between single and multiple scattering are tightly related to wave–particle duality.

Scattering Theory

Scattering theory is a framework for studying and understanding the scattering of waves and particles. Prosaically, wave scattering corresponds to the collision and scattering of a wave with some material object, for instance sunlight scattered by rain drops to form a rainbow. Scattering also includes the interaction of billiard balls on a table, the Rutherford scattering (or angle change) of alpha particles by gold nuclei, the Bragg scattering (or diffraction) of electrons and X-rays by a cluster of atoms, and the inelastic scattering of a fission fragment as it traverses a thin foil. More precisely, scattering consists of the study of how solutions of partial differential equations, propagating freely "in the distant past", come together and interact with one another or with a boundary condition, and then propagate away "to the distant future".

Electromagnetic Scattering

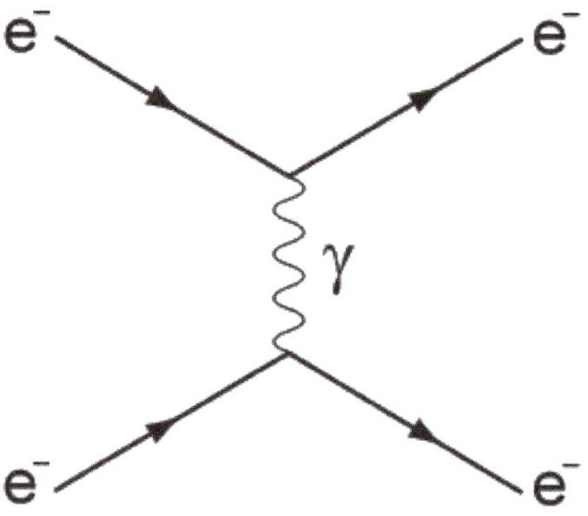

A Feynman diagram of scattering between two electrons by emission of a virtual photon.

Electromagnetic waves are one of the best known and most commonly encountered forms of radiation that undergo scattering. Scattering of light and radio waves (especially in radar) is particularly important. Several different aspects of electromagnetic scattering are distinct enough to have conventional names. Major forms of elastic light scattering (involving negligible energy transfer) are Rayleigh scattering and Mie scattering. Inelastic scattering includes Brillouin scattering, Raman scattering, inelastic X-ray scattering and Compton scattering.

Light scattering is one of the two major physical processes that contribute to the visible appearance of most objects, the other being absorption. Surfaces described as *white* owe their appearance to multiple scattering of light by internal or surface inhomogeneities in the object, for example by the boundaries of transparent microscopic crystals that make up a stone or by the microscopic fibers in a sheet of paper. More generally, the gloss (or lustre or sheen) of the surface is determined by scattering. Highly scattering surfaces are described as being dull or having a matte finish, while the absence of surface scattering leads to a glossy appearance, as with polished metal or stone.

Spectral absorption, the selective absorption of certain colors, determines the color of most objects with some modification by elastic scattering. The apparent blue color of veins in skin is a common example where both spectral absorption and scattering play important and complex roles in the coloration. Light scattering can also create color without absorption, often shades of blue, as with the sky (Rayleigh scattering), the human blue iris, and the feathers of some birds (Prum et al. 1998). However, resonant light scattering in nanoparticles can produce many different highly saturated and vibrant hues, especially when surface plasmon resonance is involved (Roqué et al. 2006).

Models of light scattering can be divided into three domains based on a dimensionless size parameter, α which is defined as:

$$\alpha = \pi D_p / \lambda,$$

where πD_p is the circumference of a particle and λ is the wavelength of incident radiation. Based on the value of α, these domains are:

$\alpha \ll 1$: Rayleigh scattering (small particle compared to wavelength of light);

$\alpha \approx 1$: Mie scattering (particle about the same size as wavelength of light, valid only for spheres);

$\alpha \gg 1$: geometric scattering (particle much larger than wavelength of light).

Rayleigh scattering is a process in which electromagnetic radiation (including light) is scattered by a small spherical volume of variant refractive index, such as a particle, bubble, droplet, or even a density fluctuation. This effect was first modeled successfully by Lord Rayleigh, from whom it gets its name. In order for Rayleigh's model to apply, the sphere must be much smaller in diameter than the wavelength (λ) of the scattered wave; typically the upper limit is taken to be about 1/10 the wavelength. In this size regime, the exact shape of the scattering center is usually not very significant and can often be treated as a sphere of equivalent volume. The inherent scattering that radiation undergoes passing through a pure gas is due to microscopic density fluctuations as the gas molecules move around, which are normally small enough in scale for Rayleigh's model to apply. This scattering mechanism is the primary cause of the blue color of the Earth's sky on a clear day, as the shorter blue wavelengths of sunlight passing overhead are more strongly scattered than the longer red wavelengths according to Rayleigh's famous $1/\lambda^4$ relation. Along with absorption, such scattering is a major cause of the attenuation of radiation by the atmosphere. The degree of scattering varies as a function of the ratio of the particle diameter to the wavelength of the radiation, along with many other factors including polarization, angle, and coherence.

For larger diameters, the problem of electromagnetic scattering by spheres was first solved by Gustav Mie, and scattering by spheres larger than the Rayleigh range is therefore usually known as Mie scattering. In the Mie regime, the shape of the scattering center becomes much more significant and the theory only applies well to spheres and, with some modification, spheroids and ellipsoids. Closed-form solutions for scattering by certain other simple shapes exist, but no general closed-form solution is known for arbitrary shapes.

Both Mie and Rayleigh scattering are considered elastic scattering processes, in which the energy (and thus wavelength and frequency) of the light is not substantially changed. However, electromagnetic radiation scattered by moving scattering centers does undergo a Doppler shift, which can be detected and used to measure the velocity of the scattering center/s in forms of techniques such as lidar and radar. This shift involves a slight change in energy.

At values of the ratio of particle diameter to wavelength more than about 10, the laws of geometric optics are mostly sufficient to describe the interaction of light with the particle, and at this point the interaction is not usually described as scattering.

For modeling of scattering in cases where the Rayleigh and Mie models do not apply such as irregularly shaped particles, there are many numerical methods that can be used. The most common are finite-element methods which solve Maxwell's equations to find the distribution of the scattered electromagnetic field. Sophisticated software packages exist which allow the user to specify the refractive index or indices of the scattering feature in space, creating a 2- or sometimes 3-dimensional model of the structure. For relatively large and complex structures, these models usually require substantial execution times on a computer.

Raman Scattering

Raman scattering or the Raman effect /ˈrɑːmən/ is the inelastic scattering of a photon upon interaction with matter. It was discovered by C. V. Raman and K. S. Krishnan (who was a student of C.V. Raman) in liquids, and independently by Grigory Landsberg and Leonid Mandelstam in crystals. The effect had been predicted theoretically by Adolf Smekal in 1923.

When photons are scattered from an atom or molecule, most photons are elastically scattered (Rayleigh scattering), such that the scattered photons have the same energy (frequency and wavelength) as the incident photons. A small fraction of the scattered photons (approximately 1 in 10 million) are scattered by an excitation, with the scattered photons having a frequency different from, and usually lower than, that of the incident photons. In a gas, Raman scattering can occur with a change in energy of a molecule due to a transition to another (usually higher) energy level. Chemists are primarily concerned with the transitional Raman effect.

History

The inelastic scattering of light was predicted by Adolf Smekal in 1923 (and in German-language literature it may be referred to as the Smekal-Raman effect). In 1922, Indian physicist C. V. Raman published his work on the "Molecular Diffraction of Light," the first of a series of investigations with his collaborators that ultimately led to his discovery (on 28 February 1928) of the radiation effect that bears his name. The Raman effect was first reported by C. V. Raman and K. S. Krishnan, and independently by Grigory Landsberg and Leonid Mandelstam, on 21 February 1928 (that is why in the former Soviet Union the priority of Raman was always disputed; thus in Russian scientific literature this effect is usually referred to as "combination scattering" or "combinatory scattering"). Raman received the Nobel Prize in 1930 for his work on the scattering of light.

In 1998 the Raman effect was designated a National Historic Chemical Landmark by the American Chemical Society in recognition of its significance as a tool for analyzing the composition of liquids, gases, and solids.

Description

Degrees of Freedom

For any given chemical compound, there are a total of 3N degrees of freedom, where N is the number of atoms in the compound. This number arises from the ability of each atom in a molecule to move in three different directions (x, y, and z). When dealing with molecules, it is more common to consider the movement of the molecule as a whole. Consequently, the 3N degrees of freedom are partitioned into molecular translational, rotational, and vibrational motion. Three of the degrees of freedom correspond to translational motion of the molecule as a whole (along each of the three spatial dimensions). Similarly, three degrees of freedom correspond to rotations of the molecule about the x, y, and z -axes. Linear molecules only have two rotations because rotations along the bond axis do not change the positions of the atoms in the molecule. The remaining degrees of freedom correspond to molecular vibrational modes. These modes include stretching and bending motions of the chemical bonds of the molecule. For a linear molecule, the number of vibrational modes is:

$$3N - 3 - 2 = 3N - 5$$

whereas for a non-linear molecule the number of vibrational modes are

$$3N - 6$$

The frequencies of molecular vibrations range from less than 10^{12} to approximately 10^{14} Hz. These frequencies correspond to radiation in the infrared (IR) region of the electromagnetic spectrum. At any given instant, each molecule in a sample has a certain amount of vibrational energy. However, the amount of vibrational energy that a molecule has continually changes due to collisions and other interactions with other molecules in the sample.

At room temperature, most of the molecules will be in the lowest energy state, which is known as the ground state. A few molecules will be in higher energy states, which are known as excited states. The fraction of molecules occupying a given vibrational mode at a given temperature can be calculated using the Boltzmann distribution. Performing such a calculation shows that, for relatively low temperatures (such as those used for most routine spectroscopy), most of the molecules occupy the ground vibrational state. Such a molecule can be excited to a higher vibrational mode through the direct absorption of a photon of the appropriate energy. This is the mechanism by which IR spectroscopy operates: infrared radiation is passed through the sample, and the intensity of the transmitted light is compared with that of the incident light. A reduction in intensity at a given wavelength of light indicates the absorption of energy by a vibrational transition. The energy, E, of a photon is

$$E = h\nu,$$

where h is Planck's constant and ν is the frequency of the radiation. Thus, the energy required for such a transition may be calculated if the frequency of the incident radiation is known.

Raman Scattering

It is also possible to observe molecular vibrations by an inelastic scattering process. In inelastic (Raman) scattering, an absorbed photon is re-emitted with lower energy; the difference in energy between the incident photons and scattered photons corresponds to the energy required to excite a molecule to a higher vibrational mode.

Typically, in Raman spectroscopy high intensity laser radiation with wavelengths in either the visible or near-infrared regions of the spectrum is passed through a sample. Photons from the laser beam produce an oscillating polarization in the molecules, exciting them to a virtual energy state. The oscillating polarization of the molecule can couple with other possible polarizations of the molecule, including vibrational and electronic excitations. If the polarization in the molecule does not couple to these other possible polarizations, then it will not change the vibrational state that the molecule started in and the scattered photon will have the same energy as the original photon. This type of scattering is known as Rayleigh scattering.

When the polarization in the molecules couples to a vibrational state that is higher in energy than the state they started in, then the original photon and the scattered photon differ in energy by the amount required to vibrationally excite the molecule. In perturbation theory, the Raman effect corresponds to the absorption and subsequent emission of a photon via an intermediate quantum state of a material. The intermediate state can be either a "real", i.e., stationary state or a virtual state.

Stokes and anti-Stokes Scattering

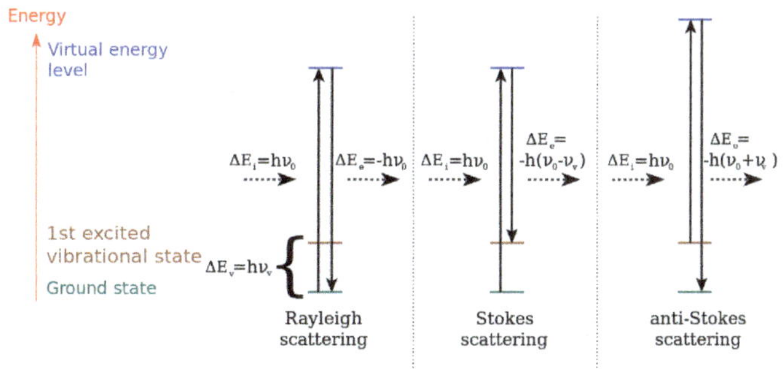

The different possibilities of light scattering: Rayleigh scattering (no exchange of energy: incident and scattered photons have the same energy), Stokes Raman scattering (atom or molecule absorbs energy: scattered photon has less energy than the incident photon) and anti-Stokes Raman scattering (atom or molecule loses energy: scattered photon has more energy than the incident photon)

The Raman interaction leads to two possible outcomes:

- the material absorbs energy and the emitted photon has a lower energy than the absorbed photon. This outcome is labeled Stokes Raman scattering in honor of George Stokes who showed in 1852 that fluorescence is due to light emission at longer wavelength (now known to correspond to lower energy) than the absorbed incident light.
- the material loses energy and the emitted photon has a higher energy than the absorbed photon. This outcome is labeled anti-Stokes Raman scattering.

The energy difference between the absorbed and emitted photon corresponds to the energy difference between two resonant states of the material and is independent of the absolute energy of the photon.

The spectrum of the scattered photons is termed the Raman spectrum. It shows the intensity of the scattered light as a function of its frequency difference $\Delta\nu$ to the incident photons. The locations of corresponding Stokes and anti-Stokes peaks form a symmetric pattern around $\Delta\nu=0$. The frequency shifts are symmetric because they correspond to the energy difference between the same upper and lower resonant states. The intensities of the pairs of features will typically differ, though. They depend on the populations of the initial states of the material, which in turn depend on the temperature. In thermodynamic equilibrium, the upper state will be less populated than the lower state.

Therefore, the rate of transitions from the lower to the upper state (Stokes transitions) will be higher than in the opposite direction (anti-Stokes transitions). Correspondingly, Stokes scattering peaks are stronger than anti-Stokes scattering peaks. Their ratio depends on the temperature, and can therefore be exploited to measure it.

Distinction from Fluorescence

The Raman effect differs from the process of fluorescence in that it is a scattering process. For fluorescence, the incident light is completely absorbed, transferring the system to an excited state. After a certain resonance lifetime, the system de-excites to lower energy states via emission of photons. The result of both processes is in essence the same: A photon with a frequency different from that of the incident photon is produced and the molecule is brought to a higher or lower energy level. But the major difference is that the Raman effect can take place for any frequency of incident light. In contrast to the fluorescence effect, the Raman effect is therefore not a resonant effect. In practice, this means that a fluorescence peak is anchored at a specific frequency, whereas a Raman peak maintains a constant separation from the excitation frequency.

Selection Rules

A Raman transition from one state to another is allowed only if the molecular polarizability of those states is different. For a vibration, this means that the derivative of the polarizability with respect to the normal coordinate associated to the vibration is non zero: $\partial\alpha / \partial Q \neq 0$. In general, a normal mode is Raman active if it transforms with the same symmetry of the quadratic forms ($x^2, y^2, z^2, xy, xz, yz$),), which can be verified from the character table of the molecule's symmetry group.

The specific selection rules state that the allowed rotational transitions are $\ddot{A}J = \pm 2$, where J is the rotational state.

The allowed vibrational transitions are $\ddot{A}\nu = \pm 1$, where ν is the vibrational state.

Stimulated Raman Scattering and Raman Amplification

The Raman-scattering process as described above takes place spontaneously; i.e., in random time intervals, one of the many incoming photons is scattered by the material. This process is thus called *spontaneous Raman scattering*.

On the other hand, *stimulated Raman scattering* can take place when some Stokes photons have previously been generated by spontaneous Raman scattering (and somehow forced to remain in the material), or when deliberately injecting Stokes photons ("signal light") together with the original light ("pump light"). In that case, the total Raman-scattering rate is increased beyond that of spontaneous Raman scattering: pump photons are converted more rapidly into additional Stokes photons. The more Stokes photons are already present, the faster more of them are added. Effectively, this *am-*

plifies the Stokes light in the presence of the pump light, which is exploited in Raman amplifiers and Raman lasers.

Stimulated Raman scattering is a nonlinear-optical effect. It can be described using a third-order nonlinear susceptibility $\chi^{(3)}$.

Need of Space-coherence

Suppose that the distance between two points A and B of an exciting beam is x. Generally, as the exciting frequency is not equal to the scattered Raman frequency, the corresponding relative wavelengths λ and λ' are not equal. Thus, a phase-shift $\Theta = 2\pi x(1/\lambda - 1/\lambda')$ appears. For $\Theta = \pi$, the scattered amplitudes are opposite, so that the Raman scattered beam remains weak.

- A crossing of the beams may limit the path x.

Several tricks may be used to get a larger amplitude:

- In an optically anisotropic crystal, a light ray may have two modes of propagation with different polarizations and different indices of refraction. If energy may be transferred between these modes by a quadrupolar (Raman) resonance, phases remain coherent along the whole path, transfer of energy may be large. It is an Optical parametric generation.

- Light may be pulsed, so that beats do not appear.

It is the Impulsive Stimulated Raman Scattering (ISRS), in which the length of the pulses must be shorter than all relevant time constants. Interference of Raman and incident lights is too short to allow beats, so that it produces a frequency shift roughly, in best conditions, inversely proportional to cube of length of pulses. In labs, femtosecond laser pulses must be used because the ISRS becomes very weak if the pulses are too long. Thus ISRS cannot be observed using nanosecond pulses making ordinary time-incoherent light.

Applications

Raman spectroscopy employs the Raman effect for substances analysis. The spectrum of the Raman-scattered light depends on the molecular constituents present and their state, allowing the spectrum to be used for material identification and analysis. Raman spectroscopy is used to analyze a wide range of materials, including gases, liquids, and solids. Highly complex materials such as biological organisms and human tissue can also be analyzed by Raman spectroscopy.

For solid materials, Raman scattering is used as a tool to detect high-frequency phonon and magnon excitations.

Raman lidar is used in atmospheric physics to measure the atmospheric extinction coefficient and the water vapour vertical distribution.

Stimulated Raman transitions are also widely used for manipulating a trapped ion's energy levels, and thus basis qubit states.

Raman spectroscopy can be used to determine the force constant and bond length for molecules that do not have an infrared absorption spectrum.

Raman amplification is used in optical amplifiers.

Supercontinuum Generation

For high-intensity continuous wave (CW) lasers, SRS can be used to produce broad bandwidth spectra. This process can also be seen as a special case of four-wave mixing, wherein the frequencies of the two incident photons are equal and the emitted spectra are found in two bands separated from the incident light by the phonon energies. The initial Raman spectrum is built up with spontaneous emission and is amplified later on. At high pumping levels in long fibers, higher-order Raman spectra can be generated by using the Raman spectrum as a new starting point, thereby building a chain of new spectra with decreasing amplitude. The disadvantage of intrinsic noise due to the initial spontaneous process can be overcome by seeding a spectrum at the beginning, or even using a feedback loop as in a resonator to stabilize the process. Since this technology easily fits into the fast evolving fiber laser field and there is demand for transversal coherent high-intensity light sources (i.e., broadband telecommunication, imaging applications), Raman amplification and spectrum generation might be widely used in the near-future.

Rayleigh Scattering

Rayleigh scattering causes the blue hue of the daytime sky and the reddening of the sun at sunset.

Rayleigh scattering (pronounced /ˈreɪli/ *RAY-LEE*), named after the British physicist Lord Rayleigh (John William Strutt), is the (dominantly) elastic scattering of light or other electromagnetic radiation by particles much smaller than the wavelength of the radiation. Rayleigh scattering does not change the state of material and is, hence, a parametric process. The particles may be individual atoms or molecules. It can occur when light travels through transparent solids and liquids, but is most prominently seen in gases. Rayleigh scattering results from the electric polarizability of the particles. The oscillating electric field of a light wave acts on the charges within a particle, causing them to move at the same frequency. The particle therefore becomes a small radiating dipole whose radiation we see as scattered light.

Rayleigh scattering of sunlight in the atmosphere causes diffuse sky radiation, which is the reason for the blue color of the sky and the yellow tone of the sun itself.

Rayleigh scattering of molecular nitrogen and oxygen in the atmosphere includes elastic scattering as well as the inelastic contribution from rotational Raman scattering in air, since the changes in wavenumber of the scattered photon are typically smaller than 50 cm^{-1}. This can lead to changes in the rotational state of the molecules. Furthermore, the inelastic contribution has the same wavelengths dependency as the elastic part.

Scattering by particles similar to, or larger than, the wavelength of light is typically treated by the Mie theory, the discrete dipole approximation and other computational techniques. Rayleigh scattering applies to particles that are small with respect to wavelengths of light, and that are optically "soft" (i.e. with a refractive index close to 1). On the other hand, Anomalous Diffraction Theory applies to optically soft but larger particles.

Small Size Parameter Approximation

The size of a scattering particle is often parameterized by the ratio

$$x = \frac{2\pi r}{\lambda},$$

where r is its characteristic length (radius) and λ is the wavelength of the light. The amplitude of light scattered from within any transparent dielectric is proportional to the inverse square of its wavelength and to the volume of material, that is to the cube of its characteristic length. The wavelength dependence is characteristic of dipole scattering and the volume dependence will apply to any scattering mechanism. Objects with $x \gg 1$ act as geometric shapes, scattering light according to their projected area. At the intermediate $x \simeq 1$ of Mie scattering, interference effects develop through phase variations over the object's surface. Rayleigh scattering applies to the case when the scattering particle is very small ($x \ll 1$, with a particle size < 1 /10 wavelength) and the whole surface re-radiates with the same phase. Because the particles are randomly positioned, the scattered light arrives at a particular point with a random collection of phases; it is incoherent and the resulting intensity is just the sum of the squares of the amplitudes from each particle and therefore proportional to the inverse fourth power of the wavelength and the sixth power of its size. In detail, the intensity I of light scattered by any one of the small spheres of diameter d and refractive index n from a beam of unpolarized light of wavelength λ and intensity I_0 is given by

$$I = I_0 \frac{1+\cos^2\theta}{2R^2}\left(\frac{2\pi}{\lambda}\right)^4\left(\frac{n^2-1}{n^2+2}\right)^2\left(\frac{d}{2}\right)^6,$$

where R is the distance to the particle and θ is the scattering angle. Averaging this over all angles gives the Rayleigh scattering cross-section

$$\sigma_s = \frac{2\pi^5}{3}\frac{d^6}{\lambda^4}\left(\frac{n^2-1}{n^2+2}\right)^2.$$ [8]

The fraction of light scattered by a group of scattering particles is the number of particles per unit volume N times the cross-section. For example, the major constituent of the atmosphere, nitrogen, has a Rayleigh cross section of 5.1×10^{-31} m² at a wavelength of 532 nm (green light). This means that at atmospheric pressure, where there are about 2×10^{25} molecules per cubic meter, about a fraction 10^{-5} of the light will be scattered for every meter of travel.

The strong wavelength dependence of the scattering ($\sim\lambda^{-4}$) means that shorter (blue) wavelengths are scattered more strongly than longer (red) wavelengths.

From Molecules

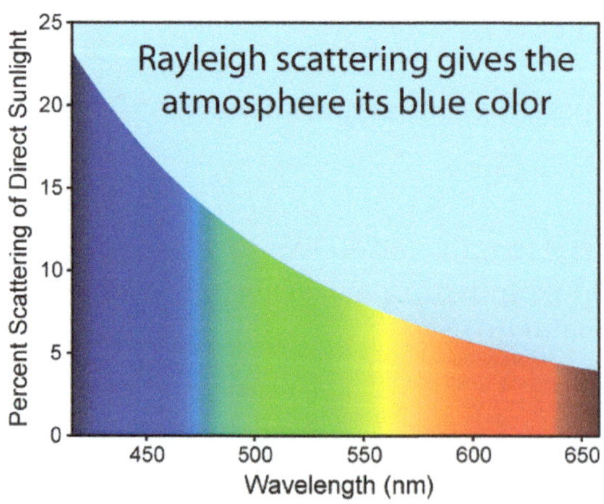

Figure showing the greater proportion of blue light scattered by the atmosphere relative to red light.

The expression above can also be written in terms of individual molecules by expressing the dependence on refractive index in terms of the molecular polarizability α, proportional to the dipole moment induced by the electric field of the light. In this case, the Rayleigh scattering intensity for a single particle is given in CGS-units by

$$I = I_0 \frac{8\pi^4\alpha^2}{\lambda^4 R^2}(1+\cos^2\theta).$$

Reason for the Blue Color of the Sky

Scattered blue light is polarized. The picture on the right is shot through a polarizing filter: the polarizer transmits light that is linearly polarized in a specific direction.

The strong wavelength dependence of the scattering ($\sim\lambda^{-4}$) means that shorter (blue) wavelengths are scattered more strongly than longer (red) wavelengths. This results in the indirect blue light coming from all regions of the sky. Rayleigh scattering is a good approximation of the manner in which light scattering occurs within various media for which scattering particles have a small size (parameter).

A portion of the beam of light coming from the sun scatters off molecules of gas and other small particles in the atmosphere. Here, Rayleigh scattering primarily occurs through sunlight's interaction with randomly located air molecules. It is this scattered light that gives the surrounding sky its brightness and its color. As previously stated, Rayleigh scattering is inversely proportional to the fourth power of wavelength, so that shorter wavelength violet and blue light will scatter more than the longer wavelengths (yellow and especially red light). However, the Sun, like any star, has its own spectrum and so I_0 in the scattering formula above is not constant but falls away in the violet. In addition the oxygen in the Earth's atmosphere absorbs wavelengths at the edge of the ultra-violet region of the spectrum. The resulting color, which appears like a pale blue, actually is a mixture of all the scattered colors, mainly blue and green. Conversely, glancing toward the sun, the colors that were not scattered away — the longer wavelengths such as red and yellow light — are directly visible, giving the sun itself a slightly yellowish hue. Viewed from space, however, the sky is black and the sun is white.

The reddening of the sun is intensified when it is near the horizon because the light being received directly from it must pass through more of the atmosphere. The effect is further increased because the sunlight must pass through a greater proportion of the atmosphere nearer the earth's surface, where it is denser. This removes a significant proportion of the shorter wavelength (blue) and medium wavelength (green) light from the direct path to the observer. The remaining unscattered light is therefore mostly of longer wavelengths and appears redder.

Some of the scattering can also be from sulfate particles. For years after large Plinian eruptions, the blue cast of the sky is notably brightened by the persistent sulfate load of the stratospheric gases. Some works of the artist J. M. W. Turner may owe their vivid red colours to the eruption of Mount Tambora in his lifetime.

In locations with little light pollution, the moonlit night sky is also blue, because moonlight is reflected sunlight, with a slightly lower color temperature due to the brownish color of the moon. The moonlit sky is not perceived as blue, however, because at low light levels human vision comes mainly from rod cells that do not produce any color perception (Purkinje effect).

In Optical Fibers

Rayleigh scattering is an important component of the scattering of optical signals in optical fibers. Silica fibers are glasses, disordered materials with microscopic variations of density and refractive index. These give rise to energy losses due to the scattered light, with the following coefficient:

$$\alpha_{\text{scat}} = \frac{8\pi^3}{3\lambda^4} n^8 p^2 k T_{\text{f}} \beta$$

where n is the refraction index, p is the photoelastic coefficient of the glass, k is the Boltzmann constant, and β is the isothermal compressibility. T_f is a *fictive temperature*, representing the temperature at which the density fluctuations are "frozen" in the material.

In Porous Materials

Rayleigh scattering in opalescent glass: it appears blue from the side, but orange light shines through.

λ^{-4} Rayleigh-type scattering can also be exhibited by porous materials. An example is the strong optical scattering by nanoporous materials. The strong contrast in refractive index between pores and solid parts of sintered alumina results in very strong scattering, with light completely changing direction each 5 micrometers on average. The λ^{-4}-type scattering is caused by the nanoporous structure (a narrow pore size distribution around ~70 nm) obtained by sintering monodispersive alumina powder.

Rutherford Scattering

Rutherford scattering is the elastic scattering of charged particles by the Coulomb interaction. It is a physical phenomenon explained by Ernest Rutherford in 1911 that led to the development of the planetary Rutherford model of the atom and eventually the Bohr model. It is now exploited by the materials analytical technique Rutherford backscattering. Rutherford scattering was first referred to as Coulomb scattering because it relies only upon static electric (Coulomb) forces, and the minimal distance between particles is set only by this potential. The classical Rutherford scattering of alpha particles against gold nuclei is an example of "elastic scattering" because the energy and velocity of the outgoing scattered particle is the same as that with which it began.

Rutherford also later analyzed inelastic scattering when he projected alpha particles against hydrogen nuclei (protons); however this latter process is not referred to as "Rutherford scattering", although Rutherford was first to observe it. At the end of such processes, non-Coulombic forces come into play. These forces, and also energy gained from the scattering particle by the lighter target, change the scattering results in fundamental ways which suggest structural information about the target. A similar process probed the insides of nuclei in the 1960s, and is called deep inelastic scattering.

The initial discovery was made by Hans Geiger and Ernest Marsden in 1909 when they performed the gold foil experiment in collaboration with Rutherford, in which they fired a beam of alpha particles (helium nuclei) at layers of gold leaf only a few atoms thick. At the time of the experiment, the atom was thought to be analogous to a plum pudding (as proposed by J. J. Thomson), with the negative charges (the plums) found throughout a positive sphere (the pudding). If the plum-pudding model were correct, the positive "pudding", being more spread out than in the current model of a concentrated nucleus, would not be able to exert such large columbic forces, and the alpha particles should only be deflected by small angles as they pass through.

However, the intriguing results showed that around 1 in 8000 alpha particles were deflected by very large angles (over 90°), while the rest passed straight through with little or no deflection. From this, Rutherford concluded that the majority of the mass was concentrated in a minute, positively charged region (the nucleus/ central charge) surrounded by electrons. When a (positive) alpha particle approached sufficiently close to the nucleus, it was repelled strongly enough to rebound at high angles. The small size of the nucleus explained the small number of alpha particles that were repelled in this way. Rutherford showed, using the method below, that the size of the nucleus was less than about 10^{-14} m (how *much* less than this size, Rutherford could not tell from this experiment alone; see more below on this problem of lowest possible size).

Derivation

The differential cross section can be derived from the equations of motion for a particle interacting with a central potential. In general, the equations of motion describing two particles interacting under a central force can be decoupled into the center of mass and the motion of the particles relative to one another. For the case of light alpha particles scattering off heavy nuclei, as in the experiment performed by Rutherford, the reduced mass is essentially the mass of the alpha particle and the nucleus off of which it scatters is essentially stationary in the lab frame.

Substituting into the Binet equation yields the equation of trajectory

$$\frac{d^2u}{d\theta^2}+u=-\frac{Z_1Z_2e^2}{4\pi\epsilon_0 m v_0^2 b^2}=-\kappa,$$

where $u=\frac{1}{r}$, v_0 is the speed at infinity, and b is the impact parameter.
The general solution of the above differential equation is

$$u=u_0\cos(\theta-\theta_0)-\kappa,$$

and the boundary condition is

$$u\to 0 \quad r\sin\theta\to b \quad (\theta\to\pi).$$

If we choose

$$\theta_0=\frac{\pi}{2}+\arctan b\kappa.$$

then the deflection angle Θ can be seen from solving $u\to 0$ as

$$\Theta=2\theta_0-\pi=2\arctan b\kappa=2\arctan\frac{Z_1Z_2e^2}{4\pi\epsilon_0 m v_0^2 b}.$$

b can be solved to give

$$b=\frac{Z_1Z_2e^2}{4\pi\epsilon_0 m v_0^2}\cot\frac{\text{È}}{2}.$$

To find the scattering cross section from this result consider its definition

$$\frac{d\sigma}{d\Omega}(\Omega)d\Omega=\frac{\text{number of particles scattered into solid angle d}\Omega\text{ per unit time}}{\text{incident intensity}}$$

Since the scattering angle is uniquely determined for a given E and b, the number of particles scattered into an angle between È and È + dÈ must be the same as the number of particles with associated impact parameters between b and $b+db$.. For an incident intensity I, this implies the following equality

$$2\pi Ib\left|db\right| = I\frac{d\sigma}{d\Omega}d\Omega$$

For a radially symmetric scattering potential, as in the case of the Coulomb potential, $d\Omega = 2\pi \sin\Theta d\Theta$, yielding the expression for the scattering cross section

$$\frac{d\sigma}{d\Omega} = \frac{b}{\sin\Theta}\left|\frac{db}{d\Theta}\right|$$

Finally, plugging in the previously derived expression for the impact parameter bÈ) we find the Rutherford scattering cross section

$$\frac{d\sigma}{d\Omega} = \left(\frac{Z_1 Z_2 e^2}{8\pi\epsilon_0 m v_0^2}\right)^2 \csc^4\left(\frac{\Theta}{2}\right).$$

Details of Calculating Maximal Nuclear Size

For head-on collisions between alpha particles and the nucleus, all the kinetic energy of the alpha particle is turned into potential energy and the particle is at rest. The distance from the centre of the alpha particle to the centre of the nucleus (b) at this point is a maximum value for the radius, if it is evident from the experiment that the particles have not hit the nucleus.

Applying the inverse-square law between the charges on the alpha particle and nucleus, one can write:

$$\frac{1}{2}mv^2 = \frac{1}{4\pi\epsilon_0}\cdot\frac{q_1 q_2}{b}$$

Rearranging:

$$b = \frac{1}{4\pi\epsilon_0}\cdot\frac{2q_1 q_2}{mv^2}$$

For an alpha particle:

- m (mass) = 6.64424×10^{-27} kg = 3.7273×10^{9} eV
- $q_1 = 2\times(1.6\times10^{-19})$ C

- q_2 (for gold) = $79\times(1.6\times10^{-19})$ C
- **v** (initial velocity) = 2×10^{7} m/s

Substituting these in gives the value of about 2.7×10^{-14} m. (The true radius is about 7.3×10^{-15} m.) The true radius of the nucleus is not recovered in these experiments because the alphas do not have enough energy to penetrate to more than 27 fm of the nuclear center, as noted, when the actual radius of gold is 7.3 fm. Rutherford realized this, and also realized that actual impact of the alphas on gold causing any force-deviation from that of the 1/r coulomb potential would change the *form* of his scattering curve at high scattering angles (the smallest impact parameters) from a hyperbola to something else. This was not seen, indicating that the gold had not been "hit" so that Rutherford also knew the gold nucleus (or the sum of the gold and alpha radii) was smaller than 27 fm (2.7×10^{-14} m)

Extension to Situations With Relativistic Particles and Target Recoil

The extension of Rutherford-type scattering to energy regions in which the incoming particle has spin and magnetic moment, and is traveling at relativistic energies, and there is enough momentum-transfer that the struck particle recoils with some of the incoming particle's energy (so the process is inelastic rather than elastic), is called Mott scattering.

Compton Scattering

Compton scattering, discovered by Arthur Holly Compton, is the inelastic scattering of a photon by a charged particle, usually an electron. It results in a decrease in energy (increase in wavelength) of the photon (which may be an X ray or gamma ray photon), called the Compton effect. Part of the energy of the photon is transferred to the recoiling electron. Inverse Compton scattering exists, in which a charged particle transfers part of its energy to a photon.

Introduction

Compton scattering is an example of inelastic scattering of light by a free charged particle, where the wavelength of the scattered light is different from that of the incident radiation. In Compton's original experiment, the energy of the X ray photon ($\approx$17 keV) was very much larger than the binding energy of the atomic electron, so the electrons could be treated as being free. The amount by which the light's wavelength changes is called the Compton shift. Although nuclear Compton scattering exists, Compton scattering usually refers to the interaction involving only the electrons of an atom. The

Compton effect was observed by Arthur Holly Compton in 1923 at Washington University in St. Louis and further verified by his graduate student Y. H. Woo in the years following. Compton earned the 1927 Nobel Prize in Physics for the discovery.

The effect is significant because it demonstrates that light cannot be explained purely as a wave phenomenon. Thomson scattering, the classical theory of an electromagnetic wave scattered by charged particles, cannot explain shifts in wavelength at low intensity: classically, light of sufficient intensity for the electric field to accelerate a charged particle to a relativistic speed will cause radiation-pressure recoil and an associated Doppler shift of the scattered light, but the effect would become arbitrarily small at sufficiently low light intensities *regardless of wavelength*. Thus, light must behave as if it consists of particles, if we are to explain low-intensity Compton scattering. Compton's experiment convinced physicists that light can behave as a stream of particle-like objects (quanta), whose energy is proportional to the light wave's frequency.

The interaction between an electron and a photon results in the electron being given part of the energy (making it recoil), and a photon of the remaining energy being emitted in a different direction from the original, so that the overall momentum of the system is conserved. If the scattered photon still has enough energy, the process may be repeated. In this scenario, the electron is treated as free or loosely bound. Experimental verification of momentum conservation in individual Compton scattering processes by Bothe and Geiger as well as by Compton and Simon has been important in disproving the BKS theory.

If the photon is of low but sufficient energy (in general a few eV to a few keV, corresponding to visible light through soft X-rays), it can eject an electron from its host atom entirely (a process known as the photoelectric effect), instead of undergoing Compton scattering. High energy photons (1.022 MeV and above) may be able to bombard the nucleus and cause an electron and a positron to be formed, a process called pair production.

Description of the Phenomenon

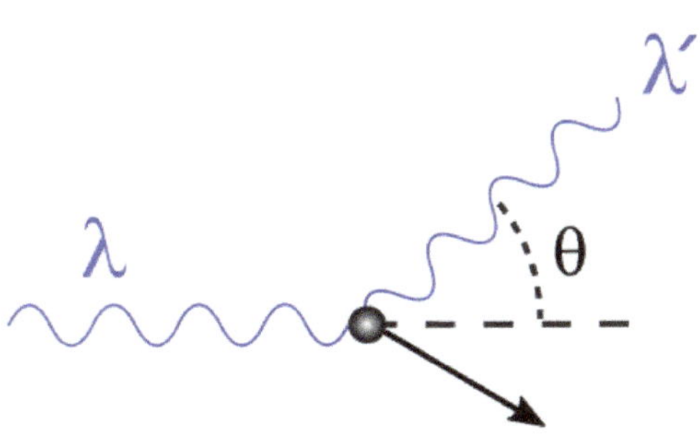

A photon of wavelength λ comes in from the left, collides with a target at rest, and a new photon of wavelength λ' emerges at an angle θ.

By the early 20th century, research into the interaction of X-rays with matter was well under way. It was observed that when X-rays of a known wavelength interact with atoms, the X-rays are scattered through an angle θ and emerge at a different wavelength related to θ. Although Classical electromagnetism predicted that the wavelength of scattered rays should be equal to the initial wavelength, multiple experiments had found that the wavelength of the scattered rays was longer (corresponding to lower energy) than the initial wavelength.

In 1923, Compton published a paper in the *Physical Review* that explained the X-ray shift by attributing particle-like momentum to light quanta (Einstein had proposed light quanta in 1905 in explaining the photo-electric effect, but Compton did not build on Einstein's work). The energy of light quanta depends only on the frequency of the light. In his paper, Compton derived the mathematical relationship between the shift in wavelength and the scattering angle of the X-rays by assuming that each scattered X-ray photon interacted with only one electron. His paper concludes by reporting on experiments which verified his derived relation:

$$\lambda' - \lambda = \frac{h}{m_e c}(1 - \cos\theta),$$

where

λ is the initial wavelength,

λ' is the wavelength after scattering,

h is the Planck constant,

m_e is the electron rest mass,

c is the speed of light, and

θ is the scattering angle.

The quantity $h/_{mec}$ is known as the Compton wavelength of the electron; it is equal to 2.43×10^{-12} m. The wavelength shift $\lambda' - \lambda$ is at least zero (for $\theta = 0°$) and at most twice the Compton wavelength of the electron (for $\theta = 180°$).

Compton found that some X-rays experienced no wavelength shift despite being scattered through large angles; in each of these cases the photon failed to eject an electron. Thus the magnitude of the shift is related not to the Compton wavelength of the electron, but to the Compton wavelength of the entire atom, which can be upwards of 10 000 times smaller.

Derivation of the Scattering Formula

A photon γ with wavelength λ collides with an electron e in an atom, which is treated as being at rest. The collision causes the electron to recoil, and a new photon γ' with wavelength λ' emerges at angle θ from the photon's incoming path. Let e' denote the electron after the collision. Compton allowed for the possibility that the interaction would sometimes accelerate the electron to speeds sufficiently close to the velocity of light and would require the application of Einstein's special relativity theory to properly describe its energy and momentum.

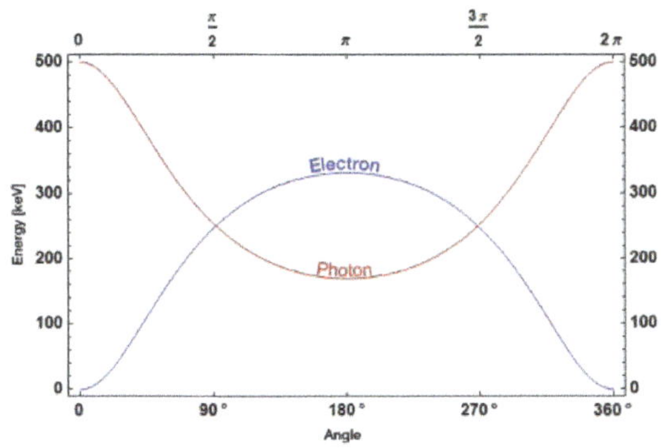

Energies of a photon at 500 keV and an electron after Compton scattering.

At the conclusion of Compton's 1923 paper, he reported results of experiments confirming the predictions of his scattering formula thus supporting the assumption that photons carry momentum as well as quantized energy. At the start of his derivation, he had postulated an expression for the momentum of a photon from equating Einstein's already established mass-energy relationship of $E = mc^2$ to the quantized photon energies of hf which Einstein has separately postulated. If $mc^2 = hf$,, the equivalent photon mass must be hf >.. The photon's momentum is then simply this effective mass times the photon's frame-invariant velocity c. For a photon, its momentum $p = hf$ >, , and thus hf can be substituted for pc for all photon momentum terms which arise in course of the derivation below. The derivation which appears in Compton's paper is more terse, but follows the same logic in the same sequence as the following derivation.

The conservation of energy E merely equates the sum of energies before and after scattering.

$$E_\gamma + E_e = E_{\gamma'} + E_{e'}.$$

Compton postulated that photons carry momentum; thus from the conservation of momentum, the momenta of the particles should be similarly related by

$$\mathbf{p}_\gamma = \mathbf{p}_{\gamma'} + \mathbf{p}_{e'},$$

in which (p_e) is omitted on the assumption it is effectively zero.

The photon energies are related to the frequencies by

$$E_\gamma = hf$$

$$E_{\gamma'} = hf'$$

where h is Planck's constant.

Before the scattering event, the electron is treated as sufficiently close to being at rest that its total energy consists entirely of the mass-energy equivalence of its rest mass m_e:

$$E_e = m_e c^2.$$

After scattering, the possibility that the electron might be accelerated to a significant fraction of the speed of light, requires that its total energy be represented using the relativistic energy–momentum relation:

$$E_{e'} = \sqrt{(p_{e'}c)^2 + (m_e c^2)^2}.$$

Substituting these quantities into the expression for the conservation of energy gives,

$$hf + m_e c^2 = hf' + \sqrt{(p_{e'}c)^2 + (m_e c^2)^2}.$$

This expression can be used to find the magnitude of the momentum of the scattered electron,

$$p_{e'}^2 c^2 = (hf - hf' + m_e c^2)^2 - m_e^2 c^4. \qquad (1)$$

Note that the magnitude of the momentum gained by the electron (formerly zero) exceeds that lost by the photon:

$$\frac{1}{c}\sqrt{(hf - hf' + m_e c^2)^2 - m_e^2 c^4} > \frac{hf - hf'}{c}.$$

Equation (1) relates the various energies associated with the collision. The electron's momentum change includes a relativistic change in the mass of the electron so it is not simply related to the change in energy in the manner that occurs in classical physics. The change in the momentum of the photon is also not simply related to the difference in energy but involves a change in direction.

Solving the conservation of momentum expression for the scattered electron's momentum gives,

$$\mathbf{p}_{e'} = \mathbf{p}_{\gamma} - \mathbf{p}_{\gamma'}.$$

Then by making use of the scalar product,

$$\begin{aligned} p_{e'}^2 &= \mathbf{p}_{e'} \cdot \mathbf{p}_{e'} = (\mathbf{p}_{\gamma} - \mathbf{p}_{\gamma'}) \cdot (\mathbf{p}_{\gamma} - \mathbf{p}_{\gamma'}) \\ &= p_{\gamma}^2 + p_{\gamma'}^2 - 2p_{\gamma} p_{\gamma'} \cos\theta. \end{aligned}$$

Anticipating that $p_\gamma c$ is replaceable with hf , multiply both sides by c^2:

$$p_{e'}^2 c^2 = p_\gamma^2 c^2 + p_{\gamma'}^2 c^2 - 2c^2 p_\gamma p_{\gamma'} \cos\theta.$$

After replacing the photon momentum terms with hf >, we get a second expression for the magnitude of the momentum of the scattered electron:

$$p_{e'}^2 c^2 = (hf)^2 + (hf')^2 - 2(hf)(hf')\cos\theta. \qquad (2)$$

Equating both expressions for this momentum gives

$$(hf - hf' + m_e c^2)^2 - m_e^2 c^4 = (hf)^2 + (hf')^2 - 2h^2 ff' \cos\theta,$$

which after evaluating the square and then canceling and rearranging terms gives

$$2hfm_e c^2 - 2hf'm_e c^2 = 2h^2 ff'(1 - \cos\theta).$$

Then dividing both sides by $2hff'm_e c$ yields

$$\frac{c}{f'} - \frac{c}{f} = \frac{h}{m_e c}(1 - \cos\theta).$$

Finally, since $f\lambda = f'\lambda' = c$,

$$\lambda' - \lambda = \frac{h}{m_e c}(1 - \cos\theta).$$

It can also be seen that the angle φ of the outgoing electron with the direction of the incoming photon is specified by

$$\cot\varphi = (1 + hf/(m_e c^2)) \tan(\theta/2).$$

Applications

Compton Scattering

Compton scattering is of prime importance to radiobiology, as it is the most probable interaction of gamma rays and high energy X-rays with atoms in living beings and is applied in radiation therapy.

In material physics, Compton scattering can be used to probe the wave function of the electrons in matter in the momentum representation.

Compton scattering is an important effect in gamma spectroscopy which gives rise to the Compton edge, as it is possible for the gamma rays to scatter out of the detectors used. Compton suppression is used to detect stray scatter gamma rays to counteract this effect.

Magnetic Compton Scattering

Magnetic Compton scattering is an extension of the previously mentioned technique which involves the magnetisation of a crystal sample hit with high energy, circularly polarised photons. By measuring the scattered photons' energy and reversing the magnetisation of the sample, two different Compton profiles are generated (one for spin up momenta and one for spin down momenta). Taking the difference between these two profiles gives the magnetic Compton profile (MCP), given by $J_{\text{mag}}(\mathbf{p}_z)$ - a one-dimensional projection of the electron spin density.

$$J_{\text{mag}}(\mathbf{p}_z) = \frac{1}{\mu} \iint_{-\infty}^{\infty} (n_{\uparrow}(\mathbf{p}) - n_{\downarrow}(\mathbf{p})) d\mathbf{p}_x d\mathbf{p}_y$$

where μ is the number of spin-unpaired electrons in the system, $n_{\uparrow}(\mathbf{p})$ and $n_{\downarrow}(\mathbf{p})$ are the three-dimensional electron momentum distributions for the majority spin and minority spin electrons respectively.

Since this scattering process is incoherent (there is no phase relationship between the scattered photons), the MCP is representative of the bulk properties of the sample and is a probe of the ground state. This means that the MCP is ideal for comparison with theoretical techniques such as density functional theory. The area under the MCP is directly proportional to the spin moment of the system and so, when combined with total moment measurements methods (such as SQUID magnetometry), can be used to isolate both the spin and orbital contributions to the total moment of a system. The shape of the MCP also yields insight into the origin of the magnetism in the system.

Inverse Compton scattering

Inverse Compton scattering is important in astrophysics. In X-ray astronomy, the accretion disc surrounding a black hole is presumed to produce a thermal spectrum. The lower energy photons produced from this spectrum are scattered to higher energies by relativistic electrons in the surrounding corona. This is surmised to cause the power law component in the X-ray spectra (0.2-10 keV) of accreting black holes.

The effect is also observed when photons from the cosmic microwave background (CMB) move through the hot gas surrounding a galaxy cluster. The CMB photons are scattered to higher energies by the electrons in this gas, resulting in the Sunyaev-Zel'dovich effect. Observations of the Sunyaev-Zel'dovich effect provide a nearly redshift-independent means of detecting galaxy clusters.

Some synchrotron radiation facilities scatter laser light off the stored electron beam. This Compton backscattering produces high energy photons in the MeV to GeV range subsequently used for nuclear physics experiments.

Small-angle Scattering

Small-angle scattering (SAS) is a scattering technique based on deflection of collimated radiation away from the straight trajectory after it interacts with structures that are much larger than the wavelength of the radiation. The deflection is small (0.1-10°) hence the name *small-angle*. SAS techniques can give information about the size, shape and orientation of structures in a sample.

Small-angle scattering (SAS) is a powerful technique for investigating large-scale structures from 10 Å up to thousands and even several tens of thousands of angstroms. The most important feature of the SAS method is its potential for analyzing the inner structure of disordered systems, and frequently the application of this method is a unique way to obtain direct structural information on systems with random arrangement of density inhomogeneities in such large-scales.

Currently, the SAS technique, with its well-developed experimental and theoretical procedures and wide range of studied objects, is a self-contained branch of the structural analysis of matter. Reflecting these situations, the international meeting on the small-angle scattering studies have been held in every three years. SAS can refer to:

- Small angle neutron scattering (SANS)
- Small angle X-ray scattering (SAXS)

Applications

Small-angle scattering is particularly useful because of the dramatic increase in forward scattering that occurs at phase transitions, known as critical opalescence, and because many materials, substances and biological systems possess interesting and complex features in their structure, which match the useful length scale ranges that these techniques probe. The technique provides valuable information over a wide variety of scientific and technological applications including chemical aggregation, defects in materials, surfactants, colloids, ferromagnetic correlations in magnetism, alloy segregation, polymers, proteins, biological membranes, viruses, ribosome and macromolecules. While analysis of the data can give information on size, shape, etc., without making any model assumptions a preliminary analysis of the data can only give information on the radius of gyration for a particle using Guinier's equation.

Theory

Continuum Description

SAXS patterns are typically represented as scattered intensity as a function of the magnitude of the *scattering vector* $q = 4\pi \sin(\theta) / \lambda$. Here 2θ is the angle between the incident X-ray beam and the detector measuring the scattered intensity, and λ is the wavelength of the X-rays. One interpretation of the scattering vector is that it is the *resolution* or *yardstick* with which the sample is observed. In the case of a two-phase sample, e.g. small particles in liquid suspension, the only contrast leading to scattering in the typical range of resolution of the SAXS is simply $\Delta\rho$, the difference in *average* electron density between the particle and the surrounding liquid, because variations in ρ due to the atomic structure only become visible at higher angles in the WAXS regime. This means that the total integrated intensity of the SAXS pattern (in 3D) is an invariant quantity proportional to the square $\Delta\rho^2$. In 1-dimensional projection, as usually recorded for an isotropic pattern this invariant quantity becomes $\int I(q)q^2\,dx$, , where the integral runs from q=0 to wherever the SAXS pattern is assumed to end and the WAXS pattern starts. It is also assumed that the density does not vary in the liquid or inside the particles, i.e. there is *binary* contrast.

SANS is described in terms of a neutron scattering length density.

Porod'S Law

At wave numbers that are relatively large on the scale of SAS, but still small when compared to wide-angle Bragg diffraction, local interface intercorrelations are probed, whereas correlations between opposite interface segments are averaged out. For smooth interfaces, one obtains Porod's law:

$$I(q) \sim Sq^{-4}$$

This allows the surface area *S* of the particles to be determined with SAS. This needs to be modified if the interface is rough on the scale q^{-1}. If the roughness can be described by a fractal dimension *d* between 2-3 then Porod's law becomes:

$$I(q) \sim S'q^{-(6-d)}$$

Scattering from Particles

Small-angle scattering from particles can be used to determine the particle shape or their size distribution. A small-angle scattering pattern can be fitted with intensities calculated from different model shapes when the size distribution is known. If the shape is known, a size distribution may be fitted to the intensity. Typically one assumes the particles to be spherical in the latter case.

If the particles are dispersed in a solution and they are known to be monodisperse, all of the same size, then a typical strategy is to measure different concentrations of particles in the solution. From the obtained SAXS patterns one can extrapolate to the intensity pattern one would get for a single particle. This is a necessary procedure that eliminates the *concentration effect*, which is a small shoulder that appears in the intensity patterns due to the proximity of neighbouring particles. The average distance between particles is then roughly the distance $2\pi/q^*$, where q^* is the position of the shoulder on the scattering vector range q. The shoulder thus comes from the structure of the solution and this contribution is called *the structure factor*. One can write for the small-angle X-ray scattering intensity:

$$I(q) = P(q)S(q),$$

where

- $I(q)$ is the intensity as a function of the magnitude q of the scattering vector
- $P(q)$ is the form factor
- and $S(q)$ is the structure factor.

When the intensities from low concentrations of particles are extrapolated to infinite dilution, the structure factor is equal to 1 and no longer disturbs the determination of the particle shape from the form factor $P(q)$. One can then easily apply the Guinier approximation (also called Guinier law, after André Guinier), which applies only at the very beginning of the scattering curve, at small q-values. According to the Guinier approximation the intensity at small q depends on the radius of gyration of the particle.

An important part of the particle shape determination is usually the distance distribution function $p(r)$, which may be calculated from the intensity using a Fourier transform

$$p(r) = \frac{r^2}{2\pi^2} \int_0^\infty I(q) \frac{\sin qr}{qr} q^2 dq.$$

The distance distribution function $p(r)$ is related to the frequency of certain distances r within the particle. Therefore it goes to zero at the largest diameter of the particle. It starts from zero at $r = 0$ due to the multiplication by r^2. The shape of the $p(r)$-function already tells something about the shape of the particle. If the function is very symmetric, the particle is also highly symmetric, like a sphere. The distance distribution function should not be confused with the size distribution.

The particle shape analysis is especially popular in biological small-angle X-ray scattering, where one determines the shapes of proteins and other natural colloidal polymers.

History

Small-angle scattering studies were initiated by André Guinier (1937). Subsequently, Peter Debye, Otto Kratky, Günther Porod, R. Hosemann and others developed the theoretical and experimental fundamentals of the method and they were established until around 1960. Later on, new progress in refining the method began in the 1970s and is continuing today.

Organisations

As a 'low resolution' diffraction technique, the worldwide interests of the small-angle scattering community are promoted and coordinated by the Commission on Small-Angle Scattering of the International Union of Crystallography. There are also a number of community-led networks and projects. One such network, CanSAS - the acronym stands for Collective Action for Nomadic Small-Angle Scatterers, emphasising the global nature of the technique, champions the development of instrumental calibration standards and data file formats. It has also created a Web Portal to a variety of SAS-related resources.

Scattering Theory

Top: the real part of a plane wave travelling upwards. Bottom: The real part of the field after inserting in the path of the plane wave a small transparent disk of index of refraction higher than the index of the surrounding medium. This object scatters part of the wave field, although at any individual point, the wave's frequency and wavelength remain intact.

In mathematics and physics, scattering theory is a framework for studying and understanding the scattering of waves and particles. Wave scattering corresponds to the collision and scattering of a wave with some material object, for instance sunlight scattered by rain drops to form a rainbow. Scattering also includes the interaction of billiard balls on a table, the Rutherford scattering (or angle change) of alpha particles by gold nuclei, the Bragg scattering (or diffraction) of electrons and X-rays by a cluster of atoms, and the inelastic scattering of a fission fragment as it traverses a thin foil. More precisely, scattering consists of the study of how solutions of partial differential equations, propagating freely "in the distant past", come together and interact with one another or with a boundary condition, and then propagate away "to the distant future". The direct scattering problem is the problem of determining the distribution of scattered radiation/particle flux basing on the characteristics of the scatterer. The inverse scattering problem is the problem of determining the characteristics of an object (e.g., its shape, internal constitution) from measurement data of radiation or particles scattered from the object.

Since its early statement for radiolocation, the problem has found vast number of applications, such as echolocation, geophysical survey, nondestructive testing, medical imaging and quantum field theory, to name just a few.

Conceptual Underpinnings

The concepts used in scattering theory go by different names in different fields. The object of this section is to point the reader to common threads.

Composite Targets and Range Equations

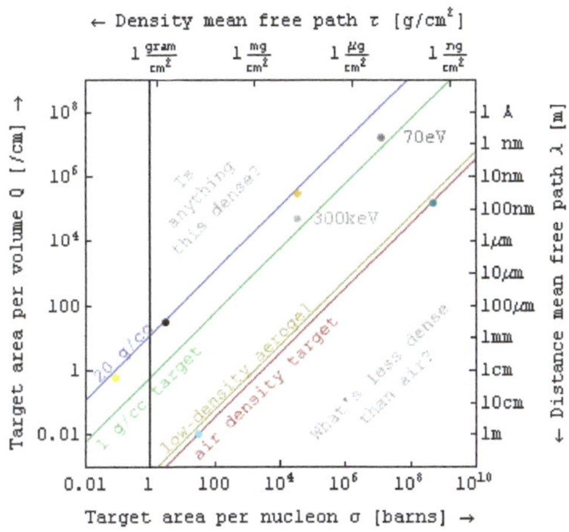

Equivalent quantities used in the theory of scattering from composite specimens, but with a variety of units.

When the target is a set of many scattering centers whose relative position varies unpredictably, it is customary to think of a range equation whose arguments take different forms in different application areas. In the simplest case consider an interaction that removes particles from the "unscattered beam" at a uniform rate that is proportional to the incident flux I of particles per unit area per unit time, i.e. that

$$\frac{dI}{dx} = -QI$$

where Q is an interaction coefficient and x is the distance traveled in the target.

The above ordinary first-order differential equation has solutions of the form:

$$I = I_o e^{-Q\text{Äx}} = I_o e^{-\frac{\text{Äx}}{\lambda}} = I_o e^{-\sigma(\eta\text{Äx})} = I_o e^{-\frac{\rho\text{Äx}}{\tau}},$$

where I_o is the initial flux, path length $\Delta x \equiv x - x_o$, the second equality defines an interaction mean free path λ, the third uses the number of targets per unit volume η to define an area cross-section σ, and the last uses the target mass density ρ to define a density mean free path τ. Hence one converts between these quantities via $Q = 1/\lambda = \eta\sigma = \rho/\tau$, as shown in the figure at left.

In electromagnetic absorption spectroscopy, for example, interaction coefficient (e.g. Q in cm^{-1}) is variously called opacity, absorption coefficient, and attenuation coefficient. In nuclear physics, area cross-sections (e.g. σ in barns or units of 10^{-24} cm^2), density mean free path (e.g. τ in grams/cm^2), and its reciprocal the mass attenuation coefficient (e.g. in cm^2/gram) or *area per nucleon* are all popular, while in electron microscopy the inelastic mean free path (e.g. λ in nanometers) is often discussed instead.

In Theoretical Physics

In mathematical physics, scattering theory is a framework for studying and understanding the interaction or scattering of solutions to partial differential equations. In acoustics, the differential equation is the wave equation, and scattering studies how its solutions, the sound waves, scatter from solid objects or propagate through non-uniform media (such as sound waves, in sea water, coming from a submarine). In the case of classical electrodynamics, the differential equation is again the wave equation, and the scattering of light or radio waves is studied. In particle physics, the equations are those of Quantum electrodynamics, Quantum chromodynamics and the Standard Model, the solutions of which correspond to fundamental particles.

In regular quantum mechanics, which includes quantum chemistry, the relevant equation is the Schrödinger equation, although equivalent formulations, such as the Lippmann-Schwinger equation and the Faddeev equations, are also largely used. The solutions of interest describe the long-term motion of free atoms, molecules, photons, electrons, and protons. The scenario is that several particles come together from an infinite distance away. These reagents then collide, optionally reacting, getting destroyed or creating new particles. The products and unused reagents then fly away to infinity again. (The atoms and molecules are effectively particles for our purposes. Also, under everyday circumstances, only photons are being created and destroyed.) The solutions reveal which directions the products are most likely to fly off to and how quickly. They also reveal the probability of various reactions, creations, and decays occurring. There are two predominant techniques of finding solutions to scattering problems: partial wave analysis, and the Born approximation.

Elastic and Inelastic Scattering

The term "elastic scattering" implies that the internal states of the scattering particles do not change, and hence they emerge unchanged from the scattering process. In inelastic scattering, by contrast, the particles' internal state is changed, which may amount to exciting some of the electrons of a scattering atom, or the complete annihilation of a scattering particle and the creation of entirely new particles.

The example of scattering in quantum chemistry is particularly instructive, as the theory is reasonably complex while still having a good foundation on which to build an intuitive understanding. When two atoms are scattered off one another, one can understand them as being the bound state solutions of some differential equation. Thus, for example, the hydrogen atom corresponds to a solution to the Schrödinger equation with a negative inverse-power (i.e., attractive Coulombic) central potential. The scattering of two hydrogen atoms will disturb the state of each atom, resulting in one or both becoming excited, or even ionized, representing an inelastic scattering process.

The term "deep inelastic scattering" refers to a special kind of scattering experiment in particle physics.

The Mathematical Framework

In mathematics, scattering theory deals with a more abstract formulation of the same set of concepts. For example, if a differential equation is known to have some simple, localized solutions, and the solutions are a function of a single parameter, that parameter can take the conceptual role of time. One then asks what might happen if two such solutions are set up far away from each other, in the "distant past", and are made to move towards each other, interact (under the constraint of the differential equation) and then move apart in the "future". The scattering matrix then pairs solutions in the "distant past" to those in the "distant future".

Solutions to differential equations are often posed on manifolds. Frequently, the means to the solution requires the study of the spectrum of an operator on the manifold. As a result, the solutions often have a spectrum that can be identified with a Hilbert space, and scattering is described by a certain map, the S matrix, on Hilbert spaces. Spaces with a discrete spectrum correspond to bound states in quantum mechanics, while a continuous spectrum is associated with scattering states. The study of inelastic scattering then asks how discrete and continuous spectra are mixed together.

An important, notable development is the inverse scattering transform, central to the solution of many exactly solvable models.

References

- Bohren, Craig F.; Donald R. Huffman (1983). Absorption and Scattering of Light by Small Particles. Wiley. ISBN 0-471-29340-7.
- Colton, David; Rainer Kress (1998). Inverse Acoustic and Electromagnetic Scattering Theory. Springer. ISBN 3-540-62838-X.
- Seinfeld, John H.; Pandis, Spyros N. (2006). Atmospheric Chemistry and Physics - From Air Pollution to Climate Change (2nd Ed.). John Wiley and Sons, Inc. ISBN 0-471-82857-2
- Ahrens, C. Donald (1994). Meteorology Today: an introduction to weather, climate, and the environment (5th ed.). St. Paul MN: West Publishing Company. pp. 88–89. ISBN 0-314-02779-3.
- Taylor, J.R.; Zafiratos, C.D.; Dubson, M.A. (2004). Modern Physics for Scientists and Engineers (2nd ed.). Prentice Hall. pp. 136–9. ISBN 0-13-805715-X.
- Feigin LA; Svergun DI (1987). Structure Analysis by Small-Angle X-Ray and Neutron Scattering (PDF). New York: Plenum Press. p. 40. ISBN 0-306-42629-3.

Permissions

All chapters in this book are published with permission under the Creative Commons Attribution Share Alike License or equivalent. Every chapter published in this book has been scrutinized by our experts. Their significance has been extensively debated. The topics covered herein carry significant information for a comprehensive understanding. They may even be implemented as practical applications or may be referred to as a beginning point for further studies.

We would like to thank the editorial team for lending their expertise to make the book truly unique. They have played a crucial role in the development of this book. Without their invaluable contributions this book wouldn't have been possible. They have made vital efforts to compile up to date information on the varied aspects of this subject to make this book a valuable addition to the collection of many professionals and students.

This book was conceptualized with the vision of imparting up-to-date and integrated information in this field. To ensure the same, a matchless editorial board was set up. Every individual on the board went through rigorous rounds of assessment to prove their worth. After which they invested a large part of their time researching and compiling the most relevant data for our readers.

The editorial board has been involved in producing this book since its inception. They have spent rigorous hours researching and exploring the diverse topics which have resulted in the successful publishing of this book. They have passed on their knowledge of decades through this book. To expedite this challenging task, the publisher supported the team at every step. A small team of assistant editors was also appointed to further simplify the editing procedure and attain best results for the readers.

Apart from the editorial board, the designing team has also invested a significant amount of their time in understanding the subject and creating the most relevant covers. They scrutinized every image to scout for the most suitable representation of the subject and create an appropriate cover for the book.

The publishing team has been an ardent support to the editorial, designing and production team. Their endless efforts to recruit the best for this project, has resulted in the accomplishment of this book. They are a veteran in the field of academics and their pool of knowledge is as vast as their experience in printing. Their expertise and guidance has proved useful at every step. Their uncompromising quality standards have made this book an exceptional effort. Their encouragement from time to time has been an inspiration for everyone.

The publisher and the editorial board hope that this book will prove to be a valuable piece of knowledge for students, practitioners and scholars across the globe.

Index

T

V

Z